FUNCTIONAL ANALYSIS
An Introduction for Physicists

FUNCTIONAL ANALYSIS

An Introduction for Physicists

NINO BOCCARA

College of Liberal Arts and Sciences
University of Illinois at Chicago
Chicago, Illinois

and

Laboratoire de Mathématique
École de Physique et de Chimie
Paris, France

ACADEMIC PRESS, INC.
Harcourt Brace Jovanovich, Publishers
Boston San Diego New York
London Sydney Tokyo Toronto

This book is printed on acid-free paper. ♾

ACADEMIC PRESS, INC.
1250 Sixth Avenue, San Diego, CA 92101

United Kingdom Edition published by
ACADEMIC PRESS LIMITED
24–28 Oval Road, London NW1 7DX

Library of Congress Cataloging-in-Publication Data

Boccara, Nino.
Functional analysis: an introduction for physicists / Nino Boccara.
p. cm.
Includes bibliographical references and index.
ISBN 0-12-108810-3 (alk. paper)
1. Functional analysis. I. Title.
QA320.B63 1990
515'.7—dc20 90-36210
CIP

Printed in the United States of America
90 91 92 93 9 8 7 6 5 4 3 2 1

Contents

Foreword vii
Notations ix
Chapter 1. Measure and Integration 1
1. Measurable Functions 1
2. Positive Measures 7
3. Integral of Measurable Functions 19
4. Lebesgue's Dominated Convergence Theorem 29
5. Fubini's Theorem 35
Appendix 48
Notes 50
Problems 54
Solutions 58
Chapter 2. Lebesgue Spaces 69
1. Convex Functions and Inequalities 69
2. $\mathcal{L}^p$ and L^p Spaces 74
3. Dense Subspaces in L^p 80
4. Fourier Transform 83
5. Laplace Transform 94
Note 104
Problems 106
Solutions 109
Chapter 3. Hilbert Spaces 118
1. Scalar Product 118
2. Hilbert Spaces 121
3. Orthogonal Systems 127
Appendix 151
Notes 153
Problems 154
Solutions 157
Chapter 4. Distributions 167
1. Test Functions and Distributions 168
2. Differentiation 173
3. Convergence in $\mathcal{D}'$ 190
4. Convolution 195

5. Fourier Series 207
6. Fourier Transform of Tempered Distributions 212
7. Laplace Transform 226
Problems 232
Solutions 236
Chapter 5. Linear Operators 255
1. Preliminaries 255
2. Bounded Linear Operators 257
3. Compact Operators 270
4. Unbounded Linear Operators 273
5. Spectral Theory 280
Appendix 301
Problems 303
Solutions 306
References 319
Index 321

Foreword

This book is intended for physicists who wish to familiarize themselves with the methods of functional analysis. These methods are extremely fruitful, but they cannot be properly used if one knows only unconnected recipes, ignoring, moreover, their conditions of applicability. That is not to say that the physicist has to be enslaved by a paralyzing rigor or indulge in a barren formalism. I have therefore sought to avoid the inflation of terminology, the introduction of concepts unnecessary to any further development, or proofs of too general a nature. Breaking away from the definition–theorem–proof style of exposition, I emphasize the practicality of the results, illustrating them with several examples. In order to assimilate a theorem, it is advisable to study the examples and to attempt to do the exercises, the solutions of which are provided, before tackling its proof.

This book is based on the course taught at the Ecole Supérieure de Physique et de Chimie in Paris. It is a modified and augmented second version of *Analyse Fonctionnelle: Une Introduction pour Physiciens*, Ellipses, Paris 1984. The prerequisite is a first course in analysis.

This book was typeset by the author using the TEX system.

Notations

The first digit refers to the number of the chapter in which the notation occurs for the first time, and the second refers to the section within the chapter. A stands for appendix, N for note, and P for problem.

$x \in A$: x is an element of the set A, 1.1
$\{x|P(x)\}$: the set of those x with property P, 1.1
$\forall$: for any, 1.1
$A \subset B$: set A is a subset of the set B, 1.1
$\cup, \cap$: union, intersection of sets, 1.1
1_A or χ_A: the characteristic function of the set A, 1.1
$\emptyset$: the empty set, 1.1
$\{a\}$: a set having a as unique element, 1.1
A^C: the complement of A, 1.1
$A - B$: the set of elements of A not in B, 1.1
$A \triangle B = (A - B) \cup (B - A)$: the symmetric difference of the sets A and B, 1.1
$\mathcal{P}(\mathrm{A})$: the set of all subsets of A, 1.1
$A \times B$: the cartesian product of the sets A and B, 1.1
(x, y): an ordered pair, 1.1
$f(x)$: the value of the mapping f at x, 1.1
$f(A)$: the image of the set A, 1.1
$f^{-1}(B)$: the preimage of the set B, 1.1
$x \mapsto f(x)$: the mapping (or the function) f, 1.1
$f^{-1}(y)$: the preimage of a one-element set $\{y\}$, 1.1
σ-algebra: a class of sets closed under countable unions and complements and containing the empty set, 1.1
$\mathcal{S}(\mathcal{E})$: the σ-algebra generated by $\mathcal{E}$, 1.1
$(X, \mathcal{A})$: the measurable space consisting of the space X and the σ-algebra $\mathcal{A}$, 1.1
$\mathbf{R}$: the set of real numbers, 1.1
$\overline{\mathbf{R}}$: the extended line, that is $\{-\infty\} \cup \mathbf{R} \cup \infty$, 1.1
$\mathbf{C}$: the set of complex numbers, 1.1
$]a, b[$, $[a, b]$: open, closed interval, 1.1
$[a, b[$, $]a, b]$: semiopen intervals, 1.1

$\sup_{x\in A} f(x)$, $\inf_{x\in A} f(x)$: supremum and infimum of f in A, 1.1
(f_n): a sequence of mappings, 1.1
$\lim_{n\to\infty} f_n$: the limit of the sequence (f_n), 1.1
$\limsup f_n$, $\liminf f_n$: the upper and lower limits of the sequence (f_n), 1.1
$f \circ g$: composed mapping of f and g, 1.1
f^+, f^-: positive, negative part of f, 1.1
$\Re$: real part, 1.1
$\Im$: imaginary part, 1.1
$x \notin A$: x is not an element of the set A, 1.2
$\mathbf{R}_+$: the set of nonnegative real numbers, 1.2
μ: positive measure, 1.2
$(X, \mathcal{A}, \mu)$: the measure space consisting of the space X, the σ-algebra $\mathcal{A}$ and the measure μ, 1.2
$\ell(I)$: the length of the interval I, 1.2
m^*: the outer measure, 1.2
m_*: the inner measure, 1.2
m: the Lebesgue measure, 1.2
$A + a = \{x + a | x \in A\}$, 1.2
xRy: equivalence relation, 1.2
$\mathbf{Q}$: the set of rational numbers, 1.2
$P \iff Q$: P is true if, and only if, Q is true, 1.2
$\mathbf{N}$: the set of positive integers, 1.2
$f(a-0)$: the left-hand limit, that is, the limit of $f(x)$ when $x \to a$ and $x < a$, 1.2
$f(a+0)$: the right-hand limit, that is, the limit of $f(x)$ when $x \to a$ and $x > a$, 1.2
a.e.: almost everywhere, that is, except on a negligible set, 1.2
$\int f\,d\mu$: integral of f with respect to μ, 1.3
$\int f\,dm$ or $\int f(x)\,dm(x)$ or $\int f(x)\,dx$: integral of f with respect to the Lebesgue measure, 1.3
$\mathbf{Z}$: the set of integers, 1.3
$\int_A f\,d\mu$: integral of f on the measurable set A, 1.3
$E(\xi)$: mathematical expectation of a random variable, 1.3
$V(\xi)$: variance of a random variable, 1.3
$|f|$: absolute value of the real or complex function f, 1.3
$\int_a^b f(x)\,dx$: Riemann integral of f on $[a,b]$, 1.3
$\mathcal{L}^1_\mu$: the vector space of integrable functions with respect to the measure μ, 1.3
J_n: Bessel function of the first kind of order n, 1.4

$\mu \otimes \nu$: product measure, 1.5
$(X \times Y, \mathcal{A} \otimes \mathcal{B})$: product of the measurable spaces $(X, \mathcal{A})$ and $(Y, \mathcal{B})$, 1,5
$(X \times Y, \mathcal{A} \otimes \mathcal{B}, \mu \otimes \nu)$: product of the measure spaces $(X, \mathcal{A}, \mu)$ and $(Y, \mathcal{B}, \nu)$, 1,5
$E_x = \{y|(x, y) \in E\}$: x-section of the set E, 1.5
$E^y = \{x|(x, y) \in E\}$: y-section of the set E, 1.5
f_x: x-section of the function f, 1.5
f^y: y-section of the function f, 1.5
$\int f \, d\mu \otimes \nu = \int\int f(x, y) \, d\mu(x) \, d\nu(y)$: integral with respect to a product measure, 1.5
$\mathcal{M}$: monotone class, 1.5
$\omega_f(x)$: oscillation of f at x, 1.A
$P \Rightarrow Q$: P implies Q, 1.N2
$\|f\|$: norm of f, 1.N4
Γ: Euler Gamma function, 2.1
$\mathcal{L}^p_\mu(X)$: the vector space of measurable functions defined on X such that $\int |f|^p \, d\mu < \infty$, 2.2
$L^p_\mu(X)$: the normed vector space of measurable functions defined almost everywhere on X such that $\int |f|^p \, d\mu < \infty$, 2.2
$\tilde{f}$: the class of functions equivalent to f, 2.2
$\tilde{0}$: the class of functions equal to 0 a.e., 2.2
$\|f\|_p$: the L^p-norm of f, 2.2
$\mathcal{L}^\infty_\mu(X)$: the vector space of essentially bounded measurable functions defined on X, 2.2
ℓ^p: the space of sequences (x_n) such that $(|x_n|^p)$ is convergent, 2.2
$d(f, g)$: distance from f to g in a metric space, 2.2
ess sup f: the essential supremum of f, 2.2
$L^\infty_\mu(X)$: the normed vector space of essentially bounded measurable functions defined almost everywhere on X, 2.2
$d(x, A)$; distance between the point x and the set A, 2.3
$\overline{A}$: the closure of the set A, 2.3
$\mathring{A}$: the interior of the set A, 2.3
B: Euler Beta function, 2.3
$\widehat{f}$: Fourier transform of the function f, 2.4
$\overline{x}$: the complex conjugate of x, 2.4
$u_n = O(n^k)$: $u_n n^k$ is bounded, 2.4
$L(f)$: Laplace transform of a function f, 2.5
θ: the Heaviside function, 2.5
$\langle x \mid y \rangle$: scalar product of x and y, 3.1

$\|x\|$: norm of the vector x, 3.1
$L^2_\mu(X)$: the Hilbert space of square integrable functions defined almost everywhere on X, 3.2
$P_F(x)$: projection of the vector x on the subspace F, 3.2
E': the topological dual of the space E, 3.2
P_n: the Legendre polynomial of degree n, 3.3
$P_n^{(\alpha,\beta)}$: Jacobi polynomial, 3.3
$L_n^{(\alpha)}$: Laguerre polynomial, 3.3
H_n: the Hermite polynomial of degree n, 3.3
$G_n^{(p)}$: Gegenbauer polynomial, 3.3
T_n, U_n: Chebyshev polynomials, 3.3
ℓ^2: the Hilbert space of sequences (x_n) such that $(|x_n|^2)$ is convergent, 3.3
D_n: the Dirichlet kernel, 3.3
F_n: the Fejér kernel, 3.3
(E, d): metric space, 3.N2
C_c^∞: the space of infinitely differentiable functions having a bounded (compact) support, 4.1
$\mathcal{D} = C_c^\infty$: the space of test functions, 4.1
φ: a test function, 4.1
$T: \varphi \mapsto T(\varphi)$: the distribution T, 4.1
$\mathcal{D}'$: the space of Schwartz distributions, 4.1
L^1_{loc}: the space of locally integrable functions, 4.1
T_f: the regular distribution associated to the function f, 4.1
δ: the Dirac distribution, 4.1
Pv: Cauchy principal value, 4.2
Pf: pseudofunction, 4.2
Fp: Hadamard's finite part, 4.2
θ_s: characteristic function of the domain $S(x_1, x_2 x_3) > 0$, 4.2
δ_s: the Dirac distribution on a surface, 4.2
$\mathbf{A}$, $\mathbf{E}$, $\mathbf{D}$: tridimensional vectors, 4.2
$\mathbf{n}$: unit tridimensional vector, 4.2
∇: nabla, 4.2
∇f: gradient of f, 4.2
$\nabla \cdot \mathbf{D}$: divergence of $\mathbf{D}$, 4.2
$\nabla \times \mathbf{E}$: curl of $\mathbf{E}$, 4.2
$S \otimes T$: the tensor product of the distributions S and T, 4.4
$S * T$: the convolution of the distributions S and T, 4.4
$\operatorname{supp} f$: the support of the function f, 4.4
B: Euler Beta function, 4.4

S^1: a circle, 4.5
$c_k(T)$: Fourier coefficient of the distribution T, 4.5
$\widehat{T}$: Fourier transform of the distribution T, 4.6
$\check{T}$: inverse Fourier transform of the distribution T, 4.6
$L(T)$: Laplace transform of a distribution T, 4.7
γ: Euler constant, 4.7
CR: canonical regularization, 4.P4
X: the position operator, 5.1
P: the momentum operator, 5.1
D_A: the domain of a linear operator A, 5.1
R_A: the range of a linear operator A, 5.1
I: the identity operator, 5.1
A^{-1} the inverse of A, 5.1
$\|A\|$: the norm of the linear operator A, 5.2
$[A, B] = AB - BA$: the commutator of the operators A and B, 5.2
$\mathcal{B}(E)$: the Banach algebra of the linear operators on E, 5.2
$\widehat{A}$: extension of the operator A, 5.2
a_{mn}: matrix element of a bounded linear operator A, 5.2
P: projection operator, 5.2
A^+: the adjoint of the operator A, 5.2
F: Fourier or Fourier–Plancherel operator, 5.2
A^+, A: creation and annihilation operators, 5.2
G_A: the graph of the operator A, 5.2
$\overline{A}$: the closure of the operator A
$R_\lambda(A)$: the resolvent of the operator A, 5.5
$\rho(A)$: the resolvent set of the operator A, 5.5
$\sigma(A)$: the spectrum of the operator A, 5.5
$P\sigma(A)$: the point spectrum of the operator A, 5.5
$C\sigma(A)$: the continuous spectrum of the operator A, 5.5
$R\sigma(A)$: the residual spectrum of the operator A, 5.5
$E_1 \oplus E_2$: direct sum, 5.5

Chapter 1

Measure and Integration

Within the framework of Riemann's theory of integration, the validity of the formula

$$\int_{-\infty}^{\infty} \lim_{n\to\infty} f_n(x)\,dx = \lim_{n\to\infty} \int_{-\infty}^{\infty} f_n(x)\,dx,$$

when (f_n) is a nonuniformly convergent sequence of Riemann-integrable functions, was extensively discussed at the end of the nineteenth century. The results obtained paved the way for Lebesgue's theory of integration.[1] This more general theory, better adapted for dealing with limit processes, allows the interchange of integral and limit in very general circumstances. The concept of measure plays a crucial role in Lebesgue's definition of the integral. Apart from Lebesgue's measure, which is a generalization of the notion of length, we shall consider different types of measures such as the Hausdorff measure. This particular measure is appropriate for Cantor-like sets found in many applications.

The most important results of this chapter are Lebesgue's dominated convergence theorem and Fubini's theorem. They are illustrated by many examples to show their wide applicability.

1. Measurable Functions

The theory of integration constructed in this chapter applies to measurable functions. Theorems 3 and 4 show that the class of measurable functions is very large; in physics, it is very unlikely that one would have to deal with nonmeasurable functions.

---

[1] On the origins and development of Lebesgue's theory, see T. Hawkins.

Definition 1. A class $\mathcal{A}$ of subsets of a set X is said to be a *σ-algebra* (sigma algebra) in X if
(1) X belongs to $\mathcal{A}$,
(2) any countable union of elements of $\mathcal{A}$ belongs to $\mathcal{A}$,
(3) the complement of any element of $\mathcal{A}$ is in $\mathcal{A}$.

From (1) and (3) it follows that the empty set $\emptyset$ is in $\mathcal{A}$; and from (2) and (3) it follows that any countable intersection of elements of $\mathcal{A}$ is in $\mathcal{A}$.

If A and B belong to $\mathcal{A}$, then $A - B = A \cap B^C$ and $A \triangle B = (A - B) \cup (B - A)$ are in $\mathcal{A}$.

The definition of a σ-algebra is similar to the definition of a topology (see Note 1). There is, however, an essential difference: the complement of an element of a σ-algebra belongs to the σ-algebra, but the the complement of an open set *is not* an open set.

Let $\mathcal{E}$ be a class of subsets of X; the set of all σ-algebras containing $\mathcal{E}$ is not empty since it contains, at least, $\mathcal{P}(X)$. The intersection of all these σ-algebras is a σ-algebra (see above), and, therefore, there exists a smallest σ-algebra containing $\mathcal{E}$. This σ-algebra is called the *σ-algebra generated by* $\mathcal{E}$; it is denoted $\mathcal{S}(\mathcal{E})$.

Theorem 1. *Let f be a mapping from X into Y, and let $\mathcal{B}$ be a σ-algebra in Y, then $\mathcal{A} = f^{-1}(\mathcal{B})$ is a σ-algebra in X.*

This result readily follows from the elementary formulas of set theory (see Note 2).

Definition 2. Let $\mathcal{A}$ be a σ-algebra in a space X, then the pair $(X, \mathcal{A})$ is called a *measurable space*, and the elements of $\mathcal{A}$ are called *measurable sets*.

Definition 3. Let $(X, \mathcal{A})$ and $(Y, \mathcal{B})$ be two measurable spaces; a mapping f from X into Y is said to be $\mathcal{A}$-$\mathcal{B}$ *measurable*[2] if, for any B in $\mathcal{B}$, $f^{-1}(B)$ belongs to $\mathcal{A}$.

In other words f is $\mathcal{A}$-$\mathcal{B}$ *measurable* if $f^{-1}(\mathcal{B}) \subset \mathcal{A}$.

Theorem 2. *Let f be a mapping from $(X, \mathcal{A})$ into $(Y, \mathcal{B})$, and let $\mathcal{E}$ be a class of subsets of Y generating the σ-algebra $\mathcal{B}$; then f is measurable if, and only if, $f^{-1}(\mathcal{E}) \subset \mathcal{A}$.*

Let us first prove the following lemma.

Lemma. *Let f be a mapping from X into Y, and let $\mathcal{E}$ be a class of subsets of Y; then $f^{-1}(\mathcal{S}(\mathcal{E})) = \mathcal{S}(f^{-1}(\mathcal{E}))$.*

---

[2] Or, simply, measurable if there is no ambiguity.

That is, the preimage of the σ-algebra generated by $\mathcal{E}$ is the σ-algebra generated by the preimage of $\mathcal{E}$.

From Theorem 1, it follows that $f^{-1}(\mathcal{S}(\mathcal{E}))$ is a σ-algebra. It contains $f^{-1}(\mathcal{E})$ and, therefore, $\mathcal{S}(f^{-1}(\mathcal{E}))$, which is the smallest σ-algebra containing $f^{-1}(\mathcal{E})$. Thus

$$f^{-1}(\mathcal{S}(\mathcal{E})) \supset \mathcal{S}(f^{-1}(\mathcal{E})).$$

But (see Note 2), the set

$$\{B | B \subset Y, f^{-1}(B) \in \mathcal{S}(f^{-1}(\mathcal{E}))\}$$

is a σ-algebra $\mathcal{S}'$ which contains $\mathcal{E}$ and, therefore, $\mathcal{S}(\mathcal{E})$. Thus

$$f^{-1}(\mathcal{S}(\mathcal{E})) \subset f^{-1}(\mathcal{S}') \subset \mathcal{S}(f^{-1}(\mathcal{E})),$$

which completes the proof.

From this lemma, it follows that

$$f^{-1}(\mathcal{B}) = f^{-1}(\mathcal{S}(\mathcal{E})) = \mathcal{S}(f^{-1}(\mathcal{E})).$$

Hence, $f^{-1}(\mathcal{B}) \subset \mathcal{A}$ if, and only if, $f^{-1}(\mathcal{E}) \subset \mathcal{A}$.

Corollary 1. *Let $(X, \mathcal{A})$, $(Y, \mathcal{B})$, $(Z, \mathcal{C})$ be three measurable spaces, and let f and g be, respectively, two measurable functions from $(X, \mathcal{A})$ into $(Y, \mathcal{B})$ and from $(Y, \mathcal{B})$ into $(Z, \mathcal{C})$; then $g \circ f$ is a measurable function from $(X, \mathcal{A})$ into $(Z, \mathcal{C})$.*

This is obvious.

Definition 4. Let $(X, \mathcal{A})$ and $(Y, \mathcal{B})$ be two measurable spaces; then $(X \times Y, \mathcal{A} \otimes \mathcal{B})$, where $\mathcal{A} \otimes \mathcal{B}$ is the σ-algebra generated by $\{A \times B | A \in \mathcal{A}, B \in \mathcal{B}\}$, is called the *product of the measurable spaces* $(X, \mathcal{A})$ and $(Y, \mathcal{B})$.

Corollary 2. *Let $(X, \mathcal{A})$, $(Y_1, \mathcal{B}_1)$ and $(Y_2, \mathcal{B}_2)$ be three measurable spaces, and let f be a mapping from X into $Y_1 \times Y_2$; f is measurable if, and only if, the mapping $p_1 \circ f$ and $p_2 \circ f$, where p_1 and p_2 are, respectively, the canonical projections of $Y_1 \times Y_2$ on Y_1 and Y_2, are measurable.*

For any ordered pair (y_1, y_2) in $Y_1 \times Y_2$, we have

$$p_1: (y_1, y_2) \mapsto y_1 \quad \text{and} \quad p_2: (y_1, y_2) \mapsto y_2.$$

Since $p_1(\mathcal{B}_1)$ and $p_2(\mathcal{B}_2)$ are both contained in $\mathcal{B}_1 \otimes \mathcal{B}_2$ (Definition 4), p_1 and p_2 are measurable. f being measurable, as a consequence of Corollary 1, $p_1 \circ f$ and $p_2 \circ f$ are, therefore, measurable.

Conversely, if $p_1 \circ f$ and $p_2 \circ f$ are measurable,

$$(\forall B_1 \in \mathcal{B}_1) \quad (\forall B_2 \in \mathcal{B}_2) \qquad f^{-1}(B_1 \times B_2) \in \mathcal{A}$$

since

$$f^{-1}(B_1 \times B_2) = (p_1 \circ f)^{-1}(B_1) \cap (p_2 \circ f)^{-1}(B_2);$$

and, therefore, f is measurable (Theorem 2).

Example 1. *Borel measurable mappings.* Let X be a topological space, and let $\mathcal{B}$ be the σ-algebra generated by the open sets of X. The elements of $\mathcal{B}$ are called the *Borel sets* of X. The closed sets of X are Borel sets, and so are all countable unions of closed sets and all countable intersections of open sets.

If $X = \mathbf{R}$, the σ-algebra $\mathcal{B}$ is generated by the set $\{]a, \infty[| a \in \mathbf{R}\}$. The intervals $[a, \infty[$, $]-\infty, a[$, $]a, b[$, and $[a, b]$ belong to $\mathcal{B}$ since

$$[a, \infty[= \cap_{n=1}^{\infty}]a - \frac{1}{n}, \infty[$$

$$]-\infty, a[= [a, \infty[^C$$

$$]a, b[=]-\infty, b[\cap]a, \infty[$$

$$[a, b] = [a, \infty[-]b, \infty[.$$

Any topological space can be regarded as a measure space with the Borel sets playing the role of the measurable sets. Let X and Y be two topological spaces; a measurable mapping $f: X \to Y$ is said to be *Borel measurable*. For instance, any continuous mapping $f: X \to Y$ is Borel measurable, since, for any open set Ω of Y, $f^{-1}(\Omega)$ is a Borel set of X (Theorem 2).

In what follows we shall deal, almost exclusively, with mappings from X into $\mathbf{R}$ (or $\mathbf{C}$), where X is any space. These particular mappings will be called *functions*, and to prove that they are measurable, we shall make use of the following criterion.

Criterion. *Let f be a function from X into $\mathbf{R}$; f is measurable if, and only if, for any real a, the set $\{x | f(x) > a\}$ is measurable.*

This result follows from Theorem 2 and from the fact that the σ-algebra of the Borel sets of $\mathbf{R}$ is generated by the set $\{]a, \infty[| a \in \mathbf{R}\}$.

Remark. Equivalent criteria could have been established by considering the sets $\{x | f(x) \geq a\}$, $\{x | f(x) < a\}$, or $\{x | f(x) \leq a\}$.

Example 2. If f is measurable, then the set $\{x | f(x) = a\}$ is a Borel set. If a is finite, this result follows from the fact that

$$\{x | f(x) = a\} = \{x | f(x) \geq a\} \cap \{x | f(x) \leq a\}.$$

If $a = \infty$, the result follows from

$$\{x|f(x) = \infty\} = \bigcap_{n=1}^{\infty} \{x|f(x) > n\}.$$

If $a = -\infty$, the proof is similar.

Example 3. The characteristic function 1_A of a set A is measurable if, and only if, the set A is measurable, since

$$\{x|1_A(x) > a\} = \begin{cases} \mathbf{R}, & \text{if } a < 0; \\ A, & \text{if } 0 < a < 1; \\ \emptyset, & \text{if } a > 1. \end{cases}$$

As a consequence of Corollary 2, a function f of X in $\mathbf{C}$ is measurable if, and only if, $\Re f$ and $\Im f$ are measurable.

Theorem 3. *Let f and g be two measurable functions from X into $\mathbf{R}$ (or $\mathbf{C}$).*

(1) *For any positive number α, $|f|^\alpha$ is measurable.*

(2) *If there is no $x \in X$ such that $f(x) = 0$, $1/f$ is measurable.*

(3) *$f + g$ and fg are measurable.*

(1) $|f|^\alpha$ is the composed mapping of f and $z \mapsto |z|^\alpha$. This last mapping is continuous and, therefore, Borel measurable. The result is then a consequence of Corollary 1.

(2) The proof is similar to the preceding one, since the mapping $z \mapsto 1/z$ from $\mathbf{R}-\{0\}$ (or from $\mathbf{C}-\{0\}$) into $\mathbf{R}$ is continuous.

(3) $f+g$ (resp. fg) is the composed mapping of $x \mapsto \big(f(x), g(x)\big)$ from X into $\mathbf{R}^2$ (or $\mathbf{C}^2$), which is measurable (Corollary 2), and of $(z_1, z_2) \mapsto z_1 + z_2$ (resp. $z_1 z_2$), which is Borel measurable (because continuous). The result is then a consequence of Corollary 1.

Example 4. Let f be a function whose range is $\{a_1, a_2, \ldots, a_n\}$, where a_i $(i = 1, 2, \ldots, n)$ are n distinct finite real numbers. If $A_i = \{x|f(x) = a_i\}$, then

$$f = \sum_{i=1}^{n} a_i 1_{A_i}.$$

If the sets A_i $(i = 1, 2, \ldots, n)$ are measurable, the characteristic functions 1_{A_i} $(i = 1, 2, \ldots, n)$ are measurable (see Example 2), and Theorem 3(3) shows that f is measurable. Finite linear combinations of characteristic functions of measurable sets will play a very important role in the construction of Lebesgue's theory (see Definition 4 and Theorem 5).

Theorem 4. *Let* (f_n) *be a sequence of measurable functions from* X *into* $\overline{\mathbf{R}}$*; then* $\sup f_n$, $\inf f_n$, $\limsup f_n$, *and* $\liminf f_n$ *are measurable functions.*

Since

$$\begin{aligned}\{x|\sup f_n(x) > a\} &= \cup_{n=1}^{\infty}\{x|f_n(x) > a\}\\ \{x|\inf f_n(x) < a\} &= \cup_{n=1}^{\infty}\{x|f_n(x) < a\}\\ \limsup f_n &= \inf_{n\geq 1}\sup_{k\geq n} f_k \qquad \text{(see Note 3)}\\ \liminf f_n &= \sup_{n\geq 1}\inf_{k\geq n} f_k \qquad \text{(see Note 3)},\end{aligned}$$

the results readily follow from the above criterion.

Remark. If f and g are two measurable functions from X into $\overline{\mathbf{R}}$, $\sup(f, g)$ and $\inf(f, g)$ are measurable. In particular, if f is measurable, $f^+ = \sup(f, 0)$ and $f^- = \inf(f, 0)$ are measurable. f^+ and f^- are respectively called the *positive* and the *negative part* of f. Note that

$$f = f^+ - f^- \qquad \text{and} \qquad |f| = f^+ + f^-.$$

Corollary. *Let* (f_n) *be a convergent sequence of measurable functions from* X *into* $\mathbf{R}$*; then* $\lim f_n$ *is measurable.*

This is an immediate consequence of the preceding theorem since, in this case, $\limsup f_n = \liminf f_n = \lim f_n$.

Definition 5. Let ϕ be a measurable function from $(X, \mathcal{A})$ into $\mathbf{R}$; ϕ is called a *simple function* if it takes only a finite number of different values.

Note that simple functions are finite-valued. If $A \subset \mathcal{A}$ is measurable, then the characteristic function 1_A is a simple function. More generally, a simple function is a finite linear combination of characteristic functions of measurable subsets of X (see Example 3). Since

$$1_A + 1_B = 1_{A\cup B} + 1_{A\cap B} \qquad \text{and} \qquad 1_A 1_B = 1_{a\cap B},$$

the sum and the product of two simple functions are simple functions.

Theorem 5. *Any measurable function is the limit of a sequence of simple functions.*

This theorem is very important. Since any simple function is, by definition, measurable, from the corollary to Theorem 4, it follows that the limit of a sequence of simple functions is also measurable. Conversely, if f is a measurable function, we now construct a convergent sequence of simple functions whose limit is f.

Let us distinguish three cases.

(1) Assume that f is positive and bounded. Then, for any x, $0 \leq f(x) \leq M$. Define

$$A_k = \{x | \frac{kM}{2^n} \leq f(x) < \frac{(k+1)M}{2^n}\},$$

where n is an integer and $k = 0, 1, 2, \ldots, 2^{n-1}$, and put

$$\phi_n = \sum_{k=0}^{2^{n-1}} \frac{kM}{2^n} 1_{A_k}.$$

Since A_k is the preimage of an interval, it is measurable. ϕ_n is, therefore, a simple function, and we have:

$$0 \leq \phi_1 \leq \phi_2 \leq \cdots \leq f$$

and, for any x,

$$\lim_{n \to \infty} \phi_n(x) = f(x).$$

Note that the convergence is uniform.

(2) Assume that f is positive but not necessarily bounded. Define $g_n = \inf(f, n)$, where n is a positive integer. g_n is positive and bounded, and there exists a simple function ψ_n such that $\sup |\psi_n - g_n| < 1/n$. It is easily seen that the sequence (ψ_n) converges to f, but we do not necessarily have $\psi_1 \leq \psi_2 \leq \cdots$. If, however, we write

$$\phi_n = \sup(\psi_1, \psi_2, \ldots, h_n),$$

then the sequence (ϕ_n) not only converges to f but we have $\phi_1 \leq \phi_2 \leq \cdots$.

(3) If f is not positive but real-valued, we have $f = f^+ - f^-$, where f^+ and f^- are positive functions, and the preceding results show that a convergent sequence can be constructed. If f is complex-valued, then $f = \Re f + i\Im f$, where $\Re f$ and $\Im f$ are real functions, and here again we reach the same conclusion.

2. Positive Measures

Apart from its utmost importance in Lebesgue's theory of integration, the material of this section has many uses in different branches of theoretical physics and probability.

Definition 6. Let $(X, \mathcal{A})$ be a measurable space; a *positive measure* on $(X, \mathcal{A})$ is a mapping $\mu: \mathcal{A} \to \overline{\mathbf{R}}_+$ such that
(1) $\mu(\emptyset) = 0$.
(2) If (A_n) is a countable collection of pairwise disjoint elements of $\mathcal{A}$, then

$$\mu\left(\bigcup_n A_n\right) = \sum_n \mu(A_n).$$

The triple $(X, \mathcal{A}, \mu)$ is called a *measure space.*

A positive measure is, therefore, a *countably* additive function defined on a σ-algebra into $[0, \infty]$. This last property, which is also called *σ-additivity*, is crucial.

Example 1. *Counting measure.* For any $A \subset X$ in $\mathcal{A}$, let $\mu(A)$ be the number of points in A if A is finite, and put $\mu(A) = \infty$ if A is infinite. This μ is called the *counting measure* on $(X, \mathcal{A})$.

Example 2. *Dirac measures .* Let $(X, \mathcal{A})$ be a measurable space, and let a be a fixed point in X; define $\mu(A) = 1$ if $a \in A$ and $\mu(A) = 0$ if $a \notin A$. This measure, usually denoted δ_a, is a *Dirac measure*. It corresponds to a *unit mass* concentrated at a.

Example 3. *Lebesgue measure.* To define the Lebesgue integral of a function from $\mathbf{R}$ into $\mathbf{R}$, it is necessary to generalize the geometric concept of length which is defined for intervals in $\mathbf{R}$. A satisfactory generalization should apply to as many subsets of $\mathbf{R}$ as possible, and fulfill certain "natural" requirements, as additivity and invariance under translation. The *Lebesgue measure* is such a generalization.

Let A be a subset of $\mathbf{R}$; the *outer measure* of A is

$$m^*(A) = \inf \sum_{j \in J} \ell(I_j),$$

where the infimum is taken over all *finite* or *countable* collections of open intervals $\{I_j | j \in J\}$ such that $A \subseteq \cup_{j \in J} I_j$. $\ell(I_j)$ denotes the *length* of the interval I_j. It is readily seen that:
(1) $m^*(\emptyset) = 0$,
(2) $m^*(A) \geq 0$,
(3) $A \subseteq B \Rightarrow m^*(A) \leq m^*(B)$.

Since, for any $x \in \mathbf{R}$, $x \in I_n =]x - 1/2n, x + 1/2n[$, and, for any positive integer n, $\ell(I_n) = 1/n$, it follows that
(4) $m^*(\{x\}) = 0$.

Note that m^* is defined for *all* subsets of $\mathbf{R}$.

The *inner measure* of A is

$$m_*(A) = \sup_{K \subset A} m^*(K),$$

where the supremum is taken over all closed subsets of A. For any $A \in \mathbf{R}$, we have $m_*(A) \leq m^*(A)$.

A subset A of $\mathbf{R}$ is said to be *Lebesgue measurable* if

$$m_*(A) = m^*(A) < \infty.$$

This finite number is called the *Lebesgue measure* of A. It is denoted by $m(A)$.

Any open set of $\mathbf{R}$, being, by definition, an union of open intervals, is Lebesgue measurable. Any bounded interval is Lebesgue measurable, and, if $b > a$, we have:

$$m(]a,b[) = m([a,b[) = m(]a,b]) = m([a,b]) = b - a.$$

All Borel sets are Lebesgue measurable, but all Lebesgue measurable sets are not Borel sets.

Since the outer measure is translation invariant, so is the Lebesgue measure. That is,

$$m(A + a) = m(A).$$

Not all subsets of $\mathbf{R}$ are Lebesgue measurable. Here is a classical example. Define in $[0,1]$ the following equivalence relation:

$$x \in [0,1] \quad y \in [0,1] \qquad xRy \iff x - y \in \mathbf{Q},$$

where $\mathbf{Q}$ is the set of rational numbers. The interval $[0,1]$ is the union of the pairwise disjoint equivalence classes. Each class can be represented by one of its elements.[3] Let V be the set of all these elements. We shall prove that V is not Lebesgue measurable.

Let $\{r_n | n \in \mathbf{N}\}$ be the set of rational numbers belonging to the interval $[-1,1]$, that is $[-1,1] \cap \mathbf{Q}$, and denote by V_n the set $V + r_n$; we have
(1) $m^*(V_n) = m^*(V)$ (invariance by translation of m^*),

---

[3] This is a consequence of the *axiom of choice*, which says that if $(A_i)_{i \in I}$ is a family of nonempty subsets of a set E, then there exists a mapping f from I into E such that, for any index i in I, we have $f(i) = A_i$.

(2) $[0,1] \subset \cup_{n=1}^{\infty} V_n \subset [-1,2]$,
(3) if $n \neq m$, then $V_n \cap V_m = \emptyset$.

If V were Lebesgue mesurable, so would be all the sets V_n ($n \in \mathbf{N}$), and we would have:

$$m(\bigcup_{n=1}^{\infty} V_n) = \sum_{n=1}^{\infty} m(V_n)$$

and

$$1 \leq \sum_{n=1}^{\infty} m(V_n) \leq 3.$$

That is,

$$1 \leq \sum_{n=1}^{\infty} m(V) \leq 3.$$

But the above sum can only be null or infinite. So V is not Lebesgue measurable.

The Lebesgue measure can be extended to subsets of $\mathbf{R}^k$. In the definition of the outer measure, open intervals of $\mathbf{R}$ have to be replaced by open intervals (or boxes) in $\mathbf{R}^k$. A point $x = (x_1, x_2, \ldots, x_k)$ of $\mathbf{R}^k$ belongs to the open interval $]a,b[$, where $a = (a_1, a_2, \ldots, a_k)$ and $b = (b_1, b_2, \ldots, b_k)$, if, for $i = 1, 2, \ldots, k$, $a_i < x_i < b_i$.

Example 4. *Stieltjes measure.* Let f be a real nondecreasing continuous from the left function[4] defined in $\mathbf{R}$. The measure μ, defined on the Borel sets of $\mathbf{R}$, such that

$$\begin{aligned}
\mu(]a,b[) &= f(b) - f(a+0) \\
\mu([a,b[) &= f(b) - f(a) \\
\mu(]a,b]) &= f(b+0) - f(a+0) \\
\mu([a,b]) &= f(b+0) - f(a)
\end{aligned}$$

is called the *Stieljes measure generated by* f. The Lebesgue measure is the Stieljes measure generated by $f\colon x \mapsto x$. The Dirac measure δ_a is the Stieljes measure generated by the function f such that

$$f(x) = \begin{cases} 0, & \text{if } x \leq a; \\ 1, & \text{otherwise.} \end{cases}$$

[4] A function f is continuous from the left at x_0 if $f(x_0 - 0) = f(x_0)$.

If f is discontinuous at x, then the measure of $\{x\}$ is nonzero. It is equal to $f(x+0)-f(x)$. If f is constant on $[a,b[$, then the measure of $[a,b[$ is equal to 0.

Let μ be a measure on X. If $\mu(X)$ is finite, then μ is said to be *bounded.* If X is the countable union of sets of finite measure, then μ is said to be *σ-finite.* The Stieljes measure on $\mathbf{R}$ associated to a bounded function is bounded. The Lebesgue measure on $\mathbf{R}$ is σ-finite.

Example 5. *Probability theory.* The axiomatic formulation of probability theory is based on measure theory. A measure space $(\Omega, \mathcal{E}, P)$ such that $P(\Omega) = 1$ is a probabilistic model. It is called a *probability space.* The elements of the σ-algebra $\mathcal{E}$ are called *events*, and the bounded measure P is a *probability measure*. The elements of Ω represent the possible outcomes of a random experiment, and, if $A \in \mathcal{E}$, $P(A)$ measures how likely the event A is to occur. The events $\emptyset$ and Ω are, respectively, said to be *impossible* $(P(\emptyset) = 0)$ and *certain* $(P(\Omega) = 1)$. If A and B are two events, the event $A \cup B$ consists of the realization of *either A or B*, and the event $A \cap B$ consists of the realization of *both A and B*. It is clear that

$$P(A \cup B) = P(A) + P(B) - P(A \cap B).$$

If $A \in \mathcal{E}$ such that $P(A) > 0$, the measure defined by

$$(\forall B \in \mathcal{E}) \qquad P(B|A) = \frac{P(A \cap B)}{P(A)}$$

is a probability measure called the *conditional probability.*[5] Since $P(A|A) = 1$, and $P(B|A) = 0$ if $A \cap B = \emptyset$, $P(B|A)$ measures how likely B is to occur knowing that A has occurred. $P(B)$ is the *a priori* probability of B, whereas $P(B|A)$ is the *a posteriori* probability of B after the occurrence of A. If $P(A \cap B) = P(A)P(B)$, that is, if $P(B|A) = P(B)$, A and B are said to be *independent.*

If (H_n) is a sequence of disjoint elements of $\mathcal{E}$ whose union is Ω, then

$$B = \bigcup_{n}^{\infty}(H_n \cap B)$$

[5] This definition is correct only if the set $\{A \cap B | B \in \mathcal{E}\}$ is a σ-algebra. It is easy to verify that this is indeed the case.

and

$$P(B) = \sum_{n=1}^{\infty} P(B|H_n)P(H_n).$$

Hence, for all $i \in \mathbf{N}$,

$$P(H_i|B) = \frac{P(H_i)P(B|H_i)}{\sum_{n=1}^{\infty} P(B|H_n)P(H_n)}.$$

This result is known as *Bayes's formula.* In statistical applications the events H_n $(n = 1, 2, \ldots)$ are called *hypotheses.*

Theorem 6. *Let $(X, \mathcal{A}, \mu)$ be a measure space.*
(1) μ is an increasing function on $\mathcal{A}$, that is, if $A \in \mathcal{A}$ and $B \in \mathcal{A}$, then

$$A \subset B \Rightarrow \mu(A) \leq \mu(B).$$

(2) Let (A_n) be an increasing sequence of measurable sets, that is, for any positive integer n, $A_n \subset A_{n+1}$, then

$$A = \bigcup_{n=1}^{\infty} A_n \Rightarrow \lim_{n\to\infty} \mu(A_n) = \mu(A).$$

(3) Let (A_n) be an decreasing sequence of measurable sets, that is, for any positive integer n, $A_n \supset A_{n+1}$, then

$$A = \bigcap_{n=1}^{\infty} A_n \Rightarrow \lim_{n\to\infty} \mu(A_n) = \mu(A),$$

provided $\mu(A_1) < \infty$.

(1) If $A \subset B$, then $B = A + (B - A)$ with $A \cap (B - A) = \emptyset$. Hence,

$$\mu(B) = \mu(A) + \mu(B - A) \geq \mu(A).$$

(2) Write $B_1 = A_1$ and $B_n = A_n - A_{n-1}$ for $n = 2, 3, \ldots$. Then $B_n \in \mathcal{A}$, $B_i \cap B_j = \emptyset$ if $i \neq j$, and $A_n = B_1 \cup B_2 \cup \cdots \cup B_n$. Hence,

$$\begin{aligned}\lim_{n\to\infty} \mu(A_n) &= \sum_{k=1}^{\infty} \mu(B_k)\\ &= \mu\Big(\bigcup_{k=1}^{\infty} B_k\Big)\\ &= \mu(A).\end{aligned}$$

(3) Write $B_n = A_1 - A_n$ for $n = 2, 3, \ldots$. Then, for all positive integers n, $B_n \in \mathcal{A}$ and $B_n \subset B_{n+1}$. Hence, from (2), it follows that

$$\begin{aligned}\mu(A_1 - A) &= \mu(A_1) - \mu(A)\\ &= \mu\Big(\bigcup_{n=1}^{\infty} B_n\Big)\\ &= \lim_{n\to\infty} \mu(B_n)\\ &= \mu(A_1) - \lim_{n\to\infty} \mu(A_n),\end{aligned}$$

which, taking into account the condition $\mu(A_1) < \infty$, completes the proof. The condition on the measure of A_1 could be replaced by a similar condition on, say, A_{n_0}. Then, we would have considered the sets $B_n = A_{n_0} - A_n$ for $n > n_0$.

Definition 7. Let $(X, \mathcal{A}, \mu)$ be a measure space; a subset N of X is said to be *μ-negligible* if there exists a set $A \in \mathcal{A}$ such that $N \subset A$ and $\mu(A) = 0$. If a property holds except on a negligible set, it is said to hold *almost everywhere*, usually abbreviated to a.e.

Note that N is not necessarily a measurable set. If all negligible sets are measurable, the measure μ is said to be *complete*. The Lebesgue measure is complete since any negligible subset N has an outer measure equal to zero, and, therefore, also an inner measure equal to zero.[6]

Example 6. The Lebesgue measure of a countable subset of $\mathbf{R}$ is equal to zero. To establish this result, it is sufficient to show that the outer measure is equal to zero since, for any subset A of $\mathbf{R}$, we have $0 \le m_*(A) \le m^*(A)$, so $m_*(A) = 0$ implies $m_*(A) = m^*(A) = m(A) = 0$. Let $\{x_n | x_n \in \mathbf{R}, n \in \mathbf{N}\}$ be the countable subset, and let ϵ be a positive real. The collection of intervals $\{I_n | n \in \mathbf{N}\}$, where

$$I_n =]x_n - \frac{\epsilon}{2^{n+1}}, x_n + \frac{\epsilon}{2^{n+1}}[,$$

is a covering of the subset. Since

$$\sum_{n=1}^{\infty} \ell(I_n) = \epsilon,$$

[6] This results from the following relations:

$$N \subset A \Rightarrow m^*(N) \le m^*(A) = 0 \quad \text{and} \quad m_*(N) \le m^*(N).$$

the result follows.

The fact that the Lebesgue measure of a countable subset of **R** is equal zero implies that
(1) any measurable subset of **R** whose Lebesgue measure is nonzero is not countable; but a noncountable subset can have a Lebesgue measure equal to zero (see the next example);
(2) any nonmeasurable subset of **R** (see Example 3) is not countable.

Example 7. *Cantor sets.* Few years ago, any physicist would probably have considered Cantor sets as rather exotic mathematical objects just useful to construct counter-examples in mathematics, but devoid of interest for his discipline. Today, many references to Cantor-like sets, called *fractals*, can be found when studying, for example, phase transitions in dilute magnetic systems or far-from-equilibrium growth phenomena.[7] Let (J_n) be the sequence of closed subsets of $[0,1]$ defined by

$$\begin{aligned} J_0 &= [0,1] \\ J_1 &= [0,\frac{1}{3}] \cup [\frac{2}{3},1] \\ J_2 &= [0,\frac{1}{9}] \cup [\frac{2}{9},\frac{1}{3}] \cup [\frac{2}{9},\frac{7}{9}] \cup [\frac{8}{9},1] \\ &\dots \end{aligned}$$

J_n is the union of 2^n pairwise disjoint closed intervals of length 3^{-n}. It is obtained from J_{n-1} by removing from each closed interval of length $3^{-(n-1)}$ an open interval of length 3^{-n} with the same center. The set

$$C = \bigcap_{n=1}^{\infty} J_n$$

is called the *Cantor ternary* or *triadic set.*[8]

C is closed since it is the intersection of closed sets. It contains no interval,[9] so its interior (see Note 1) is empty, that is, all its points are boundary points (see Note 1).

---

[7] For an attractive introductory text on fractals, presenting numerous specific applications in various fields, see B. B. Mandelbrot, *The Fractal Geometry of Nature*, W. H. Freeman, San Francisco, 1983.

[8] Note that, for all $n \in \mathbf{N}$, the end points of the closed intervals, whose union at stage n is J_n, belong to C.

[9] At stage n there is no interval of length greater than 2^{-n}.

Since at each stage of the construction of C, the open intervals which constitute the "middle thirds" of the closed intervals left at the previous stage are removed, any element x of the Cantor ternary set can be written

$$x = \sum_{n=1}^{\infty} \frac{x_n}{3^n},$$

where, for all positive integer n, $x_n = 0$ or 2. In other words, the ternary expansion (that is, the expansion to the base 3), of an element of C does not contain the digit 1.

One might think that this statement is not correct since, for example, $7/9 = 2/3 + 1/3^2$ belongs to C, and its ternary expansion (which is 0.21) contains the digit 1. There is indeed an infinite number of elements of C that have this property. The expansion of all of them contains only a finite number of terms, the last one being $1/3n_0$. If this term is replaced by the series

$$\sum_{n=n_0+1}^{\infty} \frac{2}{3^n},$$

which is equal to $1/3n_0$, the statement becomes correct. We shall always respect this convention, and write, for example, the ternary expansion of 7/9 as 0.2022.... Note that, with this convention, each point of C has a unique expansion.

The Cantor set is not countable. If C were countable, there would exist a bijection f from $\mathbf{N}$ onto C. Let $0.x_1^{(n)}x_2^{(n)}\ldots$ be the ternary expansion of $f(n)$ $(n \in \mathbf{N})$, and consider the sequence (x_n) such that

$$x_n = \begin{cases} 0, & \text{if } x_n^{(n)} = 2; \\ 2, & \text{if } x_n^{(n)} = 0; \end{cases}$$

it is clear that $x = 0.x_1x_2\ldots$ is in C but not in the set $\{f(n) | n \in \mathbf{N}\}$. Hence, there is no such bijection f.[10]

The Cantor set is indeed equipotent to $[0,1]$. To prove it, we can exhibit a bijection from [0,1] onto C. Write any $x \in [0,1]$ under the form:

$$x = \sum_{n=1}^{\infty} \frac{x_n}{2^n},$$

[10] This method of proof, called *Cantor's diagonal process*, is frequently used to demonstrate that a set is not countable.

where $x_n = 0$ or 1, and define the function ϕ by

$$\phi(x) = \sum_{n=1}^{\infty} \frac{2x_n}{3^n}.$$

ϕ is known as the *Cantor function.* Since, for all positive integers n, $2x_n = 0$ or 2, the range of ϕ lies in C, and it is readily verified that it is a bijection of $[0,1]$ onto its range. To complete the proof, it remains to show that the range of ϕ coincides with C. Let y be in C, and let its ternary expansion be $0.y_1y_2\ldots$. With the above convention, for all $n \in \mathbf{N}$, $y_n = 0$ or 2, hence $x = \phi^{-1}(y)$ exists and is unique. It is determined by its binary expansion $x = 0.x_1x_2x_3\ldots$, where $x_n = y_n/2$.

Since J_n is obtained from J_{n-1} by removing 2^n open intervals of length 3^{n+1}, we have

$$m(C) = 1 - \sum_{n=0}^{\infty} \frac{2^n}{3^{n+1}} = 0.$$

That is, *the Lebesgue measure of the Cantor set is zero.*[11]

More general Cantor sets can be defined. Let b be a positive integer; the set of the numbers whose expansion to the base b does not contain the digits $k_1, k_2, \ldots, k_\lambda$, where for $i = 1, 2, \ldots, \lambda$, $0 < k_i < b-1$ is a Cantor-like set. It is easy to show that this set is not countable, and has zero Lebesgue measure.

It is also possible to define a Cantor-like set whose Lebesgue measure is not equal to zero. Let (u_n) be a sequence of positive real numbers such that $\sum_{n=1}^{\infty} u_n = s < 1$, and let us modify the construction of the Cantor ternary set as follows: At stage 1 remove from the closed interval $[0,1]$ an open interval of length u_1 centered at $1/2$ to obtain two closed intervals. At stage 2 remove from the center of each of these two closed intervals an open interval of length $u_2/2$ to obtain 2^2 closed intervals .... At stage n remove from the center of each of the 2^{n-1} closed intervals obtained at stage $n-1$ an open interval of length $u_n/2^{n-1}$ to obtain 2^n closed intervals. The resulting closed set, obtained when $n \to \infty$, has a Lebesgue measure equal to $1 - s > 0$. Note that this set has an empty interior, but its boundary has a nonzero Lebesgue measure!

---

[11] (J_n) is a decreasing sequence of Lebesgue measurable sets, and the result also follows from Theorem 6(3).

Example 8. *Lebesgue function.* This is also a fairly exotic object recently rediscovered by physicists, and called the *Devil's staircase.* It is probably best viewed as the limit of the sequence of functions (L_n) defined as follows. For each positive integer n, let L_n be the nondecreasing continuous function on $[0, 1]$ such that $L_n(0) = 0$, $L_n(1) = 1$, and that is linear and increasing by 2^{-n} on each closed interval whose union is J_n, and constant on the removed middle third intervals, where J_n and the removed intervals refer to the inductive construction of the Cantor ternary set. If $n > m$, then, for any $x \in [0, 1]$, $|L_n(x) - L_m(x)| < 2^{-m}$, that is, for all x, the sequence $\big(L_n(x)\big)$ is a Cauchy sequence. $L(x) = \lim_{n\to\infty} L_n(x)$ is, therefore, well defined for all $x \in [0, 1]$, and from $|L(x) - L_m(x)| < 2^{-m}$ it follows that the convergence is uniform. L, which is the *Lebesgue function*, is, consequently, a nondecreasing *continuous* function on $[0, 1]$ such that $L(0) = 0$ and $L(1) = 1$. Since L is constant on each removed middle third interval, it is *constant almost everywhere.* If

$$x = \sum_{n=1}^{\infty} \frac{x_n}{3^n},$$

it can be shown that

$$L(x) = \sum_{n=1}^{m-1} \frac{x_n}{2^{n+1}} + \frac{1}{2^m},$$

where m is the smallest index n such that $x_n = 1$. If x belongs to the Cantor ternary set, the ternary expansion of x does not contain the digit 1, then m is taken equal to ∞. Note that the restriction of L to the Cantor ternary set, that is $L: C \to [0, 1]$, coincides with ϕ^{-1}, where ϕ is the Cantor function used in the previous example to show that C was equipotent to $[0, 1]$. The Lebesgue function is also frequently referred to as the Cantor function.

Example 9. *Hausdorff dimension.* This notion will help making a distinction between sets of zero Lebesgue measure. Let A be a subset of $\mathbf{R}$; the *Hausdorff outer measure* of A is

$$H_d^*(A) = \lim_{\epsilon\to 0} \inf \sum_{j\in J} \big(\ell(I_j)\big)^d, \qquad (d > 0)$$

where the infimum is taken over all *finite* or *countable* coverings of A by open intervals whose length is less than ϵ. The limit exists[12] since the infimum can only increase as ϵ decreases.

[12] It might be infinite.

The Hausdorff outer measure is a generalization of the Lebesgue outer measure m^*, which corresponds to $d = 1$.

(1) If $H^*_{d_1}(A) < \infty$, then, for $d_2 > d_1$, $H^*_{d_2}(A) = 0$.
Indeed, the condition $\ell(I_j) < \epsilon$, for all $j \in J$, implies

$$\frac{(\ell(I_j))^{d_2}}{(\ell(I_j))^{d_1}} = (\ell(I_j))^{d_2-d_1} \leq \epsilon^{d_2-d_1}.$$

Therefore,

$$\begin{aligned}\inf \sum_{j\in J} (\ell(I_j))^{d_2} &\leq \sum_{j \in J} (\ell(I_j))^{d_2} \\ &\leq \epsilon^{d_2-d_1} \sum_{j\in J} (\ell(I_j))^{d_1}.\end{aligned}$$

Letting $\epsilon \to 0$, the result follows.

(2) If $0 < H^*_{d_1}(A) < \infty$, then, for $d_2 < d_1$, $H^*_{d_2}(A) = \infty$.
This is a corollary to the preceding result.

If I is an interval such that $\ell(I) > 0$, then

$$H^*_d(I) = \begin{cases} 0, & \text{if } d > 1; \\ \infty, & \text{if } d < 1. \end{cases}$$

Let A be a subset of $\mathbf{R}$; the number $\inf\{d | H^*_d(A) = 0\}$ is called the *Hausdorff dimension* of the subset A.

Let us determine the Hausdorff dimension of the Cantor ternary set.

Note first that C is *self-similar*, that is,

$$3C \cap [0, \frac{1}{3}] = C \quad \text{and} \quad 3C \cap [\frac{2}{3}, 1] - 2 = C,$$

where, for a given real k, kA denotes the set $\{kx | x \in A \subset \mathbf{R}\}$. But

$$H^*_d(C) = H^*_d(C \cap [0, \frac{1}{3}]) + H^*_d(C \cap [\frac{2}{3}, 1]),$$

and

$$\begin{aligned}H^*_d(C \cap [0, \frac{1}{3}]) &= H^*_d(C \cap [\frac{2}{3}, 1]) \\ &= \frac{1}{3^d} H^*_d(C),\end{aligned}$$

since, if $A = kB$, $H^*_d(A) = k^d H^*_d(B)$. Finally,

$$H^*_d(C) = \frac{2}{3^d} H^*_d(C).$$

This relation shows that $H_d^*(C)$ can take a finite nonzero value if, and only if,

$$\frac{2}{3^d} = 1.$$

The Hausdorff dimension of C is, therefore, equal to $\log 2/\log 3$.

In Example 7 we considered more general Cantor sets of zero Lebesgue measure. Following the same method, it is easy to find that their Hausdorff dimension is equal to $\log(b-\lambda)/\log b$.

The Hausdorff outer measure and Hausdorff dimension of subsets of $\mathbf{R}^k$ can be defined in a similar way.

3. Integral of Measurable Functions

This section deals with the integral of measurable functions with respect to a measure. The particular case of the integral with respect to the Lebesgue measure is not, in any way, simpler than the general case, which will give us a tool of much wider applicability.

The definition of the general integral to be given below is closely modelled on Cauchy's definition of the integral (see Note 4).

In what follows, all functions are measurable and defined on the measure space $(X, \mathcal{A}, \mu)$.

Definition 8. Let $\phi = \sum_{i=1}^n \alpha_i 1_{A_i}$ be a positive simple function; then

$$\int \phi \, d\mu = \sum_{i=1}^n \alpha_i \mu(A_i)$$

is called the *integral of ϕ with respect to the measure μ.*

The integral of ϕ with respect to μ is finite if, and only if, the measure of the set $\{x \,|\, \phi(x) \neq 0\}$ is finite.

The integral of a simple function has the following properties:

Theorem 7. *Let ϕ_1 and ϕ_2 be two positive simple functions, and let λ be a positive real; then*

$$\int (\phi_1 + \phi_2) \, d\mu = \int \phi_1 \, d\mu + \int \phi_2 \, d\mu, \tag{1}$$

$$\int \lambda \phi_1 \, d\mu = \lambda \int \phi_1 \, d\mu, \tag{2}$$

$$\int \phi_1 \, d\mu \leq \int \phi_2 \, d\mu \qquad \text{if } \phi_1 \leq \phi_2. \tag{3}$$

(1) Let

$$\phi_1 = \sum_{i=1}^{n} \alpha_i 1_{A_i} \qquad \text{and} \qquad \phi_2 = \sum_{j=1}^{m} \beta_j 1_{B_j}.$$

Then

$$\phi_1 + \phi_2 = \sum_{i=1}^{n}\sum_{j=1}^{m} (\alpha_i + \beta_j) 1_{A_i \cap B_j},$$

and

$$\begin{aligned}\int (\phi_1 + \phi_2)\, d\mu &= \sum_{i=1}^{n}\sum_{j=1}^{m} (\alpha_i + \beta_j)\mu(A_i \cap B_j) \\ &= \sum_{i=1}^{n}\sum_{j=1}^{m} \alpha_i \mu(A_i \cap B_j) + \sum_{i=1}^{n}\sum_{j=1}^{m} \beta_j \mu(A_i \cap B_j) \\ &= \sum_{i=1}^{n} \alpha_i \mu(A_i) + \sum_{j=1}^{m} \beta_j \mu(B_j) \\ &= \int \phi_1\, d\mu + \int \phi_2\, d\mu.\end{aligned}$$

(2) and (3) are obvious.

Definition 9. Let f be a positive measurable function; then

$$\int f\, d\mu = \sup \int \phi\, d\mu,$$

where the supremum is taken over all positive simple functions ϕ such that $\phi \leq f$, is called the *integral of f with respect to the measure μ*. f is said to be *μ-integrable* if its integral is finite.

Notations. The integral of f is also denoted by $\int f(x)\, d\mu(x)$. If $A \subset X$ is a measurable set, the integral of f over A, that is $\int f 1_A\, d\mu$, is denoted by

$$\int_A f\, d\mu \qquad \text{or} \qquad \int_A f(x)\, d\mu(x).$$

f is integrable only if the set $\{x | f(x) = \infty\}$ is negligible; this necessary condition is, however, not sufficient.

Theorem 8. Monotone convergence theorem. *Let (f_n) be a sequence of positive measurable functions such that*

$$(\forall n \in \mathbf{N}) \quad (\forall x \in X) \qquad f_n(x) < f_{n+1}(x)$$

and

$$(\forall x \in X) \qquad \lim_{n\to\infty} f_n(x) = f(x).$$

Then

$$\lim_{n\to\infty} \int f_n \, d\mu = \int f \, d\mu.$$

This theorem is also called the Beppo Levi theorem. To prove it we need the following lemma.

Lemma. *Let ϕ be a positive simple function, and let (B_n) be an increasing sequence of measurable sets converging to X; then*

$$\lim_{n\to\infty} \int_{B_n} \phi \, d\mu = \int \phi \, d\mu.$$

Let

$$\phi = \sum_i \alpha_i 1_{A_i};$$

then

$$\int_{B_n} \phi \, d\mu = \int f 1_{B_n} \, d\mu$$
$$= \sum_i \alpha_i \mu(A_i \cap B_n).$$

The sequence $(A_i \cap B_n)_{n\geq 1}$ is increasing and converges to A_i; from Theorem 6(2) it, therefore, follows that

$$\lim_{n\to\infty} \mu(A_i \cap B_n) = \mu(A_i),$$

which completes the proof of the lemma.

We can now establish the monotone convergence theorem. The relation $f_n \leq f$ implies that $\int f_n \, d\mu \leq \int f \, d\mu$, that is,

$$\lim_{n\to\infty} \int f_n \, d\mu \leq \int f \, d\mu.$$

Consider now a positive simple function ϕ such that $\phi \leq f$, and let λ be a real number between 0 and 1. If

$$B_n = \{x | f_n(x) \geq \lambda\phi(x)\},$$

then

$$f_n \geq \lambda \phi 1_{B_n},$$

and we have

$$\int f_n \, d\mu \geq \int_{B_n} \lambda \, \phi \, d\mu.$$

The sequence (B_n) satisfies the conditions of the lemma, hence,

$$\lim_{n\to\infty} \int f_n \, d\mu \geq \lim_{n\to\infty} \int_{B_n} \lambda \, \phi \, d\mu = \lambda \int \phi \, d\mu.$$

Since λ is arbitrary, it follows that

$$\lim_{n\to\infty} \int f_n \, d\mu \geq \int \phi \, d\mu.$$

This last relation being valid, in particular, for any positive simple function ϕ such that $\phi \leq f$, we have

$$\lim_{n\to\infty} \int f_n \, d\mu \geq \int f \, d\mu,$$

and the theorem follows.

Since any positive measurable function is the limit of a sequence of positive simple functions (see Theorem 5), it follows from the monotone convergence theorem that Theorem 7 remains valid if we replace "positive simple functions" by "positive measurable functions."

Theorem 9. *Let f be a positive measurable function; its integral is equal to zero if, and only if, $f(x) = 0$ almost everywhere.*

If $f(x) = 0$ a.e., for any positive simple function ϕ such that $\phi \leq f$, we have $\int \phi \, d\mu = 0$, hence $\int f \, d\mu = 0$. The condition is, therefore, sufficient. To prove that it is necessary, we need the following lemma.

Lemma. *Let f be a positive measurable function, and let α be a positive real; then*

$$\mu(\{x | f(x) \geq \alpha\}) \leq \frac{1}{\alpha} \int f \, d\mu.$$

If $A = \{x | f(x) \geq \alpha\}$, we have $f \geq \alpha \, 1_A$, and

$$\alpha \mu(A) \leq \int_A f \, d\mu \leq \int f \, d\mu,$$

which proves the lemma.

From this lemma it follows that if $\int f \, d\mu = 0$, then, for any positive integer n,

$$\mu(\{x | f(x) \geq \frac{1}{n}\}) = 0.$$

Since

$$\{x|f(x) > 0\} = \bigcup_n^\infty \{x|f(x) \geq \frac{1}{n}\},$$

the set $\{x|f(x) > 0\}$ is negligible, and the condition is, therefore, necessary.

Corollary. *Let f and g be two positive measurable functions such that $f = g$ a.e.; then $\int f\,d\mu = \int g\,d\mu$.*

The functions f_1 and g_1 defined by

$$f_1(x) = \begin{cases} f(x) - \inf\big(f(x), g(x)\big), & \text{if } \inf\big(f(x), g(x)\big) < \infty, \\ 0, & \text{if } \inf\big(f(x), g(x)\big) = \infty, \end{cases}$$

and

$$g_1(x) = \begin{cases} g(x) - \inf\big(f(x), g(x)\big), & \text{if } \inf\big(f(x), g(x)\big) < \infty, \\ 0, & \text{if } \inf\big(f(x), g(x)\big) = \infty, \end{cases}$$

are measurable (lemma of Theorem 3(2)), nonnegative, and equal to zero almost everywhere. From the preceding theorem, it follows that $\int f_1\,d\mu = 0$ and $\int g_1\,d\mu = 0$, so $\int f\,d\mu = \int g\,d\mu$.

Definition 10. (1) A real-valued function f is said to be *μ-integrable* (or *integrable with respect to the measure μ*) if f^+ and f^- are μ-integrable, and the *integral* of f is

$$\int f\,d\mu = \int f^+\,d\mu - \int f^-\,d\mu.$$

(2) A complex-valued function f is said to be μ-integrable (or integrable with respect to the measure μ) if $\Re f$ and $\Im f$ are μ-integrable, and the integral of f is

$$\int f\,d\mu = \int \Re f\,d\mu + i \int \Im f\,d\mu.$$

If $f = g$ a.e., then $\int f\,d\mu = \int g\,d\mu$ (corollary to Theorem 9).

The relation $f = g$ a.e. is an equivalence relation on the set of integrable functions (with respect to a specified measure). As far as integration is concerned, functions belonging to the same class are indistinguishable.

Remark 1. If the range of an integrable function f is $\overline{\mathbf{R}}$, the sets $\{x|f(x) = \infty\}$ and $\{x|f(x) = -\infty\}$ are necessarily negligible. It is, therefore, unnecessary to distinguish functions whose range is $\mathbf{R}$ from functions whose range is $\overline{\mathbf{R}}$.

Remark 2. The sum of a convergent series is nothing else than an integral with respect to the counting measure of a function defined on the sets **N** or **Z**.

Example 1. *Random variables.* A measurable function ξ from $(\Omega, \mathcal{E}, P)$ into $(\mathbf{R}, \mathcal{B})$, where $\mathcal{B}$ is the σ-algebra of the Borel sets, is called a *real random variable.* Since ξ is measurable, for all B in $\mathcal{B}$, $\xi^{-1}(B)$ belongs to $\mathcal{E}$. Hence $P_\xi = P \circ \xi^{-1}$ is a measure on $(\mathbf{R}, \mathcal{B})$. It is called the *probability distribution* of the random variable ξ. The function F_ξ defined by

$$F_\xi(x) = P\big(\{\omega | \omega \in \Omega, \xi(\omega) < x\}\big)$$

is called the *distribution function.*[13] The probability that ξ lies in the interval $[a, b[$ is given by

$$P_\xi\big([a, b[\big) = F_\xi(b) - F_\xi(a).$$

P_ξ is the Stieljes measure generated by F_ξ.

A random variable ξ is said to be *uniformly distributed* in the interval $[0, 1]$ if

$$F_\xi(x) = \begin{cases} 0, & \text{if } x < 0; \\ x, & \text{if } 0 \le x < 1; \\ 1, & \text{if } x \ge 1. \end{cases}$$

If $0.x_1x_2x_3\ldots$ is the binary expansion of $x \in [0, 1]$, let f be the function defined on $[0, 1]$ by $f(x) = 0.x_1x_1x_2x_2x_3x_3\ldots$. What is the distribution function F_η of the random variable $\eta = f(\xi)$, if ξ is uniformly distributed? From the definition of f, it follows that

$$\begin{aligned} f(x) &< 0.01, \qquad \text{if } \ x < 0.1; \\ f(x) &\ge 0.11, \qquad \text{if } \ x \ge 0.1. \end{aligned}$$

Hence, $F_\eta(y)$ has the constant value 0.1 for $0.01 \ge y < 0.11$, that is,

$$F_\eta(y) = 1/2, \qquad \text{if } \ 1/4 \le y \le 3/4.$$

Similarly,

$$F_\eta(y) = \begin{cases} 1/4, & \text{if } 1/16 \le y \le 3/16; \\ 3/4, & \text{if } 13/16 \le y \le 15/16. \end{cases}$$

---

13 It is also called the *cumulative probability.*

If $y = f(x)$, then $F_\eta(y) = x$; if y is not of the form $0.x_1x_1x_2x_2x_3x_3\ldots$ then it lies in one of the intervals of constancy of F_η. It is readily seen that the sum of the lengths of all the intervals of constancy of F_η is equal to 1. F_η is continuous and constant almost everywhere. It is similar to the Lebesgue function described in Example 2.8.

The *mathematical expectation*[14] of the random variable ξ is $E(\xi) = \int \xi \, dP_\xi$. The *variance* of the random variable ξ is $V(\xi) = E\big((\xi - E(\xi))^2\big)$. If the expectation and variance of the random variable ξ are both finite, then, for any positive real λ,

$$P_\xi\big(|\xi - E(\xi)| \geq \lambda\sqrt{V(\xi)}\big) \leq \frac{1}{\lambda^2}.$$

This result, which is known as the *Chebyshev inequality*, is just the translation into the language of probability of the inequality used to prove Theorem 9.

Theorem 10. *The set of all μ-integrable functions is a vector space. It is denoted by $\mathcal{L}^1_\mu$.*

We have to prove that, if f and g are two integrable functions, and if λ is a scalar,

$$\int (f+g)\, d\mu = \int f\, d\mu + \int g\, d\mu,$$

and

$$\int (\lambda f)\, d\mu = \lambda \int f\, d\mu.$$

Let f and g be two real μ-integrable functions. The relations

$$(f+g)^+ \leq f^+ + g^+ \qquad \text{and} \qquad (f+g)^- \leq f^- + g^-$$

imply that

$$\int (f+g)^+\, d\mu \leq \int f^+\, d\mu + \int g^+\, d\mu,$$
$$\int (f+g)^-\, d\mu \leq \int f^-\, d\mu + \int g^-\, d\mu,$$

which shows that $(f+g)^+$ and $(f+g)^-$ are integrable, and so is $f+g$. Moreover

$$(f+g) = (f+g)^+ - (f+g)^- = f^+ - f^- + g^+ - g^-,$$

[14] Also known as the *mean value* or, especially in physics, the *expectation value.*

that is,

$$(f+g)^+ + f^- + g^- = (f+g)^- + f^+ + g^+,$$

therefore (extension of Theorem 7 to positive measurable functions),

$$\int (f+g)^+\,d\mu + \int f^-\,d\mu + \int g^-\,d\mu = \int (f+g)^-\,d\mu + \int f^+\,d\mu + \int g^+\,d\mu,$$

and, finally,

$$\int (f+g)\,d\mu = \int f\,d\mu + \int g\,d\mu.$$

If f and g are complex-valued, it is sufficient to consider $\Re(f+g)$ and $\Im(f+g)$ to verify that the above relation holds.

If λ is a real, we have

$$(\lambda f)^+ = \begin{cases} \lambda f^+, & \text{if } \lambda > 0, \\ -\lambda f^-, & \text{if } \lambda < 0, \end{cases}$$

and

$$(\lambda f)^- = \begin{cases} \lambda f^-, & \text{if } \lambda > 0, \\ -\lambda f^+, & \text{if } \lambda < 0; \end{cases}$$

then, if $\lambda > 0$,

$$\begin{aligned} \int (\lambda f)\,d\mu &= \lambda \int f^+\,d\mu - \lambda \int f^-\,d\mu \\ &= \lambda \int f\,d\mu, \end{aligned}$$

and, if $\lambda < 0$,

$$\begin{aligned} \int (\lambda f)\,d\mu &= (-\lambda) \int f^-\,d\mu - (-\lambda) \int f^+\,d\mu \\ &= \lambda \int f\,d\mu. \end{aligned}$$

If λ is complex and f complex-valued, we have

$$\lambda f = (\Re\lambda\,\Re f - \Im\lambda\,\Im f) + i(\Im\lambda\,\Re f + \Re\lambda\,\Im f),$$

and, from the above result, it follows that the relation

$$\int (\lambda f)\,d\mu = \lambda \int f\,d\mu$$

holds also in this case.

Since the integral of a positive function is positive, the mapping $f \mapsto \int f\, d\mu$ is a positive (but not definite positive) form on $\mathcal{L}^1_\mu$.

Theorem 11. *If f is μ-integrable, so is $|f|$, and we have*

$$\left|\int f\, d\mu\right| \leq \int |f|\, d\mu.$$

This is obvious since

$$|f| = f^+ + f^-, \qquad \text{if } f \text{ is real-valued,}$$

and

$$|f| \leq |\Re f| + |\Im f|, \qquad \text{if } f \text{ is complex-valued.}$$

A measurable function f is μ-integrable if, and only if, $|f|$ is μ-integrable. More generally, *if f and g are two measurable functions such that $|f| \leq |g|$ a.e., and g is μ-integrable, then f is μ-integrable.*

Example 2. *Riemann and Lebesgue integrals.* If a function f is Riemann integrable on $[a, b]$ $(b > a)$, then it is Lebesgue integrable (that is, integrable with respect to the Lebesgue measure) on $[a, b]$ and the two integrals are equal. Indeed, if f is Riemann integrable on $[a, b]$,[15] for any $\epsilon > 0$, there exist two step functions g_ϵ and h_ϵ such that

$$g_\epsilon \leq f \leq h_\epsilon,$$

and

$$\int_a^b h_\epsilon(x)\, dx - \int_a^b g_\epsilon(x)\, dx < \epsilon.$$

Step functions are simple functions, hence Lebesgue integrable, and their Lebesgue integral is equal to their Riemann integral. So

$$\int_{[a,b]} h_\epsilon\, dm - \int_{[a,b]} g_\epsilon\, dm < \epsilon.$$

In order to prove that f is measurable, let

$$G = \sup_\epsilon g_\epsilon \qquad \text{and} \qquad H = \inf_\epsilon h_\epsilon,$$

[15] See Note 4.

and consider the set

$$\{x|H(x) - G(x) > 0\} = \bigcup_{n=1}^{\infty} \{x|H(x) - G(x) > \frac{1}{n}\}.$$

Since G and H are measurable (Theorem 4), all the sets in the above relations are measurable. If $H(x) - G(x) > 1/n$, then, for any $\epsilon > 0$, $h_\epsilon - g_\epsilon > 1/n$, and, if m_n is the Lebesgue measure of the set $\{x|H(x) - G(x) > 1/n\}$, it follows that, for any $\epsilon > 0$,

$$\int_{[a,b]} (h_\epsilon - g_\epsilon)\, dm > \frac{m_n}{n}.$$

Hence, for any $\epsilon > 0$, $m_n/n < \epsilon$, that is, $m_n = 0$. This implies that, for all positive integers n, $H - G \leq 1/n$ almost everywhere, so $H = G$ a.e. The relation $G \leq f \leq H$ shows that f is almost everywhere equal to a measurable function. Since the Lebesgue measure is complete, it follows that f is measurable.

Moreover, f is bounded hence Lebesgue integrable, and we have

$$\int_{[a,b]} G\, dm = \int_{[a,b]} f\, dm = \int_{[a,b]} H\, dm;$$

but since, for all positive ϵ,

$$\int_{[a,b]} g_\epsilon(x)\, dx \leq \int_a^b f(x)\, dx \leq \int_{[a,b]} h_\epsilon(x)\, dx,$$

it follows that

$$\int_{[a,b]} f\, dm = \int_a^b f(x)\, dx.$$

If f is Lebesgue integrable on $[a, b]$, it is not in general Riemann integrable on $[a, b]$. Here is a classic example. Let f be a function defined on $[0, 1]$ such that

$$f(x) = \begin{cases} 0, & \text{if } x \text{ is rational;} \\ 1, & \text{if } x \text{ is irrational.} \end{cases}$$

f is called the *Dirichlet function*. Since $f = 1$ a.e., its Lebesgue integral is equal to 1. But, since it is impossible to find two step functions g_ϵ and h_ϵ such that

$$g_\epsilon \leq f \leq h_\epsilon$$

and

$$\int_0^1 h_\epsilon(x)\,dx - \int_0^1 g_\epsilon(x)\,dx < \epsilon$$

for all positive ϵ, f is not Riemann integrable.

The class of Riemann integrable functions is quite restricted. In fact, *a bounded function f defined on a finite interval $[a, b]$ is Riemann integrable if, and only if, it is continuous almost everywhere.*[16] The Dirichlet function, being everywhere discontinuous, cannot be Riemann integrable.

Example 3. The function

$$x \mapsto \frac{\sin x}{x}$$

is not integrable with respect to the Lebesgue measure on $[0, \infty]$ since

$$\begin{aligned}\int_0^\infty \left|\frac{\sin x}{x}\right| dx &= \sum_{n=0}^{\infty} \int_{n\pi}^{(n+1)\pi} \left|\frac{\sin x}{x}\right| dx \\ &\geq \sum_{n=0}^{\infty} \frac{1}{(n+1)\pi} \int_{n\pi}^{(n+1)\pi} |\sin x|\, dx \\ &= \infty.\end{aligned}$$

Similarly, the function

$$n \mapsto \frac{(-1)^n}{n}$$

is not integrable with respect to the counting measure on the set $\mathbf{N}$ of natural integers.

The notion of *conditionally convergent integral* does not exist in Lebesgue theory of integration.

4. Lebesgue's Dominated Convergence Theorem

The dominated convergence theorem is one of the most important results of Lebesgue's integration theory. It gives a general sufficient condition for the validity of proceeding to the limit of a sequence of functions under the integral sign. It is an invaluable tool to study functions defined by integrals.

To prove Lebesgue's theorem we need the following lemma.

16 The proof of this theorem, which is due to Lebesgue, can be found in the appendix.

Lemma. Fatou' lemma. *Let* (f_n) *be a sequence of nonnegative functions defined on the measure space* $(X, \mathcal{A}, \mu)$*; then*

$$\int \liminf_{n\to\infty} f_n \, d\mu \leq \liminf_{n\to\infty} \int f_n \, d\mu.$$

Let

$$g_n = \inf_{k\geq n} f_k.$$

Then, for all positive integers n, g_n is measurable (Theorem 4), and $g_n \leq f_n$. (g_n) is an increasing sequence of positive functions, converging to $\liminf_{n\to\infty} f_n$, and the monotone convergence theorem (Theorem 8) shows that

$$\lim_{n\to\infty} \int g_n \, d\mu = \int \liminf_{n\to\infty} f_n \, d\mu.$$

The inequality $g_n \leq f_n$ also implies that $\int g_n \, d\mu \leq \int f_n \, d\mu$, that is,

$$\lim_{n\to\infty} \int g_n \, d\mu \leq \liminf_{n\to\infty} \int f_n \, d\mu,$$

and the result follows.

Theorem 12. Lebesgue's dominated convergence theorem. *Let* $(X, \mathcal{A}, \mu)$ *be a measure space and let* (f_n) *be a sequence of measurable functions from* X *into* $\mathbf{R}$ *(or* $\mathbf{C}$*) that converges almost everywhere to a function* f*; if there exists a* μ*-integrable function* g *such that, for all positive integers* n*,* $|f_n| \leq g$*, then*

$$\lim_{n\to\infty} \int f_n \, d\mu = \int f \, d\mu.$$

f is measurable (Theorem 4), and, since $|f| \leq g$, f is integrable with respect to μ. Since $|f_n - f| \leq 2g$, Fatou's lemma applies to the sequence $(2g - |f_n - f|)$ and yields

$$\int \liminf_{n\to\infty}(2g - |f_n - f|) \, d\mu \leq \liminf_{n\to\infty} \int (2g - |f_n - f|) \, d\mu,$$

that is,

$$2\int g \, d\mu \leq 2\int g \, d\mu + \liminf_{n\to\infty}\Big(-\int |f_n - f| \, d\mu\Big).$$

Since $\int g \, d\mu < \infty$, we obtain

$$\limsup_{n\to\infty} \int |f_n - f| \, d\mu \leq 0.$$

$\int |f_n - f| \, d\mu$ being nonnegative, the upper limit must be zero. Therefore,

$$\lim_{n\to\infty} \int |f_n - f| \, d\mu = 0,$$

and, by Theorem 11 applied to $f_n - f$, the result follows.

The function g, which, of course, *must not depend on* n, is said to *dominate* the sequence (f_n).

Remark 1. Dominated convergence is essential. The theorem does not apply to the sequence $(\frac{1}{n} 1_{[0,n]})$ since no function dominating the sequence can be found. On the contrary, the theorem applies to the sequence $(x^n \, 1_{[0,1]})$ since it is dominated by $1_{[0,1]}$.

Remark 2. The notion of *improper integral*, which allows us to extend Riemann's theory to unbounded functions or infinite intervals, does not exist in Lebesgue's theory. Consider, for example, the case of an infinite interval. In Riemann's theory the improper integral $\int_0^\infty f(x)\, dx$ is said to be convergent if, for all positive n, $\int_0^n f(x)\, dx$ exists and tends to a limit when n tends to ∞. But, if f is Lebesgue integrable, the dominated convergence theorem applies since

$$\lim_{n\to\infty} f\, 1_{[0,n]} = f \qquad \text{and} \qquad |f\, 1_{[0,n]}| \le |f|.$$

Hence,

$$\lim_{n\to\infty} \int f\, 1_{[0,n]}\, dm = \int f\, dm = \int_0^\infty f(x)\, dx.$$

Example 1. The sequence (f_n) of functions defined on $[0, 1]$ by

$$f_n(x) = \frac{n^{3/2} x}{1 + n^2 x^2}$$

tends to 0. The convergence *is not* uniform since, for $x = 1/n$, $f_n(x)$ reaches its maximun value equal to $n^{1/2}/2$. Elementary theorems do not allow us to conclude that

$$\lim_{n\to\infty} \int_0^1 f_n(x)\, dx = 0.$$

However, for all positive reals t, the function $t \mapsto t^{3/2}x/(1 + t^2x^2)$ being bounded by $3^{3/4}/4\sqrt{x}$, the function $x \mapsto 3^{3/4}/4\sqrt{x}$ dominates the sequence (f_n), and, since it is integrable, the dominated convergence theorem applies and we have

$$\lim_{n\to\infty} \int_0^1 \frac{n^{3/2} x}{1 + n^2 x^2}\, dx = 0.$$

Example 2. To calculate

$$\lim_{n\to\infty}\int_0^n \left(1+\frac{x}{n}\right)^n e^{-2x}\,dx,$$

it is sufficient to note that, for all positive real x and all positive integers n,

$$\left(1+\frac{x}{n}\right)^n e^{-2x}\,1_{[0,n]}(x) < e^{-x}.$$

Therefore,

$$\begin{aligned}\lim_{n\to\infty}\int_0^n \left(1+\frac{x}{n}\right)^n e^{-2x}\,dx &= \int_0^\infty \lim_{n\to\infty}\left(1+\frac{x}{n}\right)^n e^{-2x}1_{[0,n]}(x)\,dx\\ &= \int_0^\infty e^{-x}\,dx\\ &= 1.\end{aligned}$$

The following results are important to study a function defined by an integral. Since a series is an integral with respect to the counting measure, these results will apply equally to a function defined by a series.

Theorem 13. *Let $(X, \mathcal{A}, \mu)$ be a measure space, and let (Y, d) be a metric space; if f is a function defined on $X \times Y$ such that*
(1) for all $y \in Y$, $x \mapsto f(x, y)$ is measurable,
(2) for almost all $x \in X$, $y \mapsto f(x, y)$ is continuous at y_0,
(3) there exists an integrable function g such that $|f(x, y)| \le g(x)$, for all $x \in X$ and all $y \in Y$,
then, for all $y \in Y$, $x \mapsto f(x, y)$ is integrable, and the function $F: y \to \int f(x, y)\,d\mu(x)$ is continuous at y_0.

Since $|f(x, y)| \le g(x)$ the function $x \mapsto f(x, y)$ is integrable. To prove that F is continuous at y_0, it is sufficient to show that if a sequence of points (y_n) in Y converges to y_0, then the sequence $\big(F(y_n)\big)$ converges to $F(y_0)$. Set $\phi_n(x) = f(x, y_n)$ and $\phi(x) = f(x, y_0)$. For all positive integers n and all $x \in X$, we have $|\phi_n(x)| \le g(x)$. Since $y \mapsto f(x, y)$ is continuous at y_0, the sequence (ϕ_n) converges to ϕ, and by Lebesgue's dominated convergence theorem,

$$\begin{aligned}F(y_0) &= \int \phi(x)\,d\mu(x)\\ &= \lim_{n\to\infty}\int \phi_n(x)\,d\mu(x)\\ &= \lim_{n\to\infty} F(y_n).\end{aligned}$$

Theorem 14. *Let $(X, \mathcal{A}, \mu)$ be a measure space and let I be an open interval of $\mathbf{R}$; if f is a function defined on $X \times I$ such that*
(1) for all $y \in I$, $x \mapsto f(x, y)$ is integrable,
(2) for almost all $x \in X$, $y \mapsto f(x, y)$ is differentiable, the derivative being denoted by $f'_y(x, y)$,
(3) there exists an integrable function g such that $|f'_y(x, y)| \leq g(x)$, for all $x \in X$ and all $y \in I$,
then, for all $y \in I$, $x \mapsto f'_y(x, y)$ is integrable, and the function $F: y \to \int f(x, y)\, d\mu(x)$ is differentiable, and we have $F'(y) = \int f'_y(x, y)\, d\mu(x)$.

Let y_0 be in I and let (h_n) be a sequence of reals different from 0, whose limit is 0, and such that, for all n, $y_0 + h_n$ is in I. Set

$$\phi_n(x) = \frac{1}{h_n}\big(f(x, y_0 + h_n) - f(x, y_0)\big).$$

For all $x \in X$, such that $y \mapsto f(x, y)$ is differentiable at y_0, we have

$$\lim_{n \to \infty} \phi_n(x) = f'_y(x, y_0),$$

and, for all positive integers n,

$$|\phi_n(x)| \leq \sup_{y \in I} |f'_y(x, y) < g(x).$$

$\phi_n(x)$ and $x \mapsto f'_y(x, y)$ are, therefore, integrable, and by Lebesgue's dominated convergence theorem:

$$\begin{aligned} F'(y_0) &= \lim_{n \to \infty} \frac{1}{h_n}\big(F(y_0 + h_n) - F(y_0)\big) \\ &= \lim_{n \to \infty} \int \phi_n(x)\, d\mu(x) \\ &= \int f'_y(x, y_0)\, d\mu(x). \end{aligned}$$

Example 3. *Bessel functions.* The function J_n defined on $\mathbf{R}$ by

$$J_n(x) = \frac{1}{\pi} \int_0^{\pi} \cos(n\theta - x \sin\theta)\, d\theta,$$

where n is an integer, is called the *Bessel function of the first kind of order* n. Since $x \mapsto \cos(n\theta - x\sin\theta)$ is continuous and $|\cos(n\theta - x \sin\theta)| \leq 1$, J_n is continuous (Theorem 13). $x \mapsto \cos(n\theta - x\sin\theta)$ being differentiable, and the absolute value of its derivative being not greater than 1, J_n is differentiable (Theorem 14). More generally, it is easily seen that J_n is infinitely differentiable, and the relations

$$J'_n(x) = \frac{1}{\pi}\int_0^\pi \sin(n\theta - x\sin\theta)\sin\theta\, d\theta$$
$$J''_n(x) = -\frac{1}{\pi}\int_0^\pi \cos(n\theta - x\sin\theta)\sin^2\theta\, d\theta$$

show that J_n is a solution of the following second-order linear differential equation:

$$y'' + \frac{1}{x}y' + (1 - \frac{n^2}{x^2})y = 0.$$

Example 4. Consider the function F defined on $\mathbf{R}$ by

$$F(x) = \int_{-\infty}^{\infty} \frac{e^{ixy}}{1+y^2}\, dy.$$

We have

$$\left|\frac{e^{ixy}}{1+y^2}\right| \le \frac{1}{1+y^2}.$$

$y \mapsto 1/(1+y^2)$ being integrable and $x \mapsto e^{ixy}/(1+y^2)$ being continuous, the function F is continuous (Theorem 13). If we differentiate under the integral sign, we obtain

$$\int_{-\infty}^{\infty} \frac{iye^{ixy}}{1+y^2}\, dy.$$

Since $y \mapsto y/(1+y^2)$ is not integrable, Theorem 14 does not apply, and $F'(x)$ cannot be represented by the above integral. Abel's criterion (see Note 5) shows that the above integral is, however, semiconvergent: $y \mapsto y/(1+y^2)$ is of bounded variation (because it is monotone in the three intervals $]-\infty, -1]$, $[-1, 1]$, $[1, \infty[$, and bounded) and tends to 0 when $y \to \pm\infty$, whereas $y \mapsto e^{ixy}$ is locally integrable and has bounded indefinite integrals if $x \neq 0$.[17] $F'(x)$ is represented by an improper integral, and, although Theorem 14 does not apply, F is differentiable on the complement

[17] We have $\int_a^b e^{ixy}\, dy = (e^{ibx} - e^{iax})/ix$; therefore, for all a and b, $\left|\int_a^b e^{ixy}\, dy\right| \le 2/|x|$. This is what is meant by saying that $y \mapsto e^{ixy}$ has bounded indefinite integrals if $x \neq 0$.

of the origin. If we differentiate a second time under the integral sign, we obtain

$$-\int_{-\infty}^{\infty} \frac{y^2 e^{ixy}}{1+y^2}\, dy,$$

which is meaningless.

Remark. Many lessons can be drawn from this example. The integral giving $F(x)$ can be easily calculated using the residue theorem. It is found that $F(x) = \pi e^{-|x|}$, which shows that F is infinitely differentiable on the complement of the origin. This result cannot be established using the integral representation of F. Within the framework of Lebesgue's theory, differentiation under the integral sign is not permitted. In fact, the two functions

$$x \mapsto \int_{-\infty}^{\infty} \frac{e^{ixy}}{1+y^2}\, dy \qquad \text{and} \qquad x \mapsto \pi e^{-|x|}$$

are not identical; the second one is an analytic continuation of the first one. The reason Theorem 14 did not apply is to be found in the fact that F is not everywhere differentiable.

Example 5. The periodic function defined on $\mathbf{R}$ by

$$F(x) = \sum_{n=1}^{\infty} (-1)^{n+1} \frac{\cos n\pi x}{n^2}$$

has similar properties. F is continuous, but we have to use Abel's criterion (see Note 5) to prove the semiconvergence of the series

$$\sum_{n=1}^{\infty} (-1)^n \frac{\sin n\pi x}{n},$$

which represents $F'(x)$. Here again, within the framework of Lebesgue's theory, differentiation under the sum sign is not permitted, that is, the function represented by the series is not differentiable. But, by summing the series (see Example 3.8, Chapter 3), we can define an analytic continuation of period 2, infinitely differentiable except for $x = 2k+1$ ($k \in \mathbf{Z}$), represented by $x \mapsto \pi^2(1-3x^2)/12$ if $x \in [-1, 1]$.

5. Fubini's Theorem

This section deals with integrals of measurable functions defined on the product of two measure spaces. Given two measure spaces $(X, \mathcal{A}, \mu)$ and $(Y, \mathcal{B}, \nu)$, their product is denoted by $(X \times Y, \mathcal{A} \otimes \mathcal{B}, \mu \otimes \nu)$, where

$X \times Y$ is the Cartesian product of the spaces X and Y, $\mathcal{A} \otimes \mathcal{B}$ the σ-algebra generated by the set of Cartesian products $\{A \times B | A \in \mathcal{A}, B \in \mathcal{B}\}$ (see Definition 4), and $\mu \otimes \nu)$ the product measure on $\mathcal{A} \otimes \mathcal{B}$ that we now have to define.

Definition 11. Let E be a subset of $X \times Y$; the sets

$$E_x = \{y | (x, y) \in E\} \qquad \text{and} \qquad E^y = \{x | (x, y) \in E\}$$

are, respectively, called the *x-section* and *y-section* of E.

The theorem below states that measurable sets have measurable sections.

Theorem 15. *Let $(X \times Y, \mathcal{A} \otimes \mathcal{B})$ be the product of the measurable spaces $(X, \mathcal{A})$ and $(Y, \mathcal{B})$; if E belongs to $\mathcal{A} \otimes \mathcal{B}$, then*

$$(\forall x \in X) \quad (\forall y \in Y) \qquad E_x \in \mathcal{B} \quad \text{and} \quad E^y \in \mathcal{A}.$$

Let Ω be the class of all sets $E \in \mathcal{A} \otimes \mathcal{B}$ such that, for all $x \in X$, $E_x \in \mathcal{B}$. We shall show that Ω is a σ-algebra which coincides with $\mathcal{A} \otimes \mathcal{B}$.

If $A \in \mathcal{A}$ and $B \in \mathcal{B}$, then $A \times B \in \mathcal{A} \otimes \mathcal{B}$ since

$$(A \times B)_x = \begin{cases} B, & \text{if } x \in A; \\ \emptyset, & \text{if } x \notin A. \end{cases}$$

Hence all measurable Cartesian products $A \times B$[18] are in Ω. In particular $X \times Y$ is in Ω.

If $E \in \Omega$, then $E^C \in \Omega$ since

$$(E^C)_x = \{y | (x, y) \in E^C\} = \{y | (x, y) \in E\}^C = (E_x)^C.$$

If, for all positive integers n, $E_n \in \Omega$, then $\cup_{n=1}^{\infty} E_n \in \Omega$, since

$$\left(\cup_{n=1}^{\infty} E_n\right)_x = \cup_{n=1}^{\infty} (E_n)_x.$$

So Ω is a σ-algebra, and since $\mathcal{A} \otimes \mathcal{B}$ is the smallest σ-algebra generated by the rectangles $A \times B$, $\Omega = \mathcal{A} \otimes \mathcal{B}$.

Similarly, for all $y \in Y$, $E^y \in \mathcal{A}$.

---

[18] Such sets are called *rectangles*.

Corollary. *Let f be a measurable function defined on $X \times Y$. For all $x \in X$, the function f_x defined on Y by $f_x(y) = f(x, y)$ is measurable. For all $y \in Y$, the function f^y defined on X by $f^y(x) = f(x, y)$ is measurable.*

f being measurable, for any open set V in $\mathbf{R}$ (or $\mathbf{C}$), the set $E = \{(x, y) | f(x, y) \in V\}$ belongs to $\mathcal{A} \otimes \mathcal{B}$, and by the above theorem, $E_x \in \mathcal{B}$ and $E^y \in \mathcal{A}$. Hence f_x and f^y are measurable. f_x and f^y are, respectively, called the x-section and y-section of f.

Theorem 16. *Let $(X, \mathcal{A}, \mu)$ and $(Y, \mathcal{B}, \nu)$ be two measure spaces, the measures μ and ν being σ-finite; if E belongs to the σ-algebra $\mathcal{A} \otimes \mathcal{B}$, then, for all $x \in X$ and $y \in Y$,*

$$\int \nu(E_x)\, d\mu(x) = \int \mu(E^y)\, d\nu(y).$$

This theorem is important, but its proof is rather lengthy, and should be skipped at first reading. It proves the existence of a product measure, denoted by $\mu \otimes \nu$, defined on the measure space $(X \times Y, \mathcal{A} \otimes \mathcal{B})$ by

$$(\forall E \in \mathcal{A} \otimes \mathcal{B}) \qquad (\mu \otimes \nu)(E) = \int \nu(E_x)\, d\mu(x) = \int \mu(E^y)\, d\nu(y).$$

Let Ω be the class of sets $E \in \mathcal{A} \otimes \mathcal{B})$ which satisfy the relation

$$\int \nu(E_x)\, d\mu(x) = \int \mu(E^y)\, d\nu(y);$$

we shall show that Ω coincides with $\mathcal{A} \otimes \mathcal{B}$.

First we need to prove a lemma providing a more convenient characterization of $\mathcal{A} \otimes \mathcal{B}$. A class $\mathcal{M}$ of subsets of a space X is said to be a *monotone class* if any increasing or decreasing sequence of elements of $\mathcal{M}$ converges to an element of $\mathcal{M}$.

Lemma. *Let $(X, \mathcal{A})$ and $(Y, \mathcal{B})$ be two measurable spaces; the σ-algebra $\mathcal{A} \otimes \mathcal{B}$ is the smallest monotone class containing all finite unions of measurable pairwise disjoint rectangles.*

Since any σ-algebra is a monotone class, if $\mathcal{M}$ is the smallest monotone class containing all finite unions of measurable pairwise disjoint rectangles we have $\mathcal{M} \subset \mathcal{A} \otimes \mathcal{B}$. To prove that, in fact, $\mathcal{M} = \mathcal{A} \otimes \mathcal{B}$, it is sufficient to show that $\mathcal{M}$ is a σ-algebra.

Denote by $\mathcal{E}$ the class of all finite unions of measurable pairwise disjoint rectangles. The identities

$$(A_1 \times B_1) \cap (A_2 \times B_2) = (A_1 \cap A_2) \times (B_1 \cap B_2)$$
$$(A_1 \times B_1) - (A_2 \times B_2) = \big((A_1 - A_2) \times B_1\big) \cup \big((A_1 \cap A_2) \times (B_1 - B_2)\big)$$

show that the intersection of two measurable rectangles is a measurable rectangle, and the difference of two measurable rectangles is the union of two disjoint measurable rectangles. Therefore, if P and Q belong to $\mathcal{E}$, so do $P \cap Q$ and $P - Q$. Moreover, the relation $P \cup Q = (P - Q) \cup Q$ shows that $P \cup Q$ also belongs to $\mathcal{E}$ since $(P - Q) \cap Q = \emptyset$.

If P and Q belong to $\mathcal{M}$, so do $P - Q$ and $P \cup Q$. Indeed, for any subset P of $X \times Y$, denote by $\Gamma(P)$ the class of all subsets Q of $X \times Y$ such that $P - Q$, $Q - P$, and $P \cup Q$ belong to $\mathcal{M}$. Taking into account the symmetrical role played by P and Q in the definition of $\Gamma(P)$, it is clear that Q belongs to $\Gamma(P)$ if, and only if, P belongs to $\Gamma(Q)$. Furthermore, $\mathcal{M}$ being a monotone class, so is $\Gamma(P)$. Therefore, if P belongs to $\mathcal{E}$, any Q in $\mathcal{E}$ belongs also to $\Gamma(P)$, that is, $\mathcal{E} \subset \Gamma(P)$ and, since $\Gamma(P)$ is a monotone class, $\mathcal{M} \subset \Gamma(P)$. Consider now a subset Q belonging to $\mathcal{M}$; if $P \in \mathcal{E}$, we have seen that $Q \in \Gamma(P)$, but this implies $P \in \Gamma(Q)$, hence $\mathcal{E} \subset \Gamma(Q)$ and, as before, $\mathcal{M} \subset \Gamma(Q)$. All this proves that

$$(\forall P \in \mathcal{M}) \quad (\forall Q \in \mathcal{M}) \qquad P - Q \in \mathcal{M} \quad \text{and} \quad P \cup Q \in \mathcal{M}.$$

It remains to show that $\mathcal{M}$ is a σ-algebra. This is indeed the case since
(1) $X \times Y$ belonging to $\mathcal{E}$ belongs also to $\mathcal{M}$.
(2) If $Q \in \mathcal{M}$, $Q^C = (X \times Y) - Q \in \mathcal{M}$.
(3) If $\{P_n | n \in \mathbf{N}\}$ is a collection of elements of $\mathcal{M}$, for all positive integers n, $Q_n = \cup_{i=1}^{n} P_i$ belongs to $\mathcal{M}$; the sequence (Q_n) being increasing, its limit $P = \cup_{n=1}^{\infty} Q_n = \cup_{n=1}^{\infty} P_n$ belongs also to $\mathcal{M}$.

This completes the proof of the lemma, and we are now in a position to establish the theorem.

If Ω is the class of sets $E \in \mathcal{A} \otimes \mathcal{B}$ that satisfy the relation

$$\int \nu(E_x)\, d\mu(x) = \int \mu(E^y)\, d\nu(y),$$

then:
(1) Any measurable rectangle belongs to Ω since, if $E = A \times B$ where $A \in \mathcal{A}$ and $B \in \mathcal{B}$,

$$\nu(E_x) = \nu(B) 1_A(x) \quad \text{and} \quad \mu(E^y) = \mu(A) 1_B(y),$$

hence,

$$\int \nu(E_x)\, d\mu(x) = \mu(A)\nu(B) = \int \mu(E^y)\, d\nu(y).$$

(2) If (E_n) is a monotone increasing sequence of elements of Ω, the limit $E = \cup_{n=1}^{\infty} E_n$ belongs to Ω. In fact, the σ-additivity of the measures μ and ν implies

$$\lim_{n\to\infty} \nu\big((E_n)_x\big) = \nu(E_x),$$
$$\lim_{n\to\infty} \mu\big((E_n)^y\big) = \mu(E^y).$$

Since the sequences of functions $\Big(\nu\big((E_n)_x\big)\Big)$ and $\Big(\mu\big((E_n)^y\big)\Big)$ are increasing, the monotone convergence theorem (Theorem 8) yields[19]

$$\lim_{n\to\infty} \int \nu\big((E_n)_x\big)\, d\mu = \int \nu(E_x)\, d\mu,$$
$$\lim_{n\to\infty} \int \mu\big((E_n)^y\big)\, d\nu = \int \mu(E^y)\, d\nu.$$

The result follows from the fact that the integrals on the left-hand sides of these two relations are equal.
(3) If $\{E_n | n \in \mathbf{N}\}$ is a countable collection of pairwise disjoint sets belonging to Ω, their union $E = \cup_{n=1}^{\infty} E_n$ belongs to Ω. To show this, consider the sequence (P_n), where $P_n = \cup_{i=1}^{n} E_i$. It is increasing and consists of elements of Ω. From the preceding result, it follows that its limit $E = \cup_{n=1}^{\infty} P_n = \cup_{n=1}^{\infty} E_n$ belongs to Ω.
(4) If(E_n) is a decreasing sequence of elements of Ω such that $E_1 \subset A \times B$, where $A \in \mathcal{A}$ and $B \in \mathcal{B}$ with $\mu(A) < \infty$ and $\nu(B) < \infty$, its limit $E = \cap_{n=1}^{\infty} E_n$ belongs to Ω. Proceed as for the proof of (2) but apply the dominated convergence theorem instead of the monotone convergence theorem.

To complete the proof of the theorem, we have to take into account the fact that μ and ν are σ-finite. This condition implies that there exist two sequences (X_n) and (Y_n) such that

$$X = \cup_{n=1}^{\infty} X_n \qquad (\forall n \in \mathbf{N} \quad \mu(X_n) < \infty),$$
$$Y = \cup_{n=1}^{\infty} Y_n \qquad (\forall n \in \mathbf{N} \quad \nu(Y_n) < \infty).$$

For all pairs (m, n), let

$$E_{mn} = E \cap (X_m \times Y_n),$$

[19] $\nu\big((E_n)_x\big)$ is a function of x, and $\mu\big((E_n)^y\big)$ is a function of y.

and denote by $\mathcal{M}$ the class of sets E such that all the sets E_{mn} belong to Ω. (2) and (4) show that $\mathcal{M}$ is a monotone class, and (1) and (3) show that the set $\mathcal{E}$ of all finite unions of measurable rectangles is contained in $\mathcal{M}$. Hence $\mathcal{M} \subset \mathcal{A} \otimes \mathcal{B}$, and the lemma implies $\mathcal{M} = \mathcal{A} \otimes \mathcal{B}$.

Therefore, for all $E \subset \mathcal{A} \otimes \mathcal{B}$, $E_{mn} \in \Omega$ for all pairs (m, n). Since the sets $E_{m,n}$ ($m \in \mathbf{N}$, $n \in \mathbf{N}$) are pairwise disjoint and their union is E, (3) shows that $E \in \Omega$. Therefore, Ω coincides with $\mathcal{A} \otimes \mathcal{B}$, and this completes the proof.

Theorem 17. Fubini's theorem. *Let $(X, \mathcal{A}, \mu)$ and $(Y, \mathcal{B}, \nu)$ be two measure spaces, the measures μ and ν being σ-finite, and let f be a $\mu \otimes \nu$-integrable function defined on $X \times Y$; then the functions*

$$x \mapsto \int f(x, y)\, d\nu(y) \quad \text{and} \quad y \mapsto \int f(x, y)\, d\mu(x)$$

are, respectively, μ-integrable and ν-integrable, and we have

$$\int f\, d\mu \otimes \nu = \int \Big(\int f(x, y)\, d\nu(y) \Big)\, d\mu(x)$$
$$= \int \Big(\int f(x, y)\, d\mu(x) \Big)\, d\nu(y).$$

We shall first prove Fubini's theorem for positive real-valued functions. For any set $E \in \mathcal{A} \otimes \mathcal{B}$, Fubini's theorem applied to the characteristic function 1_E is just the preceding theorem, and we have

$$\int \Big(\int 1_{E_x}(x, y)\, d\nu(y) \Big)\, d\mu(x) = \int \Big(\int 1_{E^y}(x, y)\, d\mu(x) \Big)\, d\nu(y)$$
$$= \int 1_E\, d\mu \otimes \nu.$$

If f is a positive simple function, Fubini's theorem holds as a consequence of Theorem 7(1). If f is a positive real-valued function, it is the limit of an increasing sequence of positive simple functions (f_n) (see Theorem 5 and the following remark). Then the monotone convergence theorem shows that

$$\int f(x, y)\, d\nu(y) = \lim_{n \to \infty} \int f_n(x, y)\, d\nu(y),$$
$$\int f(x, y)\, d\mu(x) = \lim_{n \to \infty} \int f_n(x, y)\, d\mu(x),$$
$$\int \Big(\int f(x, y)\, d\nu(y) \Big)\, d\mu(x) = \lim_{n \to \infty} \int \Big(\int f_n(x, y)\, d\nu(y) \Big)\, d\mu(x),$$
$$\int \Big(\int f(x, y)\, d\mu(x) \Big)\, d\nu(y) = \lim_{n \to \infty} \int \Big(\int f_n(x, y)\, d\mu(x) \Big)\, d\nu(y),$$

which proves Fubini's theorem for positive functions.[20]

If f is a real-valued function, write $f = f^+ - f^-$ and apply Fubini's theorem to the positive functions f^+ and f^-.

If f is a complex-valued function, write $f = \Re f + i\Im f$ and apply Fubini's theorem to the real-valued functions $\Re f$ and $\Im f$.

Remark 1. Since the two measures μ and ν play a symmetrical role, we can simply write

$$\int f \, d\mu \otimes \nu = \int \int f(x,y) \, d\mu(x) \, d\nu(y).$$

Remark 2. If the function f is $\mu \otimes \nu$-integrable, Fubini's theorem is often used to justify interchanging the order of the repeated integrals. For the sake of simplicity, parentheses are omitted, and we write

$$\int d\mu(x) \int f(x,y) \, d\nu(y) = \int d\nu(y) \int f(x,y) \, d\mu(x).$$

Remark 3. To apply Fubini's theorem, it is essential to verify that the function f is integrable with respect to the product measure $\mu \otimes \nu$. That is, f has to be measurable, and $|f|$ must be integrable. To verify this last property, it is sufficient to verify that one of these three integrals,

$$\int |f| \, d\mu \otimes \nu, \quad \int d\mu(x) \int |f(x,y)| \, d\nu(y), \quad \int d\nu(y) \int |f(x,y)| \, d\mu(x),$$

is finite, since these integrals (finite or infinite) are always equal as a consequence of Fubini's theorem applied to the positive function $|f|$.

Remark 4. If μ and ν are equal to the counting measure on $\mathbf{N}$ (which is σ-finite), Fubini's theorem states then that we can interchange the order of summation in the case of an absolutely convergent double series.

Remark 5. Fubini's theorem can be extended by induction to the product measure of n ($n > 2$) σ-finite measures.

Remark 6. If $(x,y) \mapsto f(x,y) = g(x)h(y)$ is $\mu \otimes \nu$-integrable, then

$$\int f \, d\mu \otimes \nu = \Big(\int g(x) \, d\mu(x)\Big)\Big(\int h(y) \, d\nu(y)\Big).$$

Remark 7. When it is valid, Fubini's theorem can be used to evaluate the integral of a function defined by an integral.

---

[20] This particular result is known as the *Fubini–Tonelli theorem.*

Example 1. If interchanging the order of the repeated integrals gives two different results, that proves *a posteriori* that Fubini's theorem does not apply. Consider, for example, the function f defined on $[0,1] \times [0,1]$ by

$$f(x,y) = \frac{x^2 - y^2}{(x^2+y^2)^2} \quad \text{if } (x,y) \neq (0,0).$$

$\{(0,0)\}$ being of measure zero, it is not necessary to define f for $(x,y) = (0,0)$. Since

$$\frac{d}{dx}\frac{-x}{x^2+y^2} = \frac{x^2-y^2}{(x^2+y^2)^2},$$

we have

$$\int_0^1 dy \int_0^1 f(x,y)\,dx = -\int_0^1 dx \int_0^1 f(x,y)\,dy = -\frac{\pi}{4}.$$

f is, therefore, not integrable. In fact, it is easy to verify that

$$\int_0^1 dy \int_0^1 \left|\frac{x^2-y^2}{(x^2+y^2)^2}\right| dx = \infty.$$

Example 2. If interchanging the order of the repeated integrals gives identical results, that does not prove that Fubini's theorem applies. Consider, for instance, the function f defined on $\mathbf{R}_+^2$ by $f(x,y) = \sin(x^2+y^2)$. Expanding $\sin(x^2+y^2)$ and taking into account the relations[21]

$$\int_0^\infty \sin x^2\,dx = \int_0^\infty \cos x^2\,dx = \frac{\sqrt{2\pi}}{4},$$

it is found that

$$\int_0^\infty dy \int_0^\infty \sin(x^2+y^2)\,dx = \int_0^\infty dx \int_0^\infty \sin(x^2+y^2)\,dy = \frac{\pi}{4},$$

[21] These integrals are known as *Fresnel's integrals.* They can be evaluated using the formula

$$\int_0^\infty \exp ix^2\,dx = \sqrt{2\pi}(1+i)/4,$$

derived by applying Cauchy's theorem to the function of a complex variable: $z \mapsto \exp iz^2$.

but

$$\int_0^\infty dy \int_0^\infty |\sin(x^2 + y^2)|\, dx = \infty.$$

This result is readily obtained using the change of variables: $x = \rho\cos\theta$ and $y = \rho\sin\theta$.

Example 3. For Fubini's theorem to be valid, the measures μ and ν ***have to be*** *σ-finite.* Let f be a function defined on $[0,1]\times[0,1]$ by

$$f(x,y) = \begin{cases} 1, & \text{if } x = y; \\ 0, & \text{otherwise.} \end{cases}$$

Let μ be the Lebesgue measure m, and let ν be the counting measure.[22] If μ is defined on the σ-algebra of the Lebesgue measurable sets, and ν is defined on the σ-algebra consisting of all the subsets of $[0,1]$, f is measurable. Then, taking into account that

$$\int_{[0,1]} f(x,y)\, d\mu(x) = 0 \qquad \text{and} \qquad \int_{[0,1]} f(x,y)\, d\nu(y) = 1,$$

we find

$$0 = \int_{[0,1]} d\nu(y) \int_{[0,1]} f(x,y)\, d\mu(x) \neq \int_{[0,1]} d\mu(x) \int_{[0,1]} f(x,y)\, d\nu(y) = 1.$$

Example 4. Let f and g be two real-valued integrable functions with respect to the Lebesgue measure. The *convolution* of f and g is the function, denoted by $f * g$, that is defined by

$$(f*g)(x) = \int_{-\infty}^{\infty} f(x-y)g(y)\, dy.$$

By Fubini's theorem $f * g$, is integrable since

$$\begin{aligned}\int_{-\infty}^{\infty} dx \int_{-\infty}^{\infty} |f(x-y)g(y)|\, dy &= \int_{-\infty}^{\infty} dy \int_{-\infty}^{\infty} |f(x-y)g(y)|\, dx \\ &= \int_{-\infty}^{\infty} |g(y)|\, dy \int_{-\infty}^{\infty} |f(x-y)|\, dx \\ &= \int_{-\infty}^{\infty} |g(y)|\, dy \int_{-\infty}^{\infty} |f(x)|\, dx,\end{aligned}$$

---

[22] Note that the counting measure on $[0,1]$ is not σ-finite since, for this measure, $[0,1]$ is not a countable union of sets of finite measure.

where we have taken into account the fact that the Lebesgue measure is translation invariant. Note that

$$\int |f * g|\, dm \leq \int |f|\, dm \int |g|\, dm.$$

Example 5. The integral

$$\int_0^\infty \frac{\sin x}{\mathrm{e}^x - 1}\, dx$$

is convergent, and we may write

$$\begin{aligned}\int_0^\infty \frac{\sin x}{\mathrm{e}^x - 1}\, dx &= \int_0^\infty \frac{\sin x}{\mathrm{e}^x(1 - \mathrm{e}^{-x})}\, dx \\ &= \int_0^\infty \Big(\sum_{k=1}^\infty \mathrm{e}^{-kx} \sin x\Big)\, dx \\ &= \sum_{k=1}^\infty \int_0^\infty \mathrm{e}^{-kx} \sin x\, dx.\end{aligned}$$

Since

$$\sum_{k=1}^\infty \int_0^\infty \mathrm{e}^{-kx} |\sin x|\, dx < \infty,$$

the interchange of the signs $\sum$ and $\int$ is justified.[23] Indeed,

$$\begin{aligned}\int_0^\infty \mathrm{e}^{-kx} |\sin x|\, dx &= \int_0^1 \mathrm{e}^{-kx} |\sin x|\, dx + \int_1^\infty \mathrm{e}^{-kx} |\sin x|\, dx \\ &\leq \int_0^1 x\mathrm{e}^{-kx}\, dx + \int_1^\infty \mathrm{e}^{-kx}\, dx \\ &\leq \frac{1}{k^2} - \frac{\mathrm{e}^{-k}}{k^2},\end{aligned}$$

and

$$\sum_{k=1}^\infty \Big(\frac{1}{k^2} - \frac{\mathrm{e}^{-k}}{k^2}\Big) < \infty.$$

---

[23] That is, Fubini's theorem applies to the function $(x, k) \mapsto \mathrm{e}^{-kx} \sin x$ defined on $\mathbf{N} \times \mathbf{R}_+$.

The integral $\int_0^\infty \mathrm{e}^{-kx} \sin x \, dx$ being equal to $1/(1+k^2)$, we obtain

$$\int_0^\infty \frac{\sin x}{\mathrm{e}^x - 1} \, dx = \sum_{k=1}^{\infty} \frac{1}{1+k^2}.$$

The sum of the series on the right-hand side may be found using the residue theorem. Finally,

$$\int_0^\infty \frac{\sin x}{\mathrm{e}^x - 1} \, dx = \frac{1}{2}(\pi \coth \pi - 1).$$

Example 6. We can sometimes evaluate an integral by transforming it into a double integral. The function

$$(x, y) \mapsto \mathrm{e}^{-x} \sin 2xy$$

is integrable on $[0, \infty[\times [0, 1]$ since

$$(\forall x \in [0, \infty[) \quad (\forall y \in [0, 1]) \qquad |\mathrm{e}^{-x} \sin 2xy| \leq \mathrm{e}^{-x},$$

and

$$\int_0^1 dy \int_0^\infty \mathrm{e}^{-x} \, dx = 1.$$

Fubini's theorem applies, and we can write

$$\int_0^\infty dx \int_0^1 \mathrm{e}^{-x} \sin 2xy \, dy = \int_0^1 dy \int_0^\infty \mathrm{e}^{-x} \sin 2xy \, dx.$$

Hence,

$$\begin{aligned} \int_0^\infty \frac{\mathrm{e}^{-x} \sin^2 x}{x} \, dx &= \int_0^1 \frac{2y}{1+4y^2} \, dy \\ &= \frac{1}{4} \log 5. \end{aligned}$$

Example 7. We have seen (Example 4.3) that the Bessel function J_n, defined by

$$J_n(x) = \frac{1}{\pi} \int_0^\pi \cos(n\theta - x \sin \theta) \, d\theta$$

for $x \in \mathbf{R}$, is infinitely differentiable. As a matter of fact J_n is an analytic function, and its expansion in powers of x may be found using Fubini's theorem. We have

$$
\begin{aligned}
J_n(x) &= \frac{1}{\pi}\int_0^\pi \cos n\theta \cos(x\sin\theta)\,d\theta - \frac{1}{\pi}\int_0^\pi \sin n\theta \sin(x\sin\theta)\,d\theta \\
&= \frac{1}{\pi}\int_0^\pi \Big(\sum_{k=0}^\infty \frac{(-1)^k x^{2k}\cos n\theta \sin^{2k}\theta}{(2k)!}\Big)\,d\theta \\
&\quad - \frac{1}{\pi}\int_0^\pi \Big(\sum_{k=0}^\infty \frac{(-1)^k x^{2k+1}\sin n\theta \sin^{2k+1}\theta}{(2k+1)!}\Big)\,d\theta.
\end{aligned}
$$

If $0 \le \theta \le \pi$ and $|x| < \infty$,

$$
\left|\frac{(-1)^k x^{2k}\cos n\theta \sin^{2k}\theta}{(2k)!}\right| \le \frac{|x|^{2k}}{(2k)!}
$$

and

$$
\left|\frac{(-1)^k x^{2k+1}\sin n\theta \sin^{2k+1}\theta}{(2k+1)!}\right| \le \frac{|x|^{2k+1}}{(2k+1)!},
$$

which shows that Fubini's theorem applies, since

$$
\frac{1}{\pi}\int_0^\pi \Big(\sum_{\ell=0}^\infty \frac{|x|^\ell}{\ell!}\Big)\,d\theta = \mathrm{e}^{|x|}.
$$

A term by term integration finally yields[24]

$$
J_n(x) = \sum_{r=0}^\infty \frac{(-1)^r (x/2)^{n+2r}}{r!\,(n+r)!}.
$$

J_n has the parity of n. The radius of convergence of these series is infinite; J_n is an entire function.

Bessel functions of the first kind satisfy certain recurrence relations that may be derived from the series representation of J_n. We have

$$
\begin{aligned}
\frac{d}{dx}\left(x^n J_n(x)\right) &= \frac{d}{dx}\sum_{r=0}^\infty \frac{(-1)^r (x/2)^{2n+2r}}{r!\,(n+r)!} \\
&= \sum_{r=0}^\infty \frac{(-1)^r (x/2)^{2n+2r-1}}{r!\,(n+r-1)!} \\
&= x^n J_{n-1}(x),
\end{aligned}
$$

[24] The trigonometric integrals are easily evaluated applying the residue theorem.

and similarly

$$\frac{d}{dx}\left(x^{-n} J_n(x)\right) = -x^{-n} J_{n+1}(x).$$

Expanding these formulas, we have

$$\begin{aligned} nx^{n-1} J_n(x) + x^n J_n'(x) &= x^n J_{n-1}(x), \\ -nx^{n-1} J_n(x) + x^n J_n'(x) &= -x^n J_{n+1}(x), \end{aligned}$$

where in the second one we have multiplied both sides by x^{2n}. Subtracting and dividing by x^n, we obtain

$$J_{n-1}(x) + J_{n+1}(x) = \frac{2x}{x} J_n(x).$$

Adding and dividing by x^n, we obtain

$$J_{n-1}(x) - J_{n+1}(x) = 2J_n'(x).$$

Appendix

The necessary and sufficient condition for a function f, defined on a bounded interval $[a, b]$ to be Riemann integrable has been put by Lebesgue into a compact form in terms of the concept of measure. Lebesgue's theorem, which gives a complete characterization of Riemann integrable functions, can be stated as follows:

Theorem. *A function f defined on a bounded interval $[a, b]$ $(a < b)$ is Riemann integrable if, and only if, it is bounded and continuous almost everywhere.*

If f is Riemann integrable, for all positive ϵ, there exists a partition

$$\pi = \{a = x_0, x_1, \ldots, x_{m-1}, x_m = b\}$$

such that, if, for $k = 1, 2, \ldots, m$,

$$m_k = \inf_{x_{k-1} \leq x \leq x_k} f(x) \qquad \text{and} \qquad M_k = \sup_{x_{k-1} \leq x \leq x_k} f(x),$$

then

$$\sum_{k=1}^{m} (x_k - x_{k-1})(M_k - m_k) < \epsilon.$$

Let

$$E_n = \{x | x \in [a, b], \omega_f(x) \geq \frac{1}{n}\},$$

where $\omega_f(x)$ is the oscillation of f at x,[25] and

$$I_n = \{k | E_n \cap [x_{k-1}, x_k] \neq \emptyset\}.$$

We have

$$E_n \subset \cup_{k \in I_n} [x_{k-1} - x_k],$$

[25] The *oscillation $\omega_f(x)$ of a function f at x* is defined by

$$\omega_f(x) = \limsup_{\xi \to x} f(\xi) - \liminf_{\xi \to x} f(\xi).$$

If f is bounded on $[a, b]$, the oscillation of f at any point is finite and positive. $\omega_f(x) = 0$ if, and only if, f is continuous at x.

and since $1/n \leq M_k - m_k$,

$$\sum_{k \in I_n} (x_k - x_{k-1}) \frac{1}{n} < \epsilon.$$

If, for fixed n, we let ϵ tend to 0, this last relation shows that E_n is of measure zero. Therefore the set of all points at which f is discontinuous,

$$\{x | x \in [a,b], \omega_f(x) > 0\} = E = \cup_{n=1}^{\infty} E_n,$$

being a countable union of sets of measure zero, is also of measure zero.

Conversely, assume that E has measure zero. Hence, for all positive ϵ, there exists a covering of E by a countable union of open intervals $\cup_{n=1}^{\infty}]\alpha_n, \beta_n[$ such that $\sum_{n=1}^{\infty} (\beta_n - \alpha_n) < \epsilon$. If f is continuous at x, there exists an open interval $]x - h, x + h[$ $(h > 0)$ such that the oscillation of f on this interval is less than ϵ. The set of these open intervals is a covering of the compact set $[a,b] - \cup_{n=1}^{\infty}]\alpha_n, \beta_n[$, and, therefore, it contains a finite covering. To this finite covering, associate a partition $\pi = \{a = x_0, x_1, \ldots, x_{m-1}, x_m = b\}$ such that, for any $k \in \{1, 2, \ldots, m\}$, the interval $]x_{k-1}, x_k[$ is contained in one of the intervals of the finite covering or in $\cup_{n=1}^{\infty}]\alpha_n, \beta_n[$. To this partition associate now two step functions g and h such that $g \leq f \leq h$ and

$$\begin{aligned} \int_a^b h(x)\,dx - \int_a^b g(x)\,dx &\leq \sum_{n=1}^{\infty} (\beta_n - \alpha_n) \sup_{x \in [a,b]} |f(x)| + \sum_{k=1}^{m} (x_k - x_{k-1})\epsilon \\ &\leq \epsilon \sup_{x \in [a,b]} |f(x)| + (b-a)\epsilon, \end{aligned}$$

which shows that f is Riemann integrable.

Notes

Note 1.

A collection $\mathcal{T}$ of subsets of a set X is said to be a *topology in* X if $\mathcal{T}$ has the following three properties:
(1) X and $\emptyset$ belong to $\mathcal{T}$.
(2) If $\{V_i | i \in I\}$ is an arbitrary collection of elements of $\mathcal{T}$, then $\cup_{i \in I} V_i$ belongs in $\mathcal{T}$.
(3) If V_1 and V_2 belong to $\mathcal{T}$, then $V_1 \cap V_2$ is in $\mathcal{T}$.

The ordered pair $(X, \mathcal{T})$ is called a *topological space.* The elements of $\mathcal{T}$ are called the *open sets* in X. The complement of an open set is called a *closed set.* Any intersection of closed sets is closed; the union of two closed sets is closed.

When no ambiguity is possible, one may speak of "the topological space X" instead of "the topological space $(X, \mathcal{T})$."

Let $(X, \mathcal{T})$ be a topological space; a point x is a *limit point* (or an *accumulation point*) of a set $E \subset X$ if every open set containing x contains, at least, one point of E different from x. A closed set contains all its limit points.

Let $(X, \mathcal{T})$ be a topological space; a point x is an *interior point* of a set $E \subset X$ if there exists an open set $U \in \mathcal{T}$ containing x such that $U \subset E$. The set of interior points of E is called the *interior* of A.

Let $(X, \mathcal{T})$ be a topological space; a point x is an *exterior point* of a set E, if there exists an open set $U \in \mathcal{T}$ containing x such that $U \cap E = \emptyset$. The set of exterior points of E is called the *exterior* of A.

Let $(X, \mathcal{T})$ be a topological space; a point x is a *boundary point* (or a *frontier point*) of a set E if there exists an open set $U \in \mathcal{T}$ containing x containing, at least, one point of E and one point of $X - E$. The set of boundary points of E is called the *boundary* of A.

Let X and Y be two topological spaces, and let f be a mapping of X into Y; f is said to be *continuous* if, for any open set V in Y, $F^{-1}(V)$ is an open set in X.

Note 2.

If f is a mapping of a set X into a set Y, then we have the following properties:
(1) $A \subset B \subset X \Rightarrow f(A) \subset f(B) \subset Y$.
(2) $A \subset B \subset X \Rightarrow f(A \cap B) \subset f(A) \cap f(B)$.

(3) $A \subset B \subset X \Rightarrow f(A \cup B) = f(A) \cup f(B)$.
(4) $A' \subset B' \subset Y \Rightarrow f^{-1}(A') \subset f^{-1}(B') \subset X$.
(5) $A' \subset B' \subset Y \Rightarrow f^{-1}(A' \cap B') = f^{-1}(A') \cap f^{-1}(B')$.
(6) $A' \subset B' \subset Y \Rightarrow f^{-1}(A' \cup B') = f^{-1}(A') \cup f^{-1}(B')$.
(7) $A' \subset Y \Rightarrow f^{-!}(A'^C) = \left(f^{-1}(A')\right)^C$.

Note the difference between relations (2) and (5). It shows why the preimage of a σ-algebra is a σ-algebra, whereas the image of a σ-algebra is not, in general, a σ-algebra. However, if $\mathcal{A}$ is a σ-algebra in X, then the set of all subsets A' of Y such that $f^{-1}(A') \in \mathcal{A}$ is a σ-algebra in Y is called the *induced* σ-algebra by f.

Note 3.

Let (a_n) be a sequence in $\overline{\mathbf{R}}$, and, for any k in $\mathbf{N}$, put

$$b_k = \inf(a_k, a_{k+1}, \ldots) = \inf_{n \geq k} a_n,$$
$$B_k = \sup(a_k, a_{k+1}, \ldots) = \sup_{n \geq k} a_n.$$

Then

$$\liminf a_n = \sup(b_1, b_2, \ldots) = \sup_{k \geq 1} \inf_{n \geq k} a_n,$$
$$\limsup a_n = \inf(B_1, B_2, \ldots) = \inf_{k \geq 1} \sup_{n \geq k} a_n$$

are, respectively, called the *lower limit* and the *upper limit* of the sequence (a_n). Since

$$b_1 \leq b_2 \leq \cdots \qquad \text{and} \qquad B_1 \geq B_2 \geq \cdots,$$

these two limits exist, and we have

$$\liminf a_n = \lim_{k \to \infty} b_k,$$
$$\limsup a_n = \lim_{k \to \infty} B_k.$$

For example, in the case of the sequence $(\cos n)$ we have:

$$\liminf \cos n = -1 \qquad \text{and} \qquad \limsup \cos n = 1.$$

If (a_n) is convergent, then

$$\liminf a_n = \limsup a_n = \lim a_n.$$

Note 4.

In this note we briefly summarize the elementary integration theory for functions defined on a finite interval $[a,b]$ $(b > a)$.

By a *partition* π of the closed interval $[a,b]$ we mean the finite set $\pi = \{a = x_0, x_1, \ldots, x_{m-1}, x_m = b\}$, where $x_0 < x_1 < \cdots < x_{m-1} < x_m$. The number $\delta(\pi) = \sup_k(x_k - x_{k-1})$ is called the *norm* of the partition.

Let f be a continuous function on $[a,b]$, and let π be a partition of $[a,b]$; then

$$S_\pi(f) = \sum_{k=1}^{m} f(x_{k-1})(x_k - x_{k-1})$$

is called the *Cauchy sum associated to* f *and* π.

f being continuous, it is uniformly continuous, and it can be shown that, for all positive ϵ, there exists η such that, if $\delta(\pi) < \eta$ and $\delta(\pi') < \eta$, then $|S_\pi(f) - S_{\pi'}(f)| < \epsilon$.

If (π_n) is a sequence of partitions of $[a,b]$ such that $\lim_{n\to\infty} \delta(\pi_n) = 0$, $(S_{\pi_n}(f))$ is a Cauchy sequence, and its limit is independent of the choice of the sequence (π_n). This limit is called the *Cauchy integral* of f on $[a,b]$, and it is denoted by

$$\int_a^b f(x)\,dx.$$

The class of Cauchy integrable functions is much larger than the class of continuous functions. This can be seen as follows.

Any finite linear combination of characteristic functions of open bounded intervals of $\mathbf{R}$ is called a *step function*. The set of step functions defined on $[a,b]$ is a vector space. If s is a step function, the mapping

$$s \mapsto \|s\| = \sup_{x\in[a,b]} s(x),$$

is a norm called the *norm of uniform convergence*. If a sequence of step functions converges uniformly to a function f, this function is said to be *regulated*. If f is a regulated function on $[a,b]$, there exists, therefore, a sequence of step functions (s_n) such that

$$\lim_{n\to\infty} \|f - s_n\| = 0.$$

Any step function being Cauchy integrable, it is legitimate to define the integral on the finite interval $[a,b]$ of a regulated function f by

$$\int_a^b f(x)\,dx = \lim_{n\to\infty} s_n(x)\,dx.$$

In fact, the mapping

$$s_n \mapsto \int_a^b s_n(x)\,dx$$

on the set of step functions is linear and continuous, and, therefore, $\left(\int_a^b s_n(x)\,dx\right)$ is a Cauchy sequence; moreover, it is easy to verify that the value of the integral of f does not depend upon the sequence (s_n) used to approach f.

A step function being discontinuous at a finite number of points, a regulated function is, at the most, discontinuous at a countable number of points.

A function is Cauchy integrable if, and only if, it is regulated.

Riemann integrability applies to a slightly larger class of functions. A function f, defined on a bounded interval $[a, b]$, is said to be *Riemann integrable* if, for all positive ϵ, there exist two step functions g_ϵ and h_ϵ such that

$$g_\epsilon \leq f \leq h_\epsilon,$$

and

$$\int_a^b h_\epsilon(x)\,dx - \int_a^b g_\epsilon(x)\,dx < \epsilon.$$

It is clear that Cauchy integrable functions are Riemann integrable; but there exist Riemann integrable functions which are discontinuous at an uncountable number of points, and that are, consequently, not regulated. More precisely, Lebesgue has shown that *a function is Riemann integrable if, and only if, it is continuous almost everywhere* (see appendix).

Note 5.

Abel's criterion is an important theorem concerning semiconvergent improper integrals. It can be stated as follows: *If u and v are two functions on $[a, b[$ into $\overline{\mathbf{R}}$ (or $\mathbf{C}$) such that u is of bounded variation on $[a, b[$ and tends to 0 when $x < b$ tends to b, and v is locally integrable with bounded indefinite integrals, then the integral $\int_a^b u(x)v(x)\,dx$ makes sense.*

There is a similar theorem for series which can be stated: *If (u_n) and (v_n) are two sequences of complex numbers such that (u_n) is of bounded variation and converges to 0, and (v_n) has bounded partial sums, then the series $\sum_{n=1}^{\infty} u_n v_n$ is convergent.*

For instance, the integral $\int_1^\infty \frac{\sin x}{x}\,dx$ and the series $\sum_1^\infty \frac{\sin n\theta}{n}$ are convergent.

Problems

1. Let $A_1, A_2, \dots, A_n$ be n measurable subsets of a measure space $(X, \mathcal{A}, \mu)$; if, for $1 \le p \le n$, we write

$$s_p = \sum_{i_1 < i_2 < \cdots < i_p} \mu(A_{i_1} \cap A_{i_2} \cap \cdots \cap A_{i_p}),$$

establish the so-called *Poincaré's formula*

$$\mu(A_1 \cup A_2 \cup \cdots \cup A_n) = \sum_{p=1}^{n} (-1)^{p-1} s_p.$$

2. Let A be a Lebesgue measurable subset of $\mathbf{R}$ such that $m(A) < \infty$; show that the function $f\colon x \mapsto m\big(A \cap]-\infty, x]\big)$ is continuous for all real x.

3. Show that the Hausdorff dimension of a countable subset of $\mathbf{R}$ is equal to zero.

4. Let K be a subset of $\mathbf{R}^2$ constructed recursively as follows. From a segment of unit length remove the middle third and replace it by two segments of length $1/3$ forming with the removed segment an equilateral triangle; then from each of the four segments of length $1/3$ remove the middle third and replace it by two segments of length $1/9$ forming with the removed segments equilateral triangles (Figure 1); and so on ....

This procedure yields the so-called *triadic Koch curve*. Determine its Hausdorff dimension.

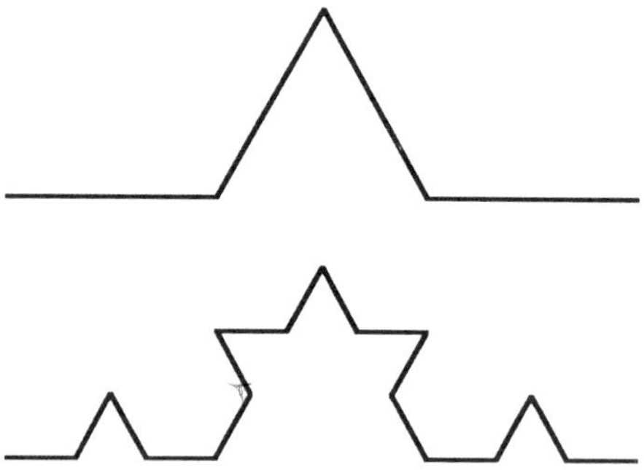

Figure 1. Construction of the Koch curve.

5. Consider a closed square of unit area. At stage 1, divide this square into nine identical squares, and remove the central open square. At stage 2, divide each of the eight remaining closed squares into nine identical squares, and remove the eight central open squares (see figure below). And so on .... If J_n denotes the closed subset of $\mathbf{R}^2$ obtained at stage n, determine the Hausdorff dimension of the so-called *Sierpiński carpet* $S = \bigcap_{n=1}^{\infty} J_n$.

Figure 2. Construction of the Sierpiński carpet.

6. Let C be the Cantor set, and let f be a function defined on $[0,1]$ by

$$f(x) = \begin{cases} 0, & \text{if } x \in C, \\ n, & \text{if } x \in J_{n-1} - J_n, \end{cases}$$

where J_n is the closed set defined in Example 2.7. Evaluate

$$\int_0^1 f(x)\,dx.$$

7. Let f_n be the function defined on $[0,1]$ by

$$f_n(x) = nx(1-x)^n.$$

Is it legitimate to write

$$\lim_{n\to\infty} \int_0^1 f_n(x)\,dx = \int_0^1 \lim_{n\to\infty} f_n(x)\,dx?$$

8. Establish the formulas

$$\lim_{n\to\infty}\int_0^n \Big(1-\frac{x}{n}\Big)^n x^{\alpha-1}\,dx = \Gamma(\alpha) \qquad (\alpha>0), \tag{1}$$

$$\lim_{n\to\infty}\frac{n!\,n^\alpha}{\alpha(\alpha-1)\cdots(\alpha+n-1)} = \Gamma(\alpha). \tag{2}$$

9. Study the continuity and the differentiability of the function defined in $\mathbf{R}$ by

$$f(x) = \int_{-\infty}^{\infty} \frac{1-\cos xt}{t^2}\,dt.$$

10. Prove that the function f defined on $]0,\infty[$ by

$$f(x) = \int_0^\infty \frac{\mathrm{e}^{-t^2 x}}{1+t^2}\,dt$$

is differentiable, and that it is the solution of a first order linear differential equation. By solving this equation, show that

$$f(x) = \frac{\pi}{2}\mathrm{e}^x\big(1-\mathrm{Erf}(\sqrt{x})\big),$$

where Erf is the *error function* defined by

$$\mathrm{Erf}(x) = \frac{2}{\sqrt{\pi}}\int_0^x \mathrm{e}^{-v^2}\,dv.$$

11. The functions K and E defined for $|x|<1$ by

$$K(x) = \int_0^{\pi/2}(1-x\sin^2\theta)^{-1/2}\,d\theta \quad \text{and} \quad E(x) = \int_0^{\pi/2}(1-x\sin^2\theta)^{1/2}\,d\theta$$

are, respectively, called *elliptic integrals* of the first and second kind. Show that K and E are infinitely differentiable, and find their series expansions in powers of x.

12. Show that

$$\lim_{n\to\infty}\int_0^n \Big(1-\frac{x}{n}\Big)^n \log x\,dx = \int_0^\infty \mathrm{e}^{-x}\log x\,dx,$$

and evaluate the integral on the right-hand side.

13. Show that

$$\int_0^\infty e^{-x}\cos\sqrt{x}\,dx = \sum_{n=0}^{\infty}(-1)^n\frac{n!}{(2n)!}.$$

14. Write the integral

$$\int_0^1 \frac{x^{p-1}}{1+x^q}\,dx,$$

where p and q are two positive numbers, as the sum of a series, and evaluate

$$\sum_{n=0}^{\infty}\frac{(-1)^n}{1+n} \quad \text{and} \quad \sum_{n=0}^{\infty}\frac{(-1)^n}{1+2n}.$$

15. Write the integral

$$\int_0^\infty \frac{x^{p-1}}{1+x}\,dx,$$

where $0 < p < 1$, as the sum of a series.

16. Evaluate the integral

$$\int_0^1 \Big(\sum_{n=1}^{\infty} a^n \lceil nx \rceil\Big)\,dx,$$

where $0 < a < 1$. $\lceil nx \rceil$ denotes the greatest integer less than nx.

17. Apply Fubini's theorem to evaluate the integral

$$\int\int\int \frac{1_{[0,1]}(x)1_{[0,1]}(y)1_{[0,\infty]}(t)}{(1+x^2t^2)(1+y^2t^2)}\,dx\,dy\,dt.$$

Prove that

$$\int_0^\infty \Big(\frac{\tan^{-1} t}{t}\Big)^2\,dt = \pi\log 2.$$

18. Evaluate the integral

$$I = \int_0^1 \frac{\log x}{1-x}\,dx.$$

19. Evaluate the integral

$$\int_0^\infty J_0(x)e^{-px}\,dx,$$

where $p > 1$. J_0 denotes the Bessel function of the first kind of order zero (see Examples 4.3 and 5.7).

Solutions

1. This formula can be established by induction. If $n = 2$, we have

$$A_1 \cup A_2 = A_1 \cup \big(A_2 - (A_1 \cap A_2)\big),$$

where A_1 and $A_2 - (A_1 \cap A_2)$ are disjoint. Hence,

$$\mu(A_1 \cup A_2) = \mu(A_1) + \mu(A_2) - \mu(A_1 \cap A_2).$$

Poincaré's formula being true for $n = 2$, assume that it is true for $n > 2$. Then

$$\mu(A_1 \cup A_2 \cup \cdots \cup A_{n+1}) = \mu(A_1 \cup A_2 \cup \cdots \cup A_n) + \mu(A_{n+1}) \\ - \mu\big((A_1 \cup A_2 \cup \cdots \cup A_n) \cap A_{n+1}\big),$$

and applying Poincaré's formula to

$$(A_1 \cap A_{n+1}) \cup (A_2 \cap A_{n+1}) \cup \cdots \cup (A_n \cap A_{n+1}) = (A_1 \cup A_2 \cup \cdots \cup A_n) \cap A_{n+1},$$

we verify that the formula is true for $n + 1$.

Remark. If μ is a probability measure, Poincaré's formula gives the probability that at least one among the events $A_1, A_2, \ldots, A_n$ occur. The following stronger result can be proved: The probability that exactly m among the n events $A_1, A_2, \ldots, A_n$ occur simultaneously is given by

$$\sum_{p=0}^{n-m} (-1)^p \binom{m+p}{m} s_{m+p}.$$

For a proof and further interesting results and applications see W. Feller, *An Introduction to Probability Theory and its Applications*, volume 1, John Wiley, New York, 1968.

2. To prove this result it is sufficient to show that, for any $x \in \mathbf{R}$, we have $f(x - 0) = f(x + 0)$. If (x_n) is an increasing sequence that tends to x, we have

$$A \cap]-\infty, x[= \cup_{n=1}^{\infty} (A \cap]-\infty, x_n]),$$

and by Theorem 6(2),

$$\begin{aligned} m(A \cap]-\infty, x]) &= m(\{x\}) + \lim_{n\to\infty} m(A \cap]-\infty, x_n]) \\ &= \lim_{n\to\infty} m(A \cap]-\infty, x_n]) \\ &= f(x - 0), \end{aligned}$$

since $m(\{x\}) = 0$. If (x_n) is a decreasing sequence that tends to x, we have

$$A\cap]-\infty, x] = \cap_{n=1}^{\infty}(A\cap]-\infty, x_n]),$$

and by Theorem 6(3),

$$\begin{aligned} m(A\cap]-\infty, x]) &= \lim_{n\to\infty} m(A\cap]-\infty, x_n]) \\ &= f(x+0), \end{aligned}$$

since $m(A) < \infty$.

3. It is sufficient to show that for all positive d the Hausdorff outer measure of any countable subset of $\mathbf{R}$ is zero. Let $\{x_n | n \in \mathbf{N}\}$ be a countable subset of $\mathbf{R}$, and let (I_n) be a sequence of intervals such that, for each $n \in \mathbf{N}$, I_n is centered at x_n and $\ell(I_n) = \epsilon u_n$, where $\epsilon > 0$, $u_n > 0$, and $\sum_{n=1}^{\infty} u_n^d < \infty$.[26] $\cup_{n=1}^{\infty} I_n$ is a covering of the countable subset of $\mathbf{R}$, and the result follows from the fact that ϵ is arbitrary.

4. Since at each stage the length of the polygonal curve is multiplied by 4/3, the length of the triadic Koch curve is infinite. To find its Hausdorff dimension, we can proceed as for the Cantor ternary set. K being self-similar, we have

$$H_d^*(K) = \frac{4}{3^d} H_d^*(K),$$

and therefore the Hausdorff dimension d of K is such that

$$\frac{4}{3^d} = 1,$$

that is,

$$d = \frac{\log 4}{\log 3}.$$

Remark. Consider an equilateral triangle and replace each side by a triadic Koch curve. The resulting figure is a bounded subset of $\mathbf{R}^2$ of finite Lebesgue measure (area) but whose boundary has an infinite length. Note that for such a figure the area is not proportional to the square of the perimeter. "Fractal geometry" differs from elementary Euclidean geometry. Coastlines are fractal objects similar to the Koch curve.

It is not difficult to construct more general Koch curves of various Hausdorff dimensions. As shown in Figure 4, we can even construct a

---

[26] Choose, for instance $u_n = \mathrm{e}^{-n}$.

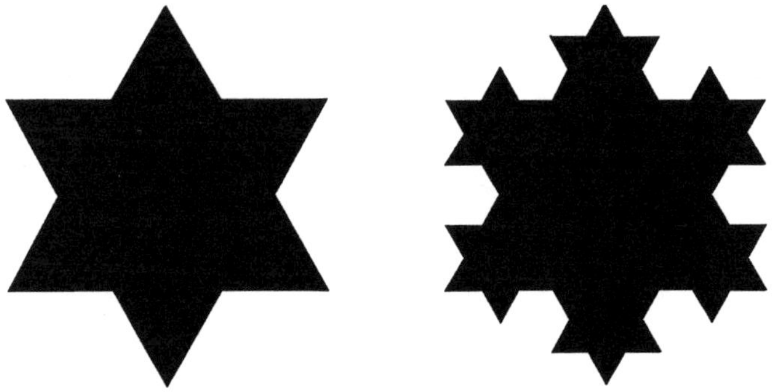

Figure 3. Construction of the triadic Koch "island."

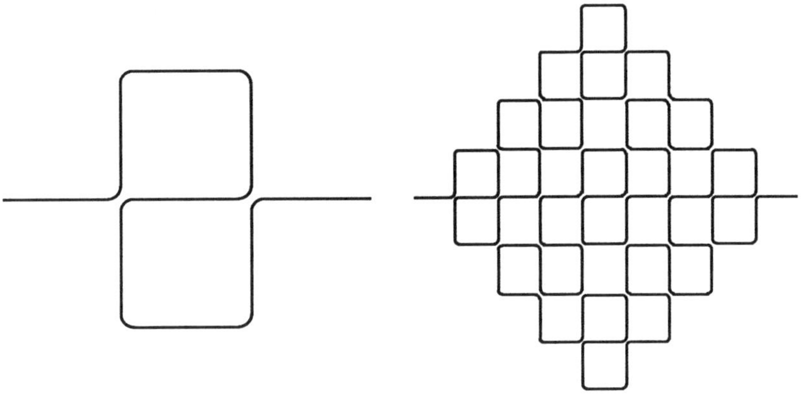

Figure 4. Construction of the Peano curve.

space-filling curve, called a *Peano curve*, whose Hausdorff dimension is 2. From a topological point of view, such an object is, however, a "curve" and not a "surface" since it is cut into two pieces if we remove a single point.

Note that the Peano curve self-contacts but does not self-intersect. A Peano curve represents the graph of a continuous mapping $f: [0,1] \to [0,1] \times [0,1]$ such that $f([0,1]) = [0,1] \times [0,1]$!

5. S is closed. It is the two-dimensional analogue of the Cantor ternary set. S is self-similar and we have

$$H_d^*(S) = \frac{8}{3^d} H_d^*(S).$$

Its Hausdorff dimension is, therefore, equal to $\log 8/\log 3$.

Remark. The area of S is equal to zero, and the length of its boundary is infinite.

It is easy to construct more general Sierpiński carpets. At the first stage we can divide the unit square into b^2 identical square and remove λ open squares. Iterating this procedure yields a closed subset of $\mathbf{R}^2$ whose Hausdorff dimension is $\log(b^2 - \lambda)/\log b$. Note that we can also remove, at each stage, the λ squares at random. The resulting closed subset is no longer self-similar, but it could be shown that it has the same Hausdorff dimension. It is an example of a nondeterministic fractal.

6. Since J_n is obtained from J_{n-1} by removing 2^{n-1} open intervals of length 3^{-n}, we have

$$\int_0^1 f(x)\,dx = \sum_{n=1}^{\infty} n 2^{n-1} 3^{-n} = 3.$$

7. For all $x \in [0,1]$, $f_n(x)$ tends to 0, but the convergence is not uniform. (The maximum of $f_n(x)$ tends to 1 when $n \to \infty$.) Since $|f_n(x)| < 1$, the dominated convergence theorem applies, and we have

$$\lim_{n\to\infty} \int_0^1 f_n(x)\,dx = \int_0^1 \lim_{n\to\infty} f_n(x)\,dx = 0.$$

This result can be verified directly. The integral of f_n is easy to evaluate. We have

$$\begin{aligned}\int_0^1 nx(1-x)^n\,dx &= \left[\frac{nx}{n+1}(1-x)^{n+1}\right]_0^1 + \frac{n}{n+1}\int_0^1 (1-x)^{n+1}\,dx \\ &= \frac{n}{(n+1)(n+2)}.\end{aligned}$$

Hence,

$$\lim_{n\to\infty} \int_0^1 nx(1-x)^n\,dx = 0.$$

8. Let f_n be the function defined in $\mathbf{R}$ by

$$f_n(x) = 1_{[0,n]}(x)\left(1 - \frac{x}{n}\right)^n x^{\alpha-1}.$$

We have

$$\lim_{n\to\infty} f_n(x) = \mathrm{e}^{-x} x^{\alpha-1},$$

and, for all positive integers n,

$$f_n(x) \leq \mathrm{e}^{-x} x^{\alpha-1}.$$

Hence, formula (1) follows by the dominated convergence theorem.

Integrating n times by parts yields

$$\begin{aligned}\int_0^n \left(1-\frac{x}{n}\right)^n x^{\alpha-1}\,dx &= \frac{n!}{n^n}\frac{1}{\alpha(\alpha-1)\cdots(\alpha+n-1)}\int_0^n x^{\alpha+n-1}\,dx\\ &= \frac{n!\,n^\alpha}{\alpha(\alpha-1)\cdots(\alpha+n-1)}.\end{aligned}$$

Letting $n \to \infty$ gives formula (2).

9. For all $t \in \mathbf{R}$, the function

$$x \mapsto \frac{1-\cos xt}{t^2}$$

is continuous. Since $t \mapsto 2/t^2$ is not integrable on $\mathbf{R}$, the continuity of f does not follow from the relation:

$$\left|\frac{1-\cos xt}{t^2}\right| \leq \frac{2}{t^2}.$$

But if we denote by g_A the function defined on $\mathbf{R}$ by

$$g_A(t) = \begin{cases} A^2/2, & \text{if } |t| \leq 2/A,\\ 2/t^2, & \text{if } |t| > 2/A,\end{cases}$$

where A is positive, then for $|x| \leq A$,

$$\left|\frac{1-\cos xt}{t^2}\right| \leq g_A(t).$$

Since g_A is integrable on $\mathbf{R}$, f is continuous for all finite x.

If we take the derivative under the integral sign, we obtain the following semi-convergent integral (Abel's criterion):

$$\int_{-\infty}^{\infty} \frac{\sin t}{t}\,dt.$$

As in Example 4.4, the differentiability of f cannot be proved by Theorem 14. In the present case, it is possible to use the residue theorem to evaluate the integral and obtain an analytic continuation of f. We find $\pi|x|$. The function $x \mapsto \pi|x|$ is infinitely differentiable in the complement of the origin.

10. f is continuous since, for $x > 0$,

$$\frac{e^{-t^2x}}{1+t^2} < \frac{1}{1+t^2}.$$

Furthermore,

$$\frac{\partial}{\partial x}\frac{e^{-t^2x}}{1+t^2} = -\frac{t^2e^{-t^2x}}{1+t^2},$$

and the relation

$$(\forall x \geq \epsilon > 0) \qquad \frac{t^2e^{-t^2x}}{1+t^2} \leq e^{-\epsilon t^2}$$

proves that f is differentiable for all positive x since $t \mapsto e^{-\epsilon t^2}$ is integrable on $\mathbf{R}_+$ when $\epsilon > 0$. We have

$$\begin{aligned} f'(x) &= -\int_0^\infty \frac{t^2e^{-t^2x}}{1+t^2}\,dt \\ &= -\int_0^\infty e^{-t^2x}\,dt + \int_0^\infty \frac{e^{-t^2x}}{1+t^2}\,dt. \end{aligned}$$

Since the first integral on the last right-hand side is equal to $-\frac{1}{2}\sqrt{\frac{\pi}{x}}$, $f(x)$ satisfies the following differential equation:

$$f'(x) - f(x) = \frac{1}{2}\sqrt{\frac{\pi}{x}}.$$

Using the traditional method to solve this equation, one finds that its general solution is

$$Ae^x + \frac{\sqrt{\pi}}{2}e^x\int_x^\infty \frac{e^{-u}}{\sqrt{u}}\,du \qquad (x > 0),$$

where A is a constant. If we put $u = v^2$, we obtain

$$e^x\left(A + \frac{\pi}{2}(1 - \operatorname{Erf}(\sqrt{x}))\right).$$

Since

$$f(0) = \int_0^\infty \frac{1}{1+t^2}\,dt = \frac{\pi}{2},$$

$A = 0$, and finally,

$$f(x) = \frac{\pi}{2} e^x \left(1 - \mathrm{Erf}(\sqrt{x})\right).$$

11. If Theorem 14 applies, the kth derivative of K is given by

$$K^{(k)}(x) = \frac{(-1)^k (2k)!}{2^{2k} k!} \int_0^{\pi/2} \sin^{2k}\theta (1 - x \sin^2\theta)^{-(2k+1/2)}\, d\theta.$$

Since, for $|x| \leq a < 1$,

$$\left| \sin^{2k}\theta (1 - x\sin^2\theta)^{-(2k+1/2)} \right| \leq (1-a)^{-(2k+1/2)},$$

Theorem 14 applies, and K is infinitely differentiable in $]-1, 1[$. Similarly, E is also infinitely differentiable in $]-1,1[$. We find

$$K(x) = \frac{\pi}{2}\left(1 + \left(\frac{1}{2}\right)^2 x + \left(\frac{1.3}{2.4}\right)^2 x^2 + \left(\frac{1.3.5}{2.4.6}\right)^2 x^3 + \cdots\right).$$

Similarly,

$$E(x) = \frac{\pi}{2}\left(1 - \left(\frac{1}{2}\right)^2 x - \left(\frac{1.3}{2.4}\right)^2 x^2 - \left(\frac{1.3.5}{2.4.6}\right)^2 x^3 - \cdots\right).$$

These two series converge for $|x| < 1$.

12. For $0 < x < n$, the inequality

$$\log\left(1 - \frac{x}{n}\right) \leq -\frac{x}{n}$$

implies that

$$\left(1 - \frac{x}{n}\right)^n = e^{n \log\left(1 - \frac{x}{n}\right)} \leq e^{-x}.$$

Hence,

$$\left| 1_{[0,n]}(x) \left(1 - \frac{x}{n}\right)^n \log x \right| \leq e^{-x} |\log x|,$$

which shows that the dominated convergence theorem applies and proves the required result.

Moreover,

$$\begin{aligned}
\int_0^n \left(1 - \frac{x}{n}\right)^n \log x \, dx &= n \int_0^1 t^n \log\big(n(1-t)\big)\, dt \\
&= n \log n \int_0^1 t^n \, dt + n \int_0^1 t^n \log(1-t)\, dt \\
&= \frac{n \log n}{n+1} + \int_0^1 \frac{t^{n+1} - 1}{1-t}\, dt.
\end{aligned}$$

Therefore,

$$\int_0^n \left(1-\frac{x}{n}\right)^n \log x\,dx = \frac{n\log n}{n+1} + \frac{n}{n+1}\int_0^1 (1+t+\cdots+t^n)\,dt$$
$$= \frac{n}{n+1}\left(\log n - 1 - \frac{1}{2} - \cdots - \frac{1}{n+1}\right).$$

Finally,

$$\int_0^\infty \mathrm{e}^{-x}\log x\,dx = -\gamma,$$

where

$$\gamma = -\lim_{n\to\infty}\left(\log n - 1 - \frac{1}{2} - \cdots - \frac{1}{n}\right)$$

is the Euler constant.

13. It is sufficient to expand $\cos\sqrt{x}$, and to verify that Fubini's theorem applies to obtain the required result.

14. x belonging to the interval $[0,1]$, we have

$$\int_0^1 \frac{x^{p-1}}{1+x^q}\,dx = \int_0^1 \left(\sum_{n=0}^\infty (-1)^n x^{p-1+nq}\right) dx$$
$$= \int_0^1 \left(\sum_{n=0}^\infty (1-x^q)x^{p-1+2nq}\right) dx.$$

Since, for all $x \in [0,1]$, $(1-x^q)x^{p-1+2nq} \geq 0$, the Fubini–Tonnelli theorem applies. Hence,

$$\int_0^1 \frac{x^{p-1}}{1+x^q}\,dx = \sum_{n=0}^\infty (-1)^n \int_0^1 x^{p-1+nq}$$
$$= \sum_{n=0}^\infty \frac{(-1)^n}{p+nq}.$$

If $p = q = 1$, we have

$$\sum_{n=0}^\infty \frac{(-1)^n}{1+n} = \int_0^1 \frac{dx}{1+x} = \log 2;$$

and, if $p = 1$ and $q = 2$, we obtain

$$\sum_{n=0}^\infty \frac{(-1)^n}{1+2n} = \int_0^1 \frac{dx}{1+x^2} = \frac{\pi}{4}.$$

15. We have

$$I = \int_0^1 \frac{x^{p-1}}{1+x}\,dx + \int_1^\infty \frac{x^{p-1}}{1+x}\,dx.$$

Changing x in $1/y$ in the second integral yields

$$I = \int_0^1 \frac{x^{p-1}}{1+x}\,dx + \int_0^1 \frac{y^{-p}}{1+y}\,dy.$$

Using the results obtained in the preceding problem, we obtain

$$\begin{aligned} I &= \sum_{n=0}^{\infty} \frac{(-1)^n}{p+n} + \sum_{n=0}^{\infty} \frac{(-1)^n}{1-p+n} \\ &= \frac{1}{p} - \frac{1}{1+p} + \frac{1}{2+p} + \cdots \\ &\quad + \frac{1}{1-p} - \frac{1}{2-p} + \frac{1}{3-p} - \cdots \\ &= \frac{1}{p} - 2p\frac{1}{p^2-1} + 2p\frac{1}{p^2-2^2} - 2p\frac{1}{p^2-3^2} + \cdots \\ &= \frac{1}{p} + 2p\sum_{n=1}^{\infty} \frac{(-1)^n}{p^2-n^2} \end{aligned}$$

Remark. The evaluation of I is a classical application of the residue theorem. It is found that

$$I = \frac{\pi}{\sin \pi p}.$$

Hence (see also Chapter 3, Problem 8),

$$\frac{\pi}{\sin \pi p} = \frac{1}{p} + 2p\sum_{n=1}^{\infty} \frac{(-1)^n}{p^2-n^2}.$$

16. For all $x \in [0,1]$, $[nx] \le n$. Hence,

$$\sum_{n=1}^{\infty} a^n [nx] \le \sum_{n=1}^{\infty} na^n = \frac{a}{(1-a)^2}.$$

The series is, therefore, convergent, and the inequality

$$\sum_{k=1}^{n} a^k [kx] \le \frac{a}{(1-a)^2}$$

shows that the dominated convergence theorem applies. We have

$$\int_0^1 \Big(\sum_{n=1}^{\infty} a^n [nx]\Big)\, dx = \lim_{n\to\infty} \int_0^1 \sum_{k=1}^{n} a^k [kx]\, dx$$
$$= \sum_{k=1}^{\infty} \int_0^1 a^k [kx]\, dx.$$

If $m/k \le x \le (m+1)/k$, then $[kx] = m$; hence,

$$\int_0^1 [kx]\, dx = \sum_{m=1}^{k-1} \frac{m}{k} = \frac{1}{2}(k-1).$$

Therefore,

$$\sum_{n=1}^{\infty} a^n [nx] = \sum_{n=1}^{\infty} \frac{n-1}{2} a^n$$
$$= \frac{a^2}{2(1-a)^2}.$$

For all positive integers n and all $x \in [0,1]$, $a^n [nx] > 0$, and we could have obtained the same result by Fubini–Tonnelli theorem.

17. Since the integrand is positive, the Fubini–Tonnelli theorem applies. Integrating first with respect to t yields

$$\int_0^{\infty} \frac{dt}{(1+x^2t^2)(1+y^2t^2)} = \frac{1}{x^2-y^2} \int_0^{\infty} \Big(\frac{x^2}{1+x^2t^2} - \frac{y^2}{1+y^2t^2}\Big)\, dt$$
$$= \frac{1}{x^2-y^2} \Big[x \tan^{-1} tx - y \tan^{-1} ty\Big]_0^1$$
$$= \frac{\pi}{2} \frac{1}{x+y}.$$

Integrating then with respect to x and y, we finally obtain

$$\frac{\pi}{2} \int_0^1 dx \int_0^1 \frac{dy}{x+y} = \frac{\pi}{2} \int_0^1 \log \frac{1+x}{x}\, dx$$
$$= \frac{\pi}{2} \Big[(1+x)\log(1+x) - (1+x) - x \log x + x\Big]_0^1$$
$$= \pi \log 2.$$

We assumed $x \neq y$, but this has no effect on the value of the integral since the Lebesgue measure of the set $\{(x,y) | (x,y) \in [0,1] \times [0,1], x = y\}$ is equal to zero.

Since

$$\int_0^1 \frac{dx}{(1+x^2t^2)} = \frac{1}{t}\Big[\tan^{-1} tx\Big]_0^1$$
$$= \frac{\tan^{-1} t}{t},$$

we obtain the required result by interchanging the order of the integrations.

18. We have

$$I = \int_0^1 \Big(\sum_{n=0}^{\infty} x^n \log x\Big)\, dx.$$

For all $n \in \mathbf{N}$ and all $x \in [0,1]$, $x^n \log x$ has a constant sign (negative). The Fubini-Tonnelli theorem applies, and we have

$$I = \sum_{n=0}^{\infty} \int_0^1 x^n \log x\, dx.$$

An integration by parts yields

$$I = -\sum_{n=0}^{\infty} \frac{1}{(n+1)^2} = \frac{\pi^2}{6}.$$

19. We have (Example 5.7)

$$J_0(x) = \sum_{r=0}^{\infty} \frac{(-1)^r}{(r!)^2} \Big(\frac{x}{2}\Big)^{2r}$$

and

$$\int_0^{\infty} x^{2r} \mathrm{e}^{-px}\, dx = \frac{(2r)!}{p^{2r+1}}.$$

Fubini's theorem applies, and we have

$$\int_0^{\infty} J_0(x)\mathrm{e}^{-px}\, dx = \sum_{r=0}^{\infty} \frac{(-1)^r}{(r!)^2} \frac{(2r)!}{2^{2r}} \frac{1}{p^{2r+1}}$$
$$= \frac{1}{p}\Big(1+\frac{1}{p^2}\Big)^{-1/2}$$
$$= (1+p^2)^{-1/2}.$$

The condition $p > 1$ is necessary to ensure the convergence of the series in powers of $1/p$.

Chapter 2

Lebesgue Spaces

The Lebesgue theory of integration developed in the first chapter makes it possible to define certain function spaces of major importance in theoretical physics. These are the so-called Lebesgue spaces introduced and studied, in fact, by F. Riesz. They consist of all the measurable functions f such that, for $1 \leq p < \infty$, $|f|^p$ is integrable. The results obtained by Riesz secured for Lebesgue's theory of integration a permanent place in functional analysis. The crucial property of the Lebesgue spaces, not shared by the corresponding spaces of Riemann integrable functions, is that of completeness for the metric

$$d: (f_1, f_2) \mapsto \Big(\int (f_1 - f_2)^p \, dm\Big)^{1/p}.$$

1. Convex Functions and Inequalities

Most inequalities have their origin in the notion of convexity.

Definition 1. A real function ϕ defined on the open interval $]a, b[$ is said to be *convex* if the inequality

$$\phi\big((1-\lambda)x + \lambda y\big) \leq (1-\lambda)\phi(x) + \lambda\phi(y)$$

holds whenever $x \in]a, b[$, $y \in]a, b[$, and $\lambda \in [0, 1]$. ϕ is said to be *concave* if $-\phi$ is convex.

Graphically, the definition says that, in $\mathbf{R}^2$, for all $t \in]x, y[$, the point $\big(t, \phi(t)\big)$ is never above the the line connecting the points $\big(x, \phi(x)\big)$ and $\big(y, \phi(y)\big)$.

It is not difficult to see that a convex function on an *open* interval $]a, b[$ is necessarily continuous (the fact that $]a, b[$ is open is essential).

Let x_1, x_2, and x_3 be such that $a < x_1 < x_2 < x_3 < b$; then, if ϕ is convex, we have

$$\frac{\phi(x_2) - \phi(x_1)}{x_2 - x_1} \leq \frac{\phi(x_3) - \phi(x_2)}{x_3 - x_2}.$$

If ϕ is convex and differentiable on $]a, b[$, the above relation implies

$$\phi'(x_1) \leq \phi'(x_2)$$

whenever $a < x_1 < x_2 < b$. If ϕ is convex and twice differentiable on $]a, b[$, then

$$(\forall x \in]a, b[) \qquad \phi''(x) \geq 0.$$

For instance, the functions

$$\begin{array}{ll} x \mapsto \exp x & x \in \mathbf{R}, \\ x \mapsto x^\alpha & \alpha < 0, \alpha > 1, x \in \mathbf{R}_+, \\ x \mapsto -\log x & x \in \mathbf{R}_+, \\ x \mapsto x \log x & x \in \mathbf{R}_+ \end{array}$$

are convex.

The following theorem has many interesting applications.

Theorem 1. Jensen's inequality. *Let $(\Omega, \mathcal{E}, P)$ be a probability space; if f is a P-integrable real-valued function in Ω into $]a, b[$, and ϕ a convex function defined on $]a, b[$, then*

$$\phi\Big(\int f\, dP\Big) \leq \int \phi \circ f\, dP.$$

That f is integrable and ϕ is continuous implies that $\phi \circ f$ is integrable, and the right-hand side makes sense. Put $t = \int f\, dP$. So $a < t < b$. Let

$$A = \sup_{u \in]a, t[} \frac{\phi(t) - \phi(u)}{t - u}.$$

Then, for all $y \in]a, b[$,

$$\phi(y) - \phi(t) \geq A(y - t).$$

If we put $y = f(x)$, for every $x \in \Omega$, we have

$$\phi\big(f(x)\big) - \phi(t) \geq A\big(f(x) - t\big).$$

The inequality follows if we integrate both sides with respect to P.
Example 1. Let $\phi(x) = \exp x$; then the Jensen's inequality yields

$$\exp\Big(\int f\,dP\Big) \leq \int \exp f\,dP.$$

If Ω is the finite set $\{x_1, x_2, \ldots, x_n\}$, and P such that $P(\{x_i\}) = 1/n$, for all $i = 1, 2, \ldots, n$, the above relation becomes

$$\exp\big(\frac{1}{n}(x_1 + x_2 + \cdots + x_n)\big) \leq \frac{1}{n}(\exp x_1 + \exp x_2 + \cdots + \exp x_n).$$

For all $i = 1, 2, \ldots, n$, put $\exp x_i = a_i$, replace in the above relation, and then change a_i into $1/a_i$ to obtain the double inequality

$$\frac{n}{\sum_{i=1}^{n} \frac{1}{a_i}} \leq \Big(\prod_{i=1}^{n} a_i\Big)^{1/n} \leq \frac{1}{n}\sum_{i=1}^{n} a_i,$$

which shows that the *harmonic mean* is less than or equal to the *geometric mean* which is itself less than or equal to the *arithmetic mean*. These three means coincide if, and only if, all the positive numbers a_i $(i = 1, 2, \ldots, n)$ are equal.
Definition 2. The two positive numbers p and q are said to be *conjugate exponents* if

$$\frac{1}{p} + \frac{1}{q} = 1.$$

Equivalently, we have $p + q = pq$. The definition implies $1 < p < \infty$ and $1 < q < \infty$. Since q tends to ∞ when p tends to 1, $(1, \infty)$ can be regarded as a pair of conjugate exponents.
Theorem 2. Hölder's inequality. *Let $(X, \mathcal{A}, \mu)$ be a measure space, and let f and g be two positive μ-measurable functions on X; then*

$$\int fg\,d\mu \leq \Big(\int f^p\,d\mu\Big)^{1/p}\Big(\int g^q\,d\mu\Big)^{1/q},$$

where p and q are two conjugate exponents.

If one of the two terms on the right-hand side is null or infinite, the inequality is trivially satisfied. So we can assume that $0 < \int f^p\,d\mu < \infty$ and $0 < \int g^q\,d\mu < \infty$. Write

$$F = \frac{f}{\left(\int f^p\,d\mu\right)^{1/p}} \quad \text{and} \quad G = \frac{g}{\left(\int g^q\,d\mu\right)^{1/q}}.$$

Then

$$\int F^p \, d\mu = 1 \quad \text{and} \quad \int G^q \, d\mu = 1.$$

If $x \in X$ is such that

$$0 < F(x) < \infty \quad \text{and} \quad 0 < G(x) < \infty,$$

there exist two reals t and u such that

$$F(x) = \exp \frac{t}{p} \quad \text{and} \quad G(x) = \exp \frac{u}{q}.$$

Since $\frac{1}{p} + \frac{1}{q} = 1$, the convexity of the exponential function implies that

$$\exp\left(\frac{t}{p} + \frac{u}{q}\right) \leq \frac{1}{p} \exp t + \frac{1}{q} \exp u.$$

For all $x \in X$, it follows that

$$F(x)G(x) \leq \frac{1}{p}(F(x))^p \frac{1}{q}(G(x))^q.$$

Integrating with respect to the measure μ yields

$$\int FG \, d\mu \leq \frac{1}{p} + \frac{1}{q} = 1.$$

Replacing F and G by their expressions, we obtain the Hölder's inequality.
Remark 1. In the next section we will show that the Hölder's inequality still holds if $p = 1$ and $q = \infty$. But we will first have to give a meaning to $q = \infty$.
Remark 2. The inequality becomes an equality if f^p and g^q are proportional a.e.
Corollary. Minkowski's inequality. *Let $(X, \mathcal{A}, \mu)$ be a measure space, and let f and g be two positive μ-measurable functions on X; then*

$$\left(\int (f+g)^p \, d\mu\right)^{1/p} \leq \left(\int f^p \, d\mu\right)^{1/p} + \left(\int g^p \, d\mu\right)^{1/p},$$

where $1 \leq p < \infty$.

The case $p = 1$ is trivial. So we need consider only the case $p > 1$. Write

$$(f+g)^p = f(f+g)^{p-1} + g(f+g)^{p-1},$$

and apply Hölder's inequality to obtain

$$\int f(f+g)^{p-1}\,d\mu \le \Big(\int f^p\,d\mu\Big)^{1/p}\Big(\int (f+g)^{(p-1)q}\,d\mu\Big)^{1/q},$$
$$\int g(f+g)^{p-1}\,d\mu \le \Big(\int g^p\,d\mu\Big)^{1/p}\Big(\int (f+g)^{(p-1)q}\,d\mu\Big)^{1/q},$$

where p and q are conjugate exponents. Since $(p-1)q = p$, addition of these two inequalities gives

$$\int (f+g)^p\,d\mu \le \Big(\big(\int f^p\,d\mu\big)^{1/p} + \big(\int g^p\,d\mu\big)^{1/p}\Big)\Big(\int (f+g)^p\,d\mu\Big)^{1/q}.$$

If we divide both sides of this relation by the last factor on the right-hand side, we obtain Minkowski's inequality.

Remark. The Minkowski's inequality can be generalized. Let $(X, \mathcal{A}, \mu)$ be a measure space, and let (f_n) be a sequence of positive μ-measurable functions on X; then

$$\Big(\int \big(\sum_{n=1}^{\infty} f_n\big)^p\,d\mu\Big)^{1/p} \le \sum_{n=1}^{\infty}\big(\int f_n^p\,d\mu\big)^{1/p},$$

where $1 \le p < \infty$.

If n is finite, Minkowski's inequality implies that

$$\Big(\int \big(\sum_{k=1}^{n} f_k\big)^p\,d\mu\Big)^{1/p} \le \Big(\int \big(\sum_{k=1}^{n-1} f_k\big)^p\,d\mu\Big)^{1/p} + \big(\int f_n^p\,d\mu\big)^{1/p},$$

which, by induction, yields

$$\Big(\int \big(\sum_{k=1}^{n} f_k\big)^p\,d\mu\Big)^{1/p} \le \sum_{k=1}^{n}\big(\int f_k^p\,d\mu\big)^{1/p}.$$

The generalized Minkowski's inequality follows by the monotone convergence theorem (Chapter 1, Theorem 8).

Example 2. Applying Hölder's inequality to the functions f and g defined on $\mathbf{R}_+$ by $f(t) = \mathrm{e}^{-at}t^{ax}$ and $g(t) = \mathrm{e}^{-(1-a)t}t^{(1-a)(x-1)}$, where $0 < a < 1$ and $0 < x < \infty$, we can prove that

$$\lim_{x\to\infty} \frac{\Gamma(x+a)}{x^a\Gamma(x)} = 1,$$

where Γ is the Euler function.

If we choose $p = 1/a$ and $q = 1/(1-a)$, we have

$$\int_0^\infty e^{-t} t^{x+a-1}\,dt \le \Big(\int_0^\infty e^{-t} t^{x}\,dt\Big)^a \Big(\int_0^\infty e^{-t} t^{x-1}\,dt\Big)^{1-a}.$$

Taking into account the relation $\Gamma(x+1) = x\Gamma(x)$, the above inequality gives

$$\Gamma(x+a) \le x^a \Gamma(x).$$

Changing first a in $1-a$ and then x in $x+a$ yields

$$\Gamma(x+1) \le (x+a)^{1-a}\Gamma(x+a).$$

Combining these two inequalities, we obtain

$$\Big(\frac{x}{x+a}\Big)^{1-a} \le \frac{\Gamma(x+a)}{x^a\Gamma(x)} \le 1,$$

and the result follows when x tends to ∞.

2. $\mathcal{L}^p$ and L^p Spaces

Definition 3. Let $(X, \mathcal{A}, \mu)$ be a measure space; the vector space of the μ-measurable functions f defined on X into $\mathbf{R}$ (or $\mathbf{C}$), such that $|f|^p$, for $1 \le p < \infty$, is integrable with respect to μ, is called $\mathcal{L}^p_\mu(X)$.

In Chapter 1 we proved that $\mathcal{L}^1_\mu(X)$ was a vector space (Theorem 10). For the present definition to make sense, we have to show that $\mathcal{L}^p_\mu(X)$ is a vector space for all $p \ge 1$. This is straightforward. If $f \in \mathcal{L}^p_\mu(X)$, for all scalar λ, λf belongs to $\mathcal{L}^p_\mu(X)$. If f and g belong to $\mathcal{L}^p_\mu(X)$, $f+g$ belongs also to $\mathcal{L}^p_\mu(X)$ since[1]

$$|f+g|^p \le 2^{p-1}(|f|^p + |g|^p).$$

If there is no ambiguity on X and μ, these spaces are denoted simply by $\mathcal{L}^p$.

---

[1] This inequality is a simple consequence of the convexity of the function $t \mapsto t^p$ defined on $\mathbf{R}_+$, where $1 \le p < \infty$. In fact, we have

$$\Big|\frac{1}{2}f + \frac{1}{2}g\Big|^p \le \frac{1}{2}|f|^p + \frac{1}{2}|g|^p.$$

If μ is the counting measure, it is customary to denote the corresponding $\mathcal{L}^p$ space by ℓ^p. An element x of ℓ^p may be regarded as sequence x_n such that

$$\sum_{n=1}^{\infty} |x_n|^p < \infty.$$

If $f \in \mathcal{L}^p$, $\left(\int |f|^p \, d\mu\right)^{1/p}$ is denoted by $\|f\|_p$. With this notation we have

$$\|fg\|_1 \leq \|f\|_p \, \|g\|_q \qquad \text{Hölder's inequality,}$$

$$\|f + g\|_p \leq \|f\|_p + \|g\|_p \qquad \text{Minkowski's inequality.}$$

Remark 1. If μ is the Lebesgue measure on $]0, \infty[$, the function $x \mapsto x^{-1/2}(1+x)^{-1}$ is in $\mathcal{L}^1$ but not in $\mathcal{L}^2$, whereas the function $x \mapsto (1+x)^{-1}$ is in $\mathcal{L}^2$ but not in $\mathcal{L}^1$. In general, if $p \neq q$, neither $\mathcal{L}^p \subset \mathcal{L}^q$ nor $\mathcal{L}^q \subset \mathcal{L}^p$ is true. It can be shown, however, that
(1) if μ is bounded, $\mathcal{L}^p \subset \mathcal{L}^q$ for $p > q$;
(2) if μ is the counting measure, $\mathcal{L}^p \supset \mathcal{L}^q$ for $p > q$.

(1) Let f be any function in $\mathcal{L}^p$, and write

$$A = \{x |\, |f| \leq 1\} \qquad \text{and} \qquad B\{x |\, |f| > 1\}.$$

$f_1 = f1_A$ and $f_2 = f1_B$ are measurable. Since f_1 is bounded, it belongs to $\mathcal{L}^q$ (because μ is bounded). $q < p$ implies that, for all $x \in X$, $|f_2(x)|^q < |f_2(x)|^p$, so f_2 is in $\mathcal{L}^q$. Hence, $f_1 + f_2 = f \in \mathcal{L}^q$, that is, $\mathcal{L}^p \subset \mathcal{L}^q$.

(2) Let f be any function in $\mathcal{L}^q$. Since $\sum_{x \in X} |f(x)|^q < \infty$, the set $A = \{x |\, |f(x)| > 1\}$ is finite, and we have $\sum_{x \in A} |f(x)|^p < \infty$. For $x \in X - A$, $|f(x)| \leq 1$, so $\sum_{x \in X-A} |f(x)|^p \leq \sum_{x \in X-A} |f(x)|^q < \infty$. Hence, $\sum_{x \in A} |f(x)|^p + \sum_{x \in X-A} |f(x)|^p < \infty$, that is, $\mathcal{L}^q \subset \mathcal{L}^p$.

Remark 2. If $f \in \mathcal{L}^p$ and $g \in \mathcal{L}^q$, from Hölder's inequality, it follows that $fg \in \mathcal{L}^1$.

Theorem 3. *The mapping $f \mapsto \|f\|_p$ from $\mathcal{L}^p$ into $\mathbf{R}_+$ is a seminorm.*

This results from Minkowski's inequality.

The mapping $f \mapsto \|f\|_p$ is a seminorm and not a norm because $\|f\|_p = 0$ implies $f = 0$ a.e. and not $f = 0$.

The relation $\|f_1 - f_2\|_p = 0$ implies that $f_1 = f_2$ a.e. It is clear that this is an equivalence relation in $\mathcal{L}^p$ that partitions $\mathcal{L}^p$ into equivalence classes.

Definition 4. If R denotes the equivalence relation

$$(f_1 \in \mathcal{L}^p)\,(f_2 \in \mathcal{L}^p) \qquad f_1 R f_2 \iff \|f_1 - f_2\|_p = 0,$$

the vector space $\mathcal{L}^p/R$ is called L^p, or $L^p_\mu(X)$ if we need specify the space X and the measure μ.

$f_1 R f_2$ is read f_1 *is equivalent to* f_2. The notation $f_1 \sim f_2$ is also used.

The elements of the space L^p *are not functions but classes of functions.* If $f \in \mathcal{L}^p$, we denote by $\tilde{f}$ the class of functions equivalent to f, and we put $\|\tilde{f}\|_p = \|f\|_p$. It is readily seen that the mapping $f \mapsto \|\tilde{f}\|_p$ from L^p into $\mathbf{R}_+$ is a norm. $\|\tilde{f}\| = 0$ implies $\tilde{f} = \tilde{0}$, where $\tilde{0}$ is the class of functions equal to zero a.e.

Since $\tilde{f}$ is a class of functions, to ask for its value at a particular point $x \in X$ is, in general, meaningless. $\tilde{f}$ can be regarded as a *function defined almost everywhere.* When there is no risk of confusion we write $f \in L^p$ (without the tilde accent) as an abbreviation for $f \in \tilde{f}$ and $\tilde{f} \in L^p$. $\|f\|_p$ is called the *L^p-norm of f.*

If f, g, and h are in L^p, from Minkowski's inequality it follows that

$$\|f-h\|_p \leq \|f-g\|_p + \|g-h\|_p.$$

This suggests that a *metric* may be defined on L^p by[2]

$$(\forall f \in L^p)\,(\forall g \in L^p) \qquad d(f,g) = \|f-g\|_p.$$

Let (f_n) be a sequence of functions in L^p. If there exists a function $f \in L^p$ such that $\lim_{n\to\infty} \|f_n - f\|_p = 0$, we say that (f_n) *converges to f in L^p* or that (f_n) *converges to f in the mean of order p.*

The following theorem is fundamental.

Theorem 4. Riesz–Fisher theorem. *L^p is a complete metric space, for $1 \leq p < \infty$.*

To prove this theorem it is sufficient to prove that, as a seminormed space, $\mathcal{L}^p$ is complete. We need the following lemma.

Lemma. *Let (f_n) be a sequence of functions in $\mathcal{L}^p$ such that the series $\sum_{n=1}^\infty \|f_n\|_p$ is convergent; then the series $\sum_{n=1}^\infty f_n(x)$ is absolutely convergent almost everywhere.*

Put $f(x) = \sum_{n=1}^\infty |f_n(x)|$. The generalized Minkowski's inequality implies

$$\|f\|_p \leq \sum_{n=1}^{\infty} \|f_n\|_p,$$

[2] See the note at the end of the chapter.

which shows that $x \mapsto f^p(x)$ is integrable. Hence, the set $\{x|f(x) = \infty\}$ is negligible, and the series $\sum_{n=1}^{\infty} f_n(x)$ is absolutely convergent on the complement of this set.

A metric space is complete if, and only if, any absolutely convergent series converges in this space.[3] It is, therefore, sufficient to prove that, if (f_n) is a sequence of functions in $\mathcal{L}^p$ such that the series $\sum_{n=1}^{\infty} \|f_n\|_p$ is convergent, then the series $\sum_{n=1}^{\infty} f_n$ converges in $\mathcal{L}^p$.

Let $E = \{x|f(x) = \infty\}$, where $f(x) = \sum_{n=1}^{\infty} |f_n(x)|$, and define the sequence (g_n) by

$$g_n(x) = \begin{cases} f_n(x), & \text{if } x \notin E; \\ 0, & \text{if } x \in E. \end{cases}$$

For all positive integers n, g_n is measurable and equal to f_n almost everywhere. The function g, such that $g(x) = \sum_{n=1}^{\infty} g_n(x)$, is defined almost everywhere and belongs to $\mathcal{L}^p$ since, for all x, $|g(x)|^p \leq f^p(x)$. As a consequence of the generalized Minkowski's inequality,

$$\begin{aligned} \|g - \sum_{k=1}^{n} g_k\|_p &= \| \sum_{k=n+1}^{\infty} g_k\|_p \\ &\leq \sum_{k=n+1}^{\infty} \|g_k\|_p \\ &\leq \sum_{k=n+1}^{\infty} \|f_k\|_p, \end{aligned}$$

which shows that the series $\sum_{n=1}^{\infty} f_n$ converges to g in $\mathcal{L}^p$, and completes the proof.

The $\mathcal{L}^p$ and L^p spaces have been defined only for $1 \leq p < \infty$. Is it possible to give a meaning to $p = \infty$, and define the corresponding spaces $\mathcal{L}^\infty$ and L^∞? To answer this question, we have to introduce the following notion.

Definition 5. Let f be a real-valued measurable function defined on the measure space $(X, \mathcal{A}, \mu)$; then $\inf\{M|f(x) \leq M \text{ a.e.}\}$ is called the *essential supremum* of f, and denoted by ess sup f.

Definition 6. Let $(X, \mathcal{A}, \mu)$ be a measure space; the vector space of the μ-measurable functions f defined on X into $\mathbf{R}$ (or $\mathbf{C}$) such that ess sup $|f| < \infty$ is called $\mathcal{L}_\mu^\infty(X)$.

[3] See the note.

It is easy to verify that $\mathcal{L}^\infty_\mu(X)$, or $\mathcal{L}^\infty$ if there is no risk of ambiguity on X and μ, is a vector space.

If ess sup $|f| < \infty$, f is said to be *essentially bounded.*

If $f \in \mathcal{L}^\infty$, ess sup $|f|$ is denoted by $\|f\|_\infty$.

Remark. This notation is justified by the following result. *If the measure μ is such that $\mu(X) < \infty$, for any essentially bounded function f we have*

$$\|f\|_\infty = \lim_{p\to\infty} \|f\|_p.$$

Indeed, the set $\{x \mid |f(x)| > \|f\|_\infty\}$ being negligible, for any $p \in [1, \infty[$, we have

$$\begin{aligned}\|f\|_p &= \Big(\int |f|^p \, d\mu\Big)^{1/p} \\ &\le \|f\|_\infty \big(\mu(X)\big)^{1/p},\end{aligned}$$

and, since $\mu(X) < \infty$,

$$\limsup_{p\to\infty} \|f\|_p \le \|f\|_\infty.$$

Conversely, if $a < \|f\|_\infty$, the set $E = \{x \mid |f(x)| > a\}$ is not negligible, and we have

$$\begin{aligned}\|f\|_p &\ge \Big(\int_E |f|^p \, d\mu\Big)^{1/p} \\ &\ge a\big(\mu(E)\big)^{1/p}.\end{aligned}$$

Hence,

$$\liminf_{p\to\infty} \|f\|_p \ge a,$$

and, since a is arbitrary,

$$\liminf_{p\to\infty} \|f\|_p \ge \|f\|_\infty.$$

Hölder's inequality remains valid for the pair of conjugate exponents $(1, \infty)$, and we may write

$$\|fg\|_1 \le \|f\|_1 \, \|g\|_\infty.$$

To prove it, consider a function h equal to g almost everywhere, and such that, for all $x \in X$, $|h(x)| \le \|g\|_\infty$. We then have

$$\|fg\|_1 = \|fh\|_1 \le \sup_{x\in X} |h(x)| \int |f| \, d\mu.$$

This is exactly Hölder's inequality.

Minkowski's inequality remains valid for $p = \infty$, that is,

$$\|f + g\|_\infty \leq \|f\|_\infty + \|g\|_\infty.$$

To prove it, consider two functions f_1 and g_1 respectively equal to f and g almost everywhere, and such that, for all $x \in X$, $|f_1(x)| \leq \|f\|_\infty$ and $|g_1(x)| \leq \|g\|_\infty$. Then

$$\begin{aligned}|f(x) + g(x)| &= |f_1(x) + g_1(x)| \text{ a.e.}\\ &\leq \|f\|_\infty + \|g\|_\infty \text{ a.e.}\end{aligned}$$

$\|f\|_\infty + \|g\|_\infty$ is therefore an essential supremum of $|f + g|$, and the result is proved.

Minkowski's inequality implies that the mapping $f \mapsto \|f\|_\infty$ from $\mathcal{L}^\infty$ into $\mathbf{R}_+$ is a seminorm since $\|f\|_\infty = 0 \Rightarrow f = 0$ a.e. Here again it is, however, possible to define a normed space.

Definition 7. If R denotes the equivalence relation

$$(f_1 \in \mathcal{L}^\infty)\,(f_2 \in \mathcal{L}^\infty) \qquad f_1 R f_2 \iff \|f_1 - f_2\|_\infty = 0,$$

the vector space $\mathcal{L}^\infty/R$ is called L^∞, or $L^\infty_\mu(X)$ if we need specify the space X and the measure μ.

The elements of the space L^∞ are not functions but classes of functions. If $f \in \mathcal{L}^\infty$, we denote by $\tilde{f}$ the class of functions equivalent to f, and we put $\|\tilde{f}\|_\infty = \|f\|_\infty$. It is readily seen that the mapping $f \mapsto \|\tilde{f}\|_\infty$ from L^∞ into $\mathbf{R}_+$ is a norm. $\|\tilde{f}\| = 0$ implies $\tilde{f} = \tilde{0}$, where $\tilde{0}$ is the class of functions equal to zero a.e.

A metric may be defined on L^∞ by

$$(\forall f \in L^\infty)\,(\forall g \in L^\infty) \qquad d(f, g) = \|f - g\|_\infty.$$

Theorem 5. Riesz–Fisher theorem. *L^∞ is a complete metric space.*

As above, to establish this important result, it is sufficient to prove that, as a seminormed space, $\mathcal{L}^\infty$ is complete.

Let (f_n) be a sequence of functions in $\mathcal{L}^\infty$ such that the series $\sum_{n=1}^{\infty} \|f_n\|_\infty$ is convergent; to prove the completeness of $\mathcal{L}^\infty$, we have to show that the series $\sum_{n=1}^{\infty} f_n(x)$ is convergent in $\mathcal{L}^\infty$ (see Theorem 4 and the note).

Define a sequence of bounded functions (g_n) such that, for all positive integers n,

$$g_n = f_n \text{ a.e.} \qquad \text{and} \qquad \sup_{x \in X} |g_n(x)| \leq \|f_n\|_\infty.$$

The series $\sum_{n=1}^{\infty} g_n(x)$ converges normally to a bounded measurable function g, which is therefore in $\mathcal{L}^\infty$. From

$$\begin{aligned} |g(x) - \sum_{k=1}^{n} g_k(x)| &\leq \sum_{k=n+1}^{\infty} |g_k(x)| \\ &\leq \sum_{k=n+1}^{\infty} \|f_k\|_\infty \end{aligned}$$

it follows that

$$\|g - \sum_{k=1}^{n} f_k\|_\infty \leq \sum_{k=n+1}^{\infty} \|f_k\|_\infty,$$

which shows that the series $\sum_{n=1}^{\infty} f_n$ converges to g in $\mathcal{L}^\infty$, and completes the proof.

3. Dense Subspaces in L^p

If μ is the Lebesgue measure m, we will show that, for $1 \leq p < \infty$, $L^p_m(\mathbf{R})$ is the completion of the space $C^0_c(\mathbf{R})$ of continuous functions with compact support[4] defined on $\mathbf{R}$ endowed with the L^p metric. For $p = 1$, this result shows that the Lebesgue integral is the "right" generalization of the Riemann integral.

Theorem 6. *Let $(X, \mathcal{A}, \mu)$ be a measure space, and let S be the space of simple functions on $(X, \mathcal{A}, \mu)$ such that*

$$(\forall s \in S) \qquad \mu(\{x | s(x) \neq 0\}) < \infty.$$

If $1 \leq p < \infty$, then S is dense in $L^p_\mu(X)$.

It is clear that $S \subset L^p_\mu(X)$. If f is a positive function belonging to $L^p_\mu(X)$, by Theorem 5, Chapter 1, there exists an increasing sequence (s_n) of positive simple functions whose limit is f. For all positive integers n, we

---

[4] The support of a real-valued function defined in X is the closure of the set $\{x | x \in X, f(x) \neq 0\}$.

have $0 \leq s_n < f$, which implies that $s_n \in L^p_\mu(X)$ and, therefore, $s_n \in S$. Since $|f - s_n|^p \leq f^p$, the dominated convergence theorem shows that

$$\lim_{n \to \infty} \|f - s_n\| = 0.$$

Thus f belongs to the closure of S (see the note). For a real or complex function the result remains valid since a real (resp. complex) function is a linear combination of two (resp. four) positive functions. Therefore, for the L^p metric, $\overline{S} = L^p$ or, equivalently, S is dense in L^p.

Remark. Since S is dense in all L^p spaces for $1 \leq p < \infty$, it is dense in $L^p \cap L^q$, where $p \neq q$. So $L^p \cap L^q$ is dense in L^p, for the L^p-metric, and dense in L^q for the L^q-metric.

In the particular, but important, case $X = \mathbf{R}$ and $\mu = m$ we have the following results.

Corollary 1. *If $1 \leq p < \infty$, then the vector space $C^0_c(\mathbf{R})$ of the continuous functions with compact support defined on $\mathbf{R}$ is dense in $L^p_m(\mathbf{R})$.*

It is clear that $C^0_c(\mathbf{R})$ is contained in $L^p_m(\mathbf{R})$. To prove that $C^0_c(\mathbf{R})$ is dense in $L^p_m(\mathbf{R})$ we have to show that any simple function with compact support is the limit in L^p of a sequence of continuous functions with compact support. Since a simple function is a linear combination of characteristic functions, we need only show that a characteristic function 1_A, where A is measurable and such that $m(A) < \infty$, is the limit in L^p of a sequence of continuous functions with compact support. Given $\epsilon > 0$, for any measurable subset A of $\mathbf{R}$, there exist two measurable subsets A_1 and A_2 such that

$$\overline{A_1} \subset A \subset \mathring{A}_2 \qquad m(A_1 - A - 2) \leq \epsilon^p,$$

where $\mathring{A}_2$ denotes the interior of the subset A_2 (see the note). Let g be the function defined by

$$g(x) = \frac{d(x, \overline{\mathring{A}_2^C})}{d(x, \overline{\mathring{A}_2^C}) + d(x, \overline{A_1})}.$$

Since $\overline{A_1}$ and $\overline{\mathring{A}_2^C}$ are two disjoint closed sets,

$$g(x) = \begin{cases} 0, & \text{if } x \in \overline{\mathring{A}_2^C}; \\ 1, & \text{if } x \in \overline{A_1}. \end{cases}$$

g is continuous, and, since

$$\begin{aligned}\|1_A - g\|_p^p &= \int_{\mathbf{R}} |1_A - g|\, dm \\ &\le \int_{A_2 - A1} dm \\ &\le \epsilon^p,\end{aligned}$$

we have $g = 1_A$ a.e.

Corollary 2. *If $1 \le p < \infty$, then the vector space $C_c^k(\mathbf{R})$ of the k times continuously differentiable functions with compact support defined on $\mathbf{R}$ is dense in $L_m^p(\mathbf{R})$.*

To establish this result it is sufficient to show that $C_c^k(\mathbf{R})$ is dense in $C_c^0(\mathbf{R})$. The proof makes use of a very general technique, which is studied in detail in Chapter 4, Section 4. Given a continuous function f with compact support, we construct a sequence (f_n) of functions in $C_c^k(\mathbf{R})$ that converges uniformly to f.

Let ρ be a function in $C_c^k(\mathbf{R})$ such that

$$\rho(x) = \begin{cases} > 0, & \text{if } x \in]-a, a[, \\ = 0, & \text{otherwise,} \end{cases}$$

and

$$\int \rho(x)\, dx = 1.$$

For all positive integers n, set

$$\begin{aligned}\rho_n(x) &= n\rho(nx) \\ f_n(x) &= \int f(t)\rho_n(x-t)\, dt.\end{aligned}$$

For example, we may choose

$$\rho(x) = \begin{cases} A(a^2 - x^2)^{k-1}, & \text{if } x \in]-a, a[, \\ 0, & \text{otherwise,} \end{cases}$$

where the constant A is such that $\int \rho(x)\, dx = 1$.[5] We have

$$\begin{aligned}f(x) - f_n(x) &= \int \big(f(x) - f(x-t)\big)\rho_n(t)\, dt \\ &= \int_{-a/n}^{a/n} \big(f(x) - f(x-t)\big)\rho_n(t)\, dt.\end{aligned}$$

[5] The integral is easily evaluated. It is found that $A^{-1} = B(k+2, 1/2)$, where B is the Euler Beta function.

f being continuous, for any positive ϵ, there exists an integer $n_0(\epsilon)$ such that, if $n \geq n_0(\epsilon)$ and $t \leq a/n$, we have $|f(x) - f(x-t)| \leq \epsilon$. $\rho(t)$ being positive, if $t \leq a/n$, it follows that

$$\left| \int_{-a/n}^{a/n} (f(x) - f(x-t))\rho_n(t)\, dt \right| \leq \int_{-a/n}^{a/n} \left|(f(x) - f(x-t))\right| \rho_n(t)\, dt$$
$$\leq \epsilon.$$

A continuous function on a bounded interval being uniformly continuous, the sequence (f_n) converges uniformly to f. Furthermore, by Theorem 14, Chapter 1, for all $n \in \mathbf{N}$, f_n is k times continuously differentiable.

To complete the proof, it is sufficient to note that uniform convergence on a bounded interval implies convergence in the mean of order p since all the functions concerned are in L^p.

Remark 1. These results are valid only for $1 \leq p < \infty$. For the L^∞-norm, the completion of $C_c^0(\mathbf{R})$ is not $L_m^\infty(\mathbf{R})$ but $C_0^0(\mathbf{R})$, the vector space of the continuous functions that *vanish at infinity*. A function f defined on $\mathbf{R}$ is said to vanish at infinity if, for every positive ϵ, there exists a compact subset K of $\mathbf{R}$ such that $|f(x)| < \epsilon$ for all $x \notin K$. Note that for a continuous function the essential supremum coincides with the ordinary supremum.

Remark 2. For $p = 1$, Corollary 1 states that if the distance between two continuous functions f and g, with compact support in $\mathbf{R}$, is defined by

$$d(f, g) = \int_{-\infty}^{\infty} |f(x) - g(x)|\, dx,$$

the completion of the resulting metric space is the class of the Lebesgue integrable functions defined almost everywhere.

4. Fourier Transform

Definition 8. Let f be a function belonging to $L_m^1(\mathbf{R})$; the function $\widehat{f}$, defined in $\mathbf{R}$ by

$$\widehat{f}(t) = \int_{-\infty}^{\infty} f(x) e^{2i\pi tx}\, dx,$$

is called the *Fourier transform* of f.

Since $f \in L_m^1(\mathbf{R})$, the integral is well defined for any real t. By Theorem 13, Chapter 1, $\widehat{f}$ is continuous.

Remark 1. When consulting a book on the Fourier transform, it is advisable to first check the definition adopted by the author, since the following definitions are also widely used.

(1) $$\widehat{f}(t) = \int_{-\infty}^{\infty} f(x)e^{-2i\pi tx}\,dx.$$

(2) $$\widehat{f}(t) = \int_{-\infty}^{\infty} f(x)e^{itx}\,dx.$$

(3) $$\widehat{f}(t) = \int_{-\infty}^{\infty} f(x)e^{-itx}\,dx.$$

If f is even, we have

$$\widehat{f}(t) = 2\int_{0}^{\infty} f(x)\cos 2\pi tx\,dx.$$

If f is odd, we have

$$\widehat{f}(t) = 2i\int_{0}^{\infty} f(x)\sin 2\pi tx\,dx.$$

Remark 2. $\widehat{f}$ is continuous and vanishes at infinity. However, any continuous function that vanishes at infinity is not necessarily the Fourier transform of a function in $L^1_m(\mathbf{R})$.

Theorem 7. *Let f be a function in $L^1_m(\mathbf{R})$, and let a be a real number.*
(1) If $g(x) = f(x)e^{2i\pi ax}$, then $\widehat{g}(t) = \widehat{f}(t+a)$.
(2) If $g(x) = f(x-a)$, then $\widehat{g}(t) = \widehat{f}(t)e^{2i\pi at}$
(3) If $g(x) = \overline{f(-x)}$, then $\widehat{g}(t) = \overline{\widehat{f}(t)}$.
(4) If $g(x) = f(x/a)$ $(a > 0)$, then $\widehat{g}(t) = a\widehat{f}(at)$.
(5) If $g(x) \in L^1_m(\mathbf{R})$ and $h(x) = (f * g)(x) = \int_{-\infty}^{\infty} f(x-y)g(y)\,dy$, then $\widehat{h}(t) = \widehat{f}(t)\widehat{g}(t)$.
(6) If $g(x) = 2i\pi x f(x)$ and $g(x) \in L^1_m(\mathbf{R})$, then $\widehat{f}$ is differentiable, and $\widehat{f}'(t) = \widehat{g}(t)$.

Proofs of (1), (2), (3), and (4) are straightforward. Since

$$\begin{aligned}\widehat{h}(t) &= \int_{-\infty}^{\infty} e^{2i\pi tx}\,dx \int_{-\infty}^{\infty} f(x-y)g(y)\,dy \\ &= \int_{-\infty}^{\infty} g(y)e^{2i\pi ty}\,dy \int_{-\infty}^{\infty} f(x-y))e^{2i\pi t(x-y)}\,dx,\end{aligned}$$

by Fubini's theorem and the invariance under translation of the Lebesgue measure, (5) follows.

To prove (6), consider the sequence (ϕ_n) defined by

$$\phi_n(t) = \frac{\widehat{f}(t+h_n) - \widehat{f}(t)}{h_n}$$
$$= \int_{-\infty}^{\infty} f(x) e^{2i\pi tx} \frac{e^{2i\pi h_n x} - 1}{h_n}\, dt,$$

where (h_n) is a seqence of positive reals that tends to zero. Since

$$\lim_{n\to\infty} \phi_n = \widehat{f'}$$

and

$$\left| \frac{\exp(2i\pi h_n x) - 1}{h_n} \right| \le 2\pi |x|,$$

the result follows.

Remark 1. The relation between the Fourier transform and the translation group is illustrated by relations (1) and (2). The translation group is isomorphic to the additive group **R**. For any real a, the mapping $x \mapsto e^{2i\pi ax}$ is a homomorphism of the additive group of reals **R** into the multiplicative group of the complex numbers of absolute value 1. Such a homomorphism is called a *character*. Properties (1) and (2) show that the Fourier transform converts multiplication by a character into translation and conversely.

Remark 2. The relation (5) shows that the Fourier transform converts the convolution of two functions into the ordinary product of their transforms. This result is frequently used (see Examples 1 and 2).

Remark 3. Under certain conditions, it is possible to establish an analogue of relation (6). If the functions f and f' both belong to $L^1_m(\mathbf{R})$ and if f is the indefinite integral of f', then the Fourier transform of f' is $-2i\pi t\widehat{f}(t)$.[6] This result is easily established by an integration by parts. We have

$$\widehat{f'}(t) = \int_{-\infty}^{\infty} f'(x) e^{2i\pi tx}\, dx$$
$$= -\int_{-\infty}^{\infty} 2i\pi t f(x) e^{2i\pi tx}\, dx,$$

since the integrated term is equal to zero. The fact that the Fourier transform converts differentiation into multiplication by it makes the Fourier

[6] The Fourier transform of the derivative of f, denoted by $\widehat{f'}$, should not be confused with the derivative of the Fourier transform, denoted by $\widehat{f}'$.

transform a particularly useful tool in the study of differential equations (see Examples 4 and 5). Furthermore, this result suggests that the derivation of a function f is probably a convolution (see Chapter 4, Theorem 3). There is, however, no function d such that

$$f'(x) = \int_{-\infty}^{\infty} d(y)f(x-y)\,dy.$$

Remark 4. Relation (6) and its analogue discussed above may be extended to higher orders. Let k be a positive integer.
(1) If the function $x \mapsto x^k f(x)$ belongs to $L^1_m(\mathbf{R})$, $\widehat{f}$ is k times continuously differentiable, and we have

$$\widehat{f}^{(k)}(t) = \int_{-\infty}^{\infty} (2i\pi x)^k f(x) \mathrm{e}^{2i\pi tx}\,dx$$

with

$$|\widehat{f}^{(k)}(t)| \le \|(2\pi x)^k f(x)\|_1.$$

(2) If f is k times continuously differentiable, and all the derivatives of $f^{(j)}$, for $j = 1, 2, \ldots, k$, are in $L^1_m(\mathbf{R})$, then

$$\widehat{f^{(k)}}(t) = (-2i\pi t)^k \widehat{f}(t)$$

and

$$|(-2i\pi t)^k \widehat{f}(t)| \le \|f^{(k)}\|_1.$$

(3) Let F be a primitive of f. If F and f are in $L^1_m(\mathbf{R})$, then

$$\widehat{F}(t) = -\frac{1}{2i\pi t}\widehat{f}(t).$$

Example 1. The Fourier transform of the function ℓ_a, defined by

$$\ell_a(x) = \frac{1}{\pi}\frac{a}{x^2+a^2},$$

where $a > 0$, is easy to determine by the residue theorem. It is found that

$$\widehat{\ell_a}(t) = \begin{cases} \mathrm{e}^{-2\pi a t}, & \text{if } t > 0; \\ \mathrm{e}^{2\pi a t}, & \text{if } t < 0. \end{cases}$$

That is,

$$\widehat{\ell_a}(t) = \mathrm{e}^{-2\pi a|t|}.$$

Hence, from Theorem 7(5), it follows that

$$\ell_a * \ell_b = \ell_{a+b}.$$

Example 2. The Fourier transform of the function g_σ, defined by

$$g_\sigma(x) = \frac{1}{\sigma\sqrt{2\pi}} \exp -\left(\frac{x^2}{2\sigma^2}\right),$$

may be determined by various methods. By Theorem 14, Chapter 1, we may differentiate under the integral sign. This yields

$$\begin{aligned}\widehat{g_\sigma}'(t) &= \frac{1}{\sigma\sqrt{2\pi}} \int_{-\infty}^{\infty} 2i\pi x \exp -\left(\frac{x^2}{2\sigma^2} - 2i\pi tx\right) dx \\ &= -4\pi^2\sigma^2 t\widehat{g_\sigma}(t).\end{aligned}$$

So $\widehat{g_\sigma}$ satisfies the differential equation

$$y' + 4\pi^2\sigma^2 ty = 0.$$

Since

$$\widehat{g_\sigma}(0) = \frac{1}{\sigma\sqrt{2\pi}} \int_{-\infty}^{\infty} \exp -\left(\frac{x^2}{2\sigma^2}\right) dx = 1,$$

we finally obtain

$$\widehat{g_\sigma}(t) = \exp -(2\pi^2\sigma^2 t^2).$$

This result may also be derived directly. In fact,

$$\begin{aligned}\widehat{g_\sigma}(t) &= \frac{1}{\sigma\sqrt{2\pi}} \int_{-\infty}^{\infty} \exp -\left(\frac{x^2}{2\sigma^2} - 2i\pi tx\right) dx \\ &= \frac{\exp -(2\pi^2\sigma^2 t^2)}{\sigma\sqrt{2\pi}} \int_{-\infty}^{\infty} \exp -\frac{1}{2}\left(\frac{x}{\sigma} - 2i\pi\sigma t\right)^2 dx,\end{aligned}$$

and, by Cauchy's theorem, the result follows.

Theorem 8(5) yields

$$g_{\sigma_1} * g_{\sigma_2} = g_{(\sigma_1^2+\sigma_2^2)^{1/2}}.$$

If $\widehat{f}$ is the Fourier transform of f, it will be shown that, under certain conditions, it is possible to establish the following inversion formula:

$$f(x) = \int_{-\infty}^{\infty} \widehat{f}(t) e^{-2i\pi tx}\, dt.$$

For instance, it is easy to verify that

$$\int_{-\infty}^{\infty} e^{-2\pi a|t|-2i\pi tx}\, dt = \frac{1}{\pi}\frac{a}{x^2+a^2}.$$

To prove the inversion formula, it is hopeless to replace $\widehat{f}(t)$ in the right-hand side by $\int_{-\infty}^{\infty} f(y)e^{2i\pi ty}\, dy$, since we would run into the integral

$$\int_{-\infty}^{\infty} e^{t(y-x)}\, dt,$$

which is meaningless.[7] It is also hopeless to expect the inverse Fourier transform of $\widehat{f}$ to be equal to f *everywhere*. As a matter of fact, $\widehat{f}$ is the Fourier transform of all the functions f belonging to the same equivalence class. So, in general, the inverse Fourier transform of $\widehat{f}$ will be, at best, a function equal to f *almost everywhere*. Finally, we have to assume that $\widehat{f}$ belongs to $L^1_m(\mathbf{R})$. This is of course not always the case. For instance, if $f = 1_{[-1,1]}$, then $\widehat{f}(t) = \sin 2\pi t/\pi t$, which is not in $L^1_m(\mathbf{R})$.

Theorem 8. *Let f be a function in $L^1_m(\mathbf{R})$ such that $\widehat{f}$ is also in $L^1_m(\mathbf{R})$. If*

$$g(x) = \int_{-\infty}^{\infty} \widehat{f}(t)e^{-2i\pi tx}\, dt,$$

then the function g belongs to $C^0_0(\mathbf{R})$, and we have $g = f$ a.e.

Let ρ be the function defined on $\mathbf{R}$ by

$$\rho(x) = \frac{1}{\pi}\frac{1}{1+x^2},$$

and let (ρ_n) be a sequence of functions such that, for all $n \in \mathbf{N}$,

$$\rho_n(x) = n\rho(nx).$$

From Example 1 ($\rho_n = \ell_{1/n}$), it follows that

$$\widehat{\rho_n}(t) = e^{-2i\pi|t|/n}.$$

We have

$$\begin{aligned}(f * \rho_n)(x) &= \int_{-\infty}^{\infty} f(x-y)\, dy \int_{-\infty}^{\infty} \widehat{\rho_n}(t)e^{-2i\pi ty}\, dt \\ &= \int_{-\infty}^{\infty} \widehat{\rho_n}(t)\, dt \int_{-\infty}^{\infty} f(x-y)e^{-2i\pi ty}\, dy,\end{aligned}$$

---

[7] See, however, Chapter 4, Examples 3.1 and 6.2 and Problem 11.

where the application of Fubini's theorem is legitimate since

$$\int_{-\infty}^{\infty}\int_{-\infty}^{\infty} |\widehat{\rho_n}(t)f(x-y)e^{-2i\pi ty}|\,dy\,dt = \int_{-\infty}^{\infty}\int_{-\infty}^{\infty} |\widehat{\rho_n}(t)f(x-y)|\,dy\,dt$$

$$\int_{-\infty}^{\infty} |\widehat{\rho_n}(t)|\,dt \int_{-\infty}^{\infty} |f(x-y)|\,dy$$

$$\leq \infty.$$

Hence,

$$(f * \rho_n)(x) = \int_{-\infty}^{\infty} \widehat{\rho_n}(t)\,dt \int_{-\infty}^{\infty} f(u)e^{-2i\pi t(x-u)}\,du,$$

$$= \int_{-\infty}^{\infty} \widehat{\rho_n}(t)\widehat{f}(t)e^{-2i\pi tx}\,dt.$$

Since

$$|\widehat{\rho_n}(t)\widehat{f}(t)e^{-2i\pi tx}| \leq |\widehat{f}(t)| \qquad \text{and} \qquad \lim_{n\to\infty} \rho_n(t) = 1,$$

by the dominated convergence theorem it follows that

$$\lim_{n\to\infty} f * \rho_n = g.$$

To prove that $g = f$ a.e., it is sufficient to show that

$$\lim_{n\to\infty} \|f * \rho_n - f\|_1 = 0.$$

Since there exists in $L^1_m(\mathbf{R})$ a sequence (f_m) of continuous functions that converges to f, we may write

$$\|f * \rho_n - f\|_1 \leq |(f - f_m) * \rho_n\|_1 + \|f_m * \rho_n - f_m\|_1 + \|f_m - f\|_1.$$

f_m being continuous, $\lim_{n\to\infty} f_m * \rho_n = f_m$, and the result follows. The fact that g belongs to $C^0_0(\mathbf{R})$ is obvious.

Corollary. *Let f and g be two functions in $L^1_m(\mathbf{R})$; if $\widehat{f} = \widehat{g}$ then $f = g$* a.e.

Indeed, $h = f - g$ belongs to $L^1_m(\mathbf{R})$, and its Fourier transform is equal to zero. By the inversion theorem, $h = 0$ a.e., that is $f = g$ a.e.

Remark. The function ρ used to establish the inversion theorem is not unique. Any positive function that has a positive Fourier transform and whose integral can be calculated could have been used. For instance, the function defined by

$$\rho(x) = \begin{cases} \frac{1}{a}\left(1 - \frac{|x|}{a}\right), & \text{if } |x| \leq a, \\ 0, & \text{otherwise,} \end{cases}$$

where $a > 0$, could have been used. Its Fourier transform,

$$\widehat{\rho}: t \mapsto \frac{\sin^2 \pi a t}{\pi^2 a^2 t^2},$$

belongs to $L^1_m(\mathbf{R})$, and the inversion formula is easy to verify applying the residue theorem. $rho = g_1$ (Example 2) is another possible choice.

If f is in $L^1_m(\mathbf{R})$, its Fourier transform is well defined and completely characterizes the equivalence class of f. However, the inversion theorem applies if, and only if, $\widehat{f}$ is in $L^1_m(\mathbf{R})$. If this is not the case, how is it possible to determine f from $\widehat{f}$?

If $\widehat{f}$ is not integrable, the relation

$$\lim_{n\to\infty} \int_{-\infty}^{\infty} \widehat{\rho_n}(t)\widehat{f}(t)\mathrm{e}^{-2i\pi tx}\, dt = \int_{-\infty}^{\infty} \widehat{f}(t)\mathrm{e}^{-2i\pi tx}\, dt$$

does not hold anymore, but if g_n denotes the function

$$x \mapsto \int_{-\infty}^{\infty} \widehat{\rho_n}(t)\widehat{f}(t)\mathrm{e}^{-2i\pi tx}\, dt,$$

we have

$$\lim_{n\to\infty} \|g_n - f\|_1 = 0.$$

We can, therefore, state

Theorem 9. *Let f be a function in $L^1_m(\mathbf{R})$; the sequence (g_n) defined by*

$$g_n(x) = \int_{-\infty}^{\infty} \widehat{\rho_n}(t)\widehat{f}(t)\mathrm{e}^{-2i\pi tx}\, dt,$$

where $\rho_n(t) = \exp(-2\pi|t|/n)$, converges to f in $L^1_m(\mathbf{R})$.

The function ρ_n is called a *convergence factor*. As mentioned in the preceding remark, it is not unique.

Example 3. Let ξ be a real random variable. The function

$$\varphi: t \mapsto \int_{-\infty}^{\infty} \mathrm{e}^{itx}\, dP_\xi(x)$$

is called the *characteristic function*[8] of the random variable ξ.

[8] This characteristic function should not be confused with the characteristic function of a subset A denoted by 1_A. To avoid confusion, 1_A is sometimes called the *indicator* of A.

From the corollary to the inversion theorem, it follows that φ_ξ "characterizes" the probability distribution of the random variable ξ.

The following three properties are readily verified.

(1) φ is continuous.

(2) $\varphi(0) = 1$.

(3) For all $t \in \mathbf{R}$, $|\varphi(t)| \leq 1$.

If $E(\xi^k) < \infty$,[9] then $\varphi^{(k)}$ exists and is continuous. By Theorem 14, Chapter 1, we have

$$\varphi^{(k)}(t) = i^k \int_{-\infty}^{\infty} x^k e^{itx}\, dP_\xi(x),$$

and $\varphi(0) = i^k E(\xi^k)$.

If $\eta = a\xi + b$, then $\varphi_\eta(t) = e^{itb}\varphi_\xi(at)$.

If ξ_1 and ξ_2 are two independent random variables, their sum $\xi_1 + \xi_2$ is also a random variable. Since

$$e^{it(x_1+x_2)} = e^{itx_1}e^{itx_2},$$

we have

$$\varphi_{\xi_1+\xi_2} = \varphi_{\xi_1}\varphi_{\xi_2},$$

that is, the characteristic function of $\xi_1 + \xi_2$ is the product of the characteristic functions of ξ_1 and ξ_2.

A random variable ξ is said to be *normally distributed* with mean value m and variance σ^2 if

$$dP_\xi(x) = \frac{1}{\sqrt{2\pi}\sigma} \exp\left(\frac{(x-m)^2}{2\sigma^2}\right) dx.$$

In this case $\varphi_\xi(t) = e^{imt}e^{-\sigma^2 t^2/2}$ (Example 2).

Let $\xi_1, \xi_2, \ldots, \xi_n$ be n independent random variables all having the same probability distribution. If, for $j = 1, 2, \ldots, n$, $E(\xi_j) = 0$, $E(\xi_j^2) = 1$,[10] and $E(|\xi|^3) < \infty$, then the characteristic function of the random variable $\zeta_n = (\xi_1 + \xi_2 + \cdots + \xi_n)/\sqrt{n}$ tends uniformly on any finite interval to $(2\pi)^{-1/2}e^{-t^2/2}$ when n tends to ∞.

---

[9] $E(\xi^k)$ is called the *moment of order k* of the random variable ξ.

[10] If $E(\xi)$ and $E(\xi^2)$ exist, we can always change the origin and the scale of ξ in order to have $E(\xi) = 0$ and $E(\xi^2) = 1$.

This important result in probability theory is known as the *central limit theorem.* It states that as n tends to ∞, the probability distribution of the random variable ζ_n tends to a *universal* probability distribution, namely the normal distribution with zero mean value and unit variance.

Let P_ξ be the probability distribution of the random variables ξ_j ($j = 1, 2, \ldots, n$). We have

$$\varphi_{\zeta_n}(t) = \left(\varphi_\xi\left(\frac{t}{\sqrt{n}}\right)\right)^n.$$

Moreover,

$$\begin{aligned}\varphi(t) &= \int_{-\infty}^{\infty} \mathrm{e}^{itx}\, dP_\xi(x) \\ &= \int_{-\infty}^{\infty} \left(1 + itx - \frac{1}{2}t^2x^2 + O(t^3x^3)\right) dP_\xi(x) \\ &= 1 - \frac{1}{2}t^2 + O(t^3).\end{aligned}$$

Therefore,

$$\varphi_\xi\left(\frac{t}{\sqrt{n}}\right) = \left(1 - \frac{t^2}{2n}\right)\left(1 + O\left(\frac{t^3}{n^{3/2}}\right)\right);$$

hence,

$$\varphi_{\zeta_n}(t) = \left(1 - \frac{t^2}{2n}\right)^n \left(1 + O\left(\frac{t^3}{n^{1/2}}\right)\right).$$

That is,

$$\lim_{n\to\infty} \varphi_{\zeta_n}(t) = \mathrm{e}^{-t^2/2}.$$

Example 4. If ϕ is harmonic in the half-plane $\mathbf{R} \times \mathbf{R}_+$,[11] and satisfies the boundary conditions

$$\phi(x, 0) = f(x) \qquad \text{and} \qquad \lim_{y\to\infty} \phi(x, y) = 0,$$

then, for all $(x, y) \in \mathbf{R} \times \mathbf{R}_+$,

$$\phi(x, y) = \frac{y}{\pi} \int_{-\infty}^{\infty} \frac{f(u)\, du}{(x-u)^2 + y^2}.$$

[11] A function ϕ is said to be *harmonic* in a domain Ω if $\nabla^2\phi = 0$ in Ω. The differential equation $\nabla^2\phi = 0$ is known as *Laplace's equation.*

This well-known formula in potential theory is due to Poisson. It may be derived as follows. Write

$$\widehat{\phi}(\xi, y) = \int_{-\infty}^{\infty} \phi(x, y) e^{2i\pi\xi x}\, dx.$$

ϕ being a solution of Laplace's equation satisfying the boundary conditions given above, $\widehat{\phi}$ is the solution of the differential equation

$$\frac{d^2\widehat{\phi}}{dy^2} - 4\pi\xi^2\widehat{\phi} = 0$$

such that

$$\widehat{\phi}(\xi, 0) = \widehat{f}(\xi) \qquad \text{and} \qquad \lim_{y\to\infty} \widehat{\phi}(\xi, y) = 0.$$

Hence,

$$\widehat{\phi}(\xi, y) = \widehat{f}(\xi) e^{-2\pi|\xi|y},$$

and, by Theorem 7(5) and Example 1, this yields Poisson's formula.

Note that the existence of $\widehat{f}$ has been implicitly assumed.

Remark. This example illustrates the use of the Fourier transform in solving certain partial differential equations. The general procedure is to transform the equation with respect to one of the variables. This gives a simpler partial differential equation in the remaining variables. This new problem is solved, with the transform variable treated as a parameter, taking into account the boundary conditions or initial conditions imposed on the transform. The inverse Fourier transform of its solution is the solution of the original differential equation. Of course, when using such a technique, each step should be justified. It is, however, much easier to check directly that the solution found in this way satisfies the equation and all the conditions imposed on it. If the solution to the problem is unique, as is usually the case in physical applications, the solution obtained by the transform technique is the "right" one.

Example 5. The *heat equation*

$$\frac{\partial\phi}{\partial t} = \frac{\partial^2\phi}{\partial x^2}$$

describes the conduction of heat in a one-dimensional medium. $\phi(x, t)$ represents the temperature at point x and time t.[12] If the medium is

---

[12] This equation is also known as the *diffusion equation*. For other solutions, see Example 5.5 and Chapter 4, Example 4.3.

infinite, $(x,t) \in \mathbf{R} \times \mathbf{R}_+$. This partial differential equation may be solved by the method used in the preceding example. If $\phi(x,t)$ satisfies the initial condition $\phi(x,0) = f(x)$, and the boundary condition $\lim_{|x|\to\infty} \phi(x,t) = 0$, we write

$$\widehat{\phi}(\xi,t) = \int_{-\infty}^{\infty} \phi(x,t) e^{2i\pi\xi x}\, dx.$$

Then $\widehat{\phi}$ is the solution of the differential equation

$$\frac{d\widehat{\phi}}{dt} + 4\pi^2\xi^2\widehat{\phi} = 0$$

such that $\widehat{\phi}(\xi,0) = \widehat{f}(\xi)$. Therefore,

$$\widehat{\phi}(\xi,t) = \widehat{f}(\xi) e^{-4\pi^2\xi^2 t},$$

and, by Theorem 7(5) and Example 2, this yields

$$\phi(x,t) = \int_{-\infty}^{\infty} \frac{f(u)}{2\sqrt{\pi t}} \exp\left(-\frac{(x-u)^2}{4t}\right) du.$$

Note that, as in the preceding example, the existence of $\widehat{f}$ has been implicitly assumed.

5. Laplace Transform

The Laplace transform is closely related to the Fourier transform.

Definition 9. If f is a locally integrable function on $\mathbf{R}_+$, the function $z \mapsto L(z,f)$, defined in $\mathbf{C}$ by

$$L(z,f) = \int_0^{\infty} e^{-zx} f(x)\, dx,$$

is called the *Laplace transform* of f.

The Laplace transform of f is well defined if the function $x \mapsto e^{-\xi x} f(x)$, where $\xi = \Re z$, belongs to $L^1_m(\mathbf{R}_+)$. If the above integral is convergent for all $z \in \mathbf{C}$ such that $\Re z = \xi$, it is also convergent for all z such that $\Re z > \xi$. The greatest lower bound of all real numbers ξ for which $x \mapsto e^{-\xi x} f(x)$ is in $L^1_m(\mathbf{R}_+)$ is called the *abscissa of convergence*. It is denoted by ξ_0. ξ_0 depends upon f.[13]

[13] ξ_0 can be equal to ∞.

Theorem 10. *The Laplace transform of a locally integrable function f on $\mathbf{R}_+$ is an analytic function in the half-plane $\{z|\Re z > \xi_0\}$, and, for all positive integers k, we have*

$$\frac{d^k L(z,f)}{dz^k} = \int_0^\infty e^{-zx}(-x)^k f(x)\,dx.$$

f being locally integrable, only the large values of x have to be taken into consideration. The existence of the Laplace transform of $x \mapsto x^k f(x)$ implies the existence of the Laplace transform of f. The abscissa of convergence ξ_k of $\int_0^\infty e^{-zx}(-x)^k f(x)\,dx$ is, therefore, greater than or equal to ξ_0. But, for any positive ϵ, $x^k \le e^{\epsilon x}$ if x is large enough. Hence,

$$|e^{-zx}x^k f(x)| \le e^{(\xi_0-\epsilon)x}|f(x)|,$$

which implies that $\xi_k \le \xi_0$. Combining these two results, we see that, for all positive integers k, the abscissa of convergence of the integral $\int_0^\infty e^{-zx}(-x)^k f(x)\,dx$ is ξ_0. Therefore, differentiation under the integral sign is legitimate if $z > \xi_0$.

In what follows, the Laplace transform of f is sometimes denoted by $L\,(z, f(x))$.

Theorem 11. *Let f be a locally integrable function on $\mathbf{R}_+$; then*

(1) $$L(z, f^{(k)}) = z^k L(z,f) - \sum_{p=0}^{k-1} z^p f^{(k-p-1)}(0^+),$$

(2) $$L\left(z, \int_a^x f(u)\,du\right) = \frac{1}{z}L(z,f) - \frac{1}{z}\int_0^a f(u)\,du.$$

k integrations by parts yield (1), and (2) follows from (1).

Example 1. Let α be a complex number such that $\Re\alpha > -1$; if θ denotes the *Heaviside function*[14] defined by

$$\theta(x) = \begin{cases} 0, & \text{if } x < 0, \\ 1, & \text{if } x \ge 0, \end{cases}$$

[14] Also called *unit step function*, and often denoted H or Υ.

the Laplace transform of the function $x \mapsto \theta(x)x^\alpha$ is given by

$$\begin{aligned} L(z, \theta(x)x^\alpha) &= \int_0^\infty \mathrm{e}^{-zx} x^\alpha \, dx \\ &= z^{-(\alpha+1)} \int_0^\infty \mathrm{e}^{-u} u^\alpha \, du \\ &= \frac{\Gamma(\alpha+1)}{z^{\alpha+1}}. \end{aligned}$$

The abscissa of convergence is equal to zero. The function

$$z \mapsto \frac{\Gamma(\alpha+1)}{z^{\alpha+1}}$$

is analytic in the half-plane $\{z \mid \Re z > 0\}$. In particular,

$$L(z, \theta) = \frac{1}{z}.$$

Example 2. The Laplace transform of the function $x \mapsto \theta(x)\mathrm{e}^{ax}$, where a is a complex number, is given by

$$L(z, \theta(x)\mathrm{e}^{ax}) = \frac{1}{z-a}.$$

The abscissa of convergence is equal to $\Re a$. From this result, it immediately follows that, if ω and γ are two positive numbers,

$$\begin{aligned} L(z, \theta(x) \sin \omega x) &= \frac{\omega}{z^2 + \omega^2} \\ L(z, \theta(x) \cos \omega x) &= \frac{z}{z^2 + \omega^2} \\ L(z, \theta(x) \sinh \omega x) &= \frac{\omega}{z^2 - \omega^2} \\ L(z, \theta(x) \cosh \omega x) &= \frac{z}{z^2 - \omega^2} \\ L(z, \theta(x)\mathrm{e}^{-\gamma x} \sin \omega x) &= \frac{\omega}{(z+\gamma)^2 + \omega^2} \\ L(z, \theta(x)\mathrm{e}^{-\gamma x} \cos \omega x) &= \frac{z+\gamma}{(z+\gamma)^2 + \omega^2}. \end{aligned}$$

Example 3. The Bessel function J_0 (see Example 4.3, Chapter 1) is defined by

$$J_0(x) = \frac{1}{\pi} \int_0^\pi \cos(x \sin u) \, du.$$

Its Laplace transform may be obtained as follows. If $\Re z > 0$, we have

$$\begin{aligned} L(z, \theta J_0) &= \frac{1}{\pi}\int_0^\infty \mathrm{e}^{-zx}\,dx \int_0^\pi \cos(x\sin u)\,du \\ &= \frac{1}{\pi}\int_0^\pi \frac{z\,du}{z^2+\sin^2 u} \\ &= \frac{1}{\sqrt{z^2+1}}. \end{aligned}$$

The application of Fubini's theorem is legitimate since, if $x > 0$,

$$|\mathrm{e}^{-zx}\cos(x\sin u)| \le \mathrm{e}^{-\xi x},$$

where $\xi = \Re z$.

Example 4. If the integral

$$\int_z^\infty d\zeta \int_0^\infty \mathrm{e}^{-\zeta x} f(x)\,dx$$

is convergent, Fubini's theorem yields

$$\int_0^\infty \mathrm{e}^{-zx}\frac{f(x)}{x}\,dx = \int_z^\infty L(\zeta, f)\,d\zeta,$$

which gives for $z = 0$

$$\int_0^\infty \frac{f(x)}{x}\,dx = \int_0^\infty L(\zeta, f)\,d\zeta.$$

Since the Laplace transform of the function $x \mapsto \theta(x)\sin x$ is $1/(z^2+1)$, from the above relation it follows that

$$\begin{aligned} L\left(z, \theta(x)\frac{\sin x}{x}\right) &= \int_z^\infty \frac{d\zeta}{\zeta^2+1} \\ &= \frac{\pi}{2} - \tan^{-1} z. \end{aligned}$$

In particular, for $z = 0$, we obtain the well-known result

$$\int_0^\infty \frac{\sin x}{x}\,dx = \frac{\pi}{2}.$$

Theorem 12. *Let f and g be two locally integrable functions defined on $\mathbf{R}_+$; then*

$$L(z, f * g) = L(z, f)L(z, g).$$

Since f and g are defined on $\mathbf{R}_+$, we have

$$(f * g)(t) = \int_0^t f(u)g(t-u)\,du.$$

Hence,

$$L(z, f * g) = \int_0^\infty \mathrm{e}^{-zt}\,dt \int_0^t f(u)g(t-u)\,du.$$

In order to apply the Fubini's theorem, the integral

$$\iint_\Omega \mathrm{e}^{-zt} f(u)g(t-u)\,du\,dt,$$

where $\Omega = \{(u,t) | t \geq u \geq 0\}$, has to be convergent. The Jacobian of the transformation

$$u = x \qquad t - u = y$$

being equal to 1, the integral over Ω may be written

$$\int_0^\infty \int_0^\infty \mathrm{e}^{-z(x+y)} f(x)g(y)\,dx\,dy.$$

If $L(f)$ and $L(g)$ exist, the double integral is convergent, and the required result follows.

Example 5. Consider the pair of functions

$$f\colon t \mapsto \theta(t)t^{a-1}, \qquad (a > 0),$$
$$g\colon t \mapsto \theta(t)t^{b-1}, \qquad (b > 0),$$

whose Laplace transforms are (Example 1)

$$L(z, f) = \frac{\Gamma(a)}{z^a} \quad \text{and} \quad L(z, g) = \frac{\Gamma(b)}{z^b}.$$

From the convolution theorem, it follows that

$$L(z, f * g) = \frac{\Gamma(a)\Gamma(b)}{z^{a+b}}.$$

Since two continuous functions that have the same Laplace transform are equal, we have

$$\int_0^t u^{a-1}(t-u)^{b-1}\,du = \frac{\Gamma(a)\Gamma(b)}{\Gamma(a+b)} t^{a+b}.$$

If we now let $t = 1$, we find that

$$B(a,b) = \int_0^1 u^{a-1}(t-u)^{b-1}\,du = \frac{\Gamma(a)\Gamma(b)}{\Gamma(a+b)},$$

which is the well-known relation between the Euler Beta and Gamma functions.

If we write $z = \xi + 2i\pi\eta$, we have

$$L(\xi + 2i\pi\eta, f) = \int_0^\infty f(x)\mathrm{e}^{-\xi x}\mathrm{e}^{-2i\pi\eta x}\,dx,$$

which shows the relation between Fourier and Laplace transforms. The integral on the right-hand side is convergent if $\xi > \xi_0$, where ξ_0 is the abscissa of convergence. If the function $\eta \mapsto L(\xi + 2i\pi\eta, f)$ belongs to $L^1_m(\mathbf{R})$, the inversion theorem applies, and we have

$$\mathrm{e}^{-\xi x} f(x) = \int_{-\infty}^{\infty} L(\xi + 2i\pi\eta, f)\mathrm{e}^{2i\pi\eta x}\,d\eta \quad \text{a.e.}$$

Multiplying both sides by $\mathrm{e}^{\xi x}$ and replacing $\xi + 2i\pi\eta$ by z gives

$$f(x) = \frac{1}{2i\pi}\int_\gamma \mathrm{e}^{zx}L(z,f)\,dz \quad \text{a.e.},$$

where, in the contour integral, the path of integration $\gamma\colon [0,1] \to \mathbf{C}$ is defined by $\gamma(t) = \xi - i\cot \pi t$ $(\xi > \xi_0)$.

This inversion formula for the Laplace transform is usually written

$$f(x) = \frac{1}{2i\pi}\int_{\xi - i\infty}^{\xi + i\infty} \mathrm{e}^{zx}L(z,f)\,dz,$$

where $\xi > \xi_0$. That is, all the singularities of the function $z \mapsto L(z,f)$ have to be in the half-plane $\{z | \Re z < \xi\}$.

Finally, we can state

Theorem 13. *Let F be an analytic function in the half-plane $\{z|\Re z > \xi_0\}$; if, as a function of $\eta = \Im z$, F is integrable, it is the Laplace transform of a continuous function f, defined on $\mathbf{R}_+$, such that*

$$f(x) = \frac{1}{2i\pi}\int_{\xi - i\infty}^{\xi + i\infty} \mathrm{e}^{zx}F(z)\,dz,$$

where $\xi > \xi_0$.

The continuity of f follows from Theorem 13, Chapter 1. By Cauchy's theorem, the integral that defines f does not depend upon ξ.

Remark. The evaluation of the inverse Laplace transform of F is a standard application of the residue theorem. It can be shown that F, as a function of $\Im z$, does not have to be integrable, an allowance that is equivalent to the condition

$$\lim_{\substack{|z|\to\infty \\ \Re z > \xi_0}} zF(z) = 0,$$

but has to satisfy the less stringent condition

$$\lim_{\substack{|z|\to\infty \\ \Re z > \xi_0}} F(z) = 0.$$

This result is the analogue, for the inverse Laplace transform, of Theorem 10.

Example 6. The results of Theorems 10 and 11(1) may be used to obtain the solution of certain types of ordinary differential equations. A linear differential equation whose coefficients are polynomials is transformed into an equation of the same type. However, the order of the transformed equation will be less than the order of the original one, if the degree of the coefficients of the latter are less than its order. This is, of course, always the case for a linear differential equation with constant coefficients.

Consider, for instance, the equation

$$xy'' + 2y' + xy = 0.$$

The degree of the coefficients is 1, whereas the order of the equation is 2. The transformed equation may be written

$$-\frac{d}{dz}\left(z^2 L(z,y) - zy(0^+) - y'(0^+)\right) + 2\left(zL(z,y) - y(0^+)\right) - \frac{d}{dz}L(z,y) = 0,$$

that is, assuming that $y(0^+)$ and $y'(0^+)$ are finite,

$$(1+z^2)\frac{d}{dz}L(z,y) + y(0^+) = 0.$$

The solution of this equation is

$$L(z,y) = y(0^+)(A - \tan^{-1} z),$$

where A is a constant. Since the inversion formula is valid[15] only if $\lim_{|z|\to\infty} L(z,y) = 0$, it follows that A must be equal to $\pi/2$. Then, from Example 4, we find

$$y(x) = y(0^+)\frac{\sin x}{x}.$$

It is clear that this method applies only if the solution of the equation has a Laplace transform. In the present case, we may obtain the solution of the equation by substituting the power series[16]

$$y(x) = \sum_0^\infty c_n x^n.$$

It is found that any solution is a linear combination of the two following particular solutions:

$$y_1(x) = \frac{\sin x}{x} \qquad \text{and} \qquad y_2(x) = \frac{\cos x}{x}.$$

The method based on the Laplace transform cannot give the solution y_2 since the Laplace transform of y_2 does not exist.

Example 7. The Laplace transform provides a useful tool for solving integral equations of the type

$$f(x) - \int_0^x k(x-u)f(u)\,du = g(x),$$

where k and g are given functions and f is an unknown function to be determined. If we denote, respectively, by F, K, and G the Laplace transforms of f, k, and g, we have

$$F(z)(1 - K(z)) = G(z),$$

whose solution is

$$F(z) = \frac{G(z)}{1 - K(z)}.$$

For instance, the Laplace transform F of the solution f of the equation

$$f(x) + 4\int_0^x (x-u)f(u)\,du = x^2$$

---

15 See the remark that follows Theorem 13.

16 See J. W. Dettman.

is given by

$$F(z) = \frac{2}{z(z^2+4)}.$$

Since the expression of $F(z)$ is particularly simple, it is not necessary, in this case, to use the inversion theorem to determine f. In fact, from the relation

$$\frac{2}{z(z^2+4)} = \frac{1}{2}\left(\frac{1}{z} - \frac{z}{z^2+4}\right),$$

it follows that

$$f(x) = \frac{1}{2}\theta(x)(1-\cos 2x) = \theta(x)\sin^2 x.$$

Example 8. Consider the one-dimensional heat equation

$$\frac{\partial \phi}{\partial t} = \frac{\partial^2 \phi}{\partial x^2}.$$

If the medium is semi-infinite, $(x,t) \in \mathbf{R}_+ \times \mathbf{R}_+$. This partial differential equation may be solved by using the general procedure we described in the remark following Example 4.4. Here, however, instead of the Fourier transform with respect to the space variable x, we make use of the Laplace transform with respect to the time variable t. If $\phi(x,t)$ satisfies the initial condition $\phi(x,0)=0$ and the two boundary conditions $\phi(0,t)=g(t)$ and $\lim_{x\to\infty}\phi(x,t)=0$, we write

$$zL\left(z,\phi(x,t)\right) = \frac{\partial^2 L\left(z,\phi(x,t)\right)}{\partial x^2},$$

where $L\left(z,\phi(0,t)\right) = L\left(z,g(t)\right)$. The appropriate solution of this differential equation is

$$L\left(z,\phi(x,t)\right) = L\left(z,g(t)\right)\mathrm{e}^{-x\sqrt{z}}.$$

To obtain the expression of $\phi(x,t)$, we have to find the inverse Laplace transform of the function $z \mapsto \mathrm{e}^{-x\sqrt{z}}$. We could evaluate directly the contour integral. It is simpler to use the following method. Consider the convergent integral

$$I(a,b) = \int_0^\infty \mathrm{e}^{-a^2u^2-b^2u^{-2}}\,du.$$

If $v = au$, then

$$\begin{aligned} I(a,b) &= a^{-1}\int_0^\infty \mathrm{e}^{-v^2-a^2b^2v^{-2}}\,dv \\ &= a^{-1}I(1,ab). \end{aligned}$$

Similarly, if we make the substitution $v = bu^{-1}$ in

$$\frac{\partial I(a,b)}{\partial b} = -2b \int_0^\infty u^{-2} \mathrm{e}^{-a^2u^2-b^2u^{-2}}\, du,$$

we obtain

$$\frac{\partial I(a,b)}{\partial b} = -2I(1,ab).$$

Eliminating $I(1,ab)$, we find that $I(a,b)$ satisfies the differential equation

$$\frac{\partial I(a,b)}{\partial b} = -2aI(a,b),$$

whose solution is

$$I(a,b) = I(a,0)\mathrm{e}^{-2ab}.$$

Since

$$I(a,0) = \int_0^\infty \mathrm{e}^{-a^2u^2}\, du = \frac{\sqrt{\pi}}{2a},$$

we finally obtain

$$\int_0^\infty \mathrm{e}^{-a^2u^2-b^2u^{-2}}\, du = \frac{\sqrt{\pi}}{2a}\mathrm{e}^{-2ab}.$$

If we replace, in this relation, u^2 by t, a^2 by z, and b^2 by c we find

$$L\left(z, \theta(t)t^{-1/2}e^{-c/t}\right) = \sqrt{\frac{\pi}{z}}\mathrm{e}^{-2\sqrt{cz}}.$$

If we now differentiate both sides of this relation with respect to c, we obtain

$$L\left(z, \theta(t)t^{-3/2}e^{-c/t}\right) = \sqrt{\frac{\pi}{c}}\mathrm{e}^{-2\sqrt{cz}}.$$

The inverse Laplace transform of the function $z \mapsto \mathrm{e}^{-x\sqrt{z}}$ is, therefore, the function $t \mapsto x\theta(t)t^{-3/2}\mathrm{e}^{-x^2/4t}/\sqrt{4\pi}$, and the required solution of the heat equation is

$$\phi(x,t) = \frac{x}{\sqrt{4\pi}} \int_0^t g(\tau)\frac{1}{(t-\tau)^{3/2}} \exp -\left(\frac{x^2}{4(t-\tau)}\right) d\tau.$$

Remark. We found two new Laplace transforms, namely

$$L\left(z, \theta(t)t^{-1/2}e^{-c/t}\right) = \sqrt{\frac{\pi}{z}}\mathrm{e}^{-2\sqrt{cz}}$$

$$L\left(z, \theta(t)t^{-3/2}e^{-c/t}\right) = \sqrt{\frac{\pi}{c}}\mathrm{e}^{-2\sqrt{cz}},$$

where c is a positive constant. Their abscissas of convergence are both equal to zero. More generally, we could derive the Laplace transforms of the family of functions $\{t \mapsto \theta(t)t^{-(2p+1)/2}\mathrm{e}^{-c/t} | p \in \mathbf{N}\}$ by differentiating p times the first relation with respect to c.

Note

A mapping $d: X \times X \to \mathbf{R}_+$ is a *metric* (or a *distance*) on the space X if it satisfies the following three conditions:
(1) $d(x, y) = 0 \iff x = y$.
(2) $d(x, y) = d(y, x)$.
(3) $d(x, z) \le d(x, y) + d(y, z)$.

The pair (X, d) is called a *metric space.*

A subset A of X is said to be *open* if, for any $a \in A$, there exists a positive number ϵ such that $\{x | d(x, a) < \epsilon\} \subset A$. The class of open sets on X forms a topology (see Chapter 1, Note 1). A set A is closed if A^C is open.

The set A is a *neighborhood* of $a \in A$ if A contains an open set containing a.

The union of all open sets included in a set A is called the *interior* of A. It is denoted by $\mathring{A}$. The interior of A is, therefore, the greatest open set contained in A. The *exterior* of A is the interior of its complement. It is the greatest open set having an empty intersection with A. The *boundary* of A is the complement of the union of the interior and the exterior.

The closure of a set A is the intersection of all the closed sets containing A; it is denoted by $\overline{A}$. A closed set coincides with its closure.

A subset A of X is *dense* in X (or simply dense) if $\overline{A} = X$. The set $\mathbf{Q}$ of rational numbers is dense in $\mathbf{R}$ for the metric

$$(x, y) \in \mathbf{R} \times \mathbf{R} \qquad d(x, y) = |x - y|.$$

A metric space is said to be *separable* if it contains a dense countable subset. $\mathbf{R}$ is separable since it contains $\mathbf{Q}$ which is dense in $\mathbf{R}$ and countable.

A sequence (x_n) in a metric space (X, d) is said to *converge* to a point x in (X, d) if

$$\lim_{n \to \infty} d(x_n, x) = 0.$$

A sequence (x_n) in a metric space (X, d) is said to be a *Cauchy sequence* if, for any positive ϵ, there exists a positive integer $n_0(\epsilon)$ such that $d(x_n, x_{n+k}) < \epsilon$ for all $n > n_0(\epsilon)$ and $k > 0$.

All convergent sequences are Cauchy sequences.

A metric space is said to be *complete* if every Cauchy sequence is convergent. For the metric defined above, $\mathbf{R}$ is complete.

Let (X, d) be a metric space; the metric space $(\widehat{X}, d)$ is called a *completion* of (X, d) if $\overline{X} = \widehat{X}$. That is, X is dense in $\widehat{X}$. Every metric space has a completion.

A mapping $x \mapsto \|x\|$ defined on a vector space X into $\mathbf{R}_+$ is a *norm* if it satisfies the following three conditions:

(1) $\|x\| = 0 \iff x = 0$.

(2) For all $x \in X$ and all scalar λ, $\|\lambda x\| = |\lambda| \, \|x\|$.

(3) For all x and y in X, $\|x + y\| \leq \|x\| + \|y\|$.

In $\mathbf{R}$ the mapping $x \mapsto |x|$ is a norm.

A vector space equipped with a norm is called a *normed vector space.* Every normed vector space becomes a metric space if, for all x and y in X, we set $d(x, y) = \|x - y\|$. A complete normed vector space relative to the above metric is called a *Banach space.*

Problems

1. Let f and g be two functions belonging, respectively, to $L_\mu^p(X)$ and $L_\mu^q(X)$, where p and q are two positive numbers; if r is such that

$$\frac{1}{r} = \frac{1}{p} + \frac{1}{q},$$

show that we have

$$fg \in L_\mu^r(X) \qquad \text{and} \qquad \|fg\|_r \leq \|f\|_p \|g\|_q.$$

2. Let p, q, and r be three positive numbers such that

$$\frac{1}{p} + \frac{1}{q} + \frac{1}{r} = 1,$$

and let f, g, and h be three functions belonging, respectively, to $L_\mu^p(X)$, $L_\mu^q(X)$, and $L_\mu^r(X)$; show that we have

$$fgh \in L_\mu^1(X) \qquad \text{and} \qquad \|fgh\|_1 \leq \|f\|_p \|g\|_q \|h\|_r.$$

3. Hölder's and Minkowski's inequalities hold if the conjugate exponents p and q are greater than 1. If $0 < p < 1$, and, therefore, $q < 0$, establish the following relations:

$$\int fg\, d\mu \geq \left(\int f^p\, d\mu\right)^{1/p} \left(\int g^q\, d\mu\right)^{1/q}, \tag{1}$$

$$\left(\int (f+g)^p\, d\mu\right)^{1/p} \geq \left(\int f^p\, d\mu\right)^{1/p} + \left(\int g^p\, d\mu\right)^{1/p}, \tag{2}$$

where f and g are measurable positive functions defined on the measure space $(X, \mathcal{A}, \mu)$. Assume that all integrals make sense.

4. Let $f: \mathbf{R} \to \mathbf{R}$ be an integrable function with respect to the Lebesgue measure; show that

$$\lim_{n\to\infty} \int_{-\infty}^{\infty} f(x) \cos nx\, dx = \lim_{n\to\infty} \int_{-\infty}^{\infty} f(x) \sin nx\, dx = 0.$$

This result is known as the *Riemann–Lebesgue lemma*.

5. The vector space $L^p_\mu(X) \cap L^q_\mu(X)$ may be equipped with either the norm $f \mapsto \|f\|_p$ or the norm $f \mapsto \|f\|_q$. Are the normed spaces so defined complete?

6. Let E be the set of functions $f: \mathbf{R} \to \mathbf{R}$ such that, for all positive numbers ϵ, the function f_ϵ defined by $f_\epsilon(x) = e^{\epsilon|x|} f(x)$ is Lebesgue integrable.
(1) Show that, for $1 \le p < \infty$, $L^p_m(\mathbf{R})$ is contained in E.
(2) Show that if f is in $L^p_m(\mathbf{R})$, so is f_ϵ, and that f_ϵ tends to f when ϵ tends to zero in the mean of order p.

7. Let f be a function belonging to $L^p_m(\mathbf{R}_+)$, where $1 < p < \infty$, and let F be the function defined on $\mathbf{R}_+$ by

$$F(x) = \frac{1}{x} \int_0^x f(t)\, dt.$$

Show that

$$\|F\|_p \le \frac{p}{p-1} \|f\|_p.$$

8. Find the Fourier transform of the function $f: x \mapsto e^{-2\pi a|x|}|x|^\alpha$, where $a > 0$ and $\alpha > -1$.

9. Show that the solution of the one-dimensional wave equation

$$\frac{\partial^2 \phi}{\partial x^2} - \frac{\partial^2 \phi}{\partial t^2} = 0,$$

which satisfies the boundary conditions

$$\phi(x, 0) = f_1(x) \qquad \text{and} \qquad \frac{\partial \phi}{\partial t}(x, 0) = f_2(x),$$

is

$$\phi(x, t) = \frac{1}{2}(f_1(x+t) + f_1(x-t)) + \frac{1}{2} \int_{x-t}^{x+t} f_2(\xi)\, d\xi.$$

10. Let f be the function defined by

$$f(x) = \theta(1 - |x|)(1 - x^2)^{-1/2}.$$

Show that $\widehat{f}(t) = \pi J_0(2\pi t)$, where J_0 is the Bessel function of the first kind of order zero.

11. For $x > 0$, the *sine integral* Si and the *cosine integral* Ci are defined by

$$\mathrm{Si}(x) = \int_0^x \frac{\sin u}{u}\, du$$

$$\mathrm{Ci}(x) = -\int_x^\infty \frac{\cos u}{u}\, du.$$

Determine their Laplace transforms.

12. Find the Laplace transform of the function $x \mapsto \theta(x)\mathrm{e}^{-ax^2}$, where $a > 0$.

13. Find the Laplace transform of the *error function* Erf defined by

$$\mathrm{Erf}(x) = \frac{2}{\sqrt{\pi}} \int_0^x \mathrm{e}^{-u^2}\, du.$$

14. L_n denotes the *Laguerre polynomial* of degree n defined by

$$L_n(x) = \mathrm{e}^x \frac{d^n}{dx^n}(x^n \mathrm{e}^{-x}).$$

Find the Laplace transform of the function θL_n, and obtain the expression of the *generating function* of the Laguerre polynomials $(x,t) \mapsto G(x,t)$ such that

$$G(x,t) = \sum_0^\infty L_n(x) \frac{t^n}{n!}.$$

15. Show that, for $x > 0$,

$$\int_0^x J_0(u) J_0(x-u)\, du = \sin x,$$

where J_0 is the Bessel function of order 0.

16. Find the solution $\phi: \mathbf{R}_+ \times \mathbf{R}_+ \to \mathbf{R}$ of the one-dimensional wave equation

$$\frac{\partial^2 \phi}{\partial x^2} - \frac{\partial^2 \phi}{\partial t^2} = 0,$$

which satisfies the conditions

$$\begin{aligned} \phi(x,0) &= 0 \\ \frac{\partial \phi}{\partial t}(x,0) &= 0 \\ \phi(0,t) &= g(t) \\ \lim_{x \to \infty} \phi(x,t) &= 0. \end{aligned}$$

Solutions

1. r is positive, and we have

$$\frac{r}{p} + \frac{r}{q} = 1.$$

So r/p and r/q are conjugate exponents. Furthermore, f and g being measurable, $|fg|^r$ is measurable, and Hölder's inequality applied to the measurable functions $|f|^r$ and $|g|^r$ yields

$$\int |fg|^r \, d\mu = \int (|f|^p)^{r/p} (|g|^q)^{r/q} \, d\mu$$
$$\leq \left(\int |f|^p \, d\mu \right)^{r/p} \left(\int |g|^q \, d\mu \right)^{r/q}.$$

Hence

$$\|fg\|_r \leq \|f\|_p \|g\|_q.$$

This result shows that $\|fg\|_r$ is finite, therefore, $fg \in L^r_\mu(X)$.

2. If s is such that

$$\frac{1}{s} = \frac{1}{p} + \frac{1}{q},$$

it follows from problem 1 that

$$fg \in L^s_\mu(X) \qquad \text{and} \qquad \|fg\|_s \leq \|f\|_p \|g\|_q.$$

Furthermore,

$$\frac{1}{s} + \frac{1}{r} = 1,$$

and Hölder's inequality applied to the functions $fg \in L^s_\mu(X)$, and $h \in L^r_\mu(X)$ yields

$$\|fgh\|_1 \leq \|fg\|_s \|h\|_r.$$

This result shows that fgh is μ-integrable, and, taking into account that $\|fg\|_s \leq \|f\|_p \|g\|_q$, we finally obtain

$$\|fgh\|_1 \leq \|f\|_p \|g\|_q \|h\|_r.$$

3. (1) Let $\phi = (fg)^p$ and $\psi = g^{-p}$. Since we assumed that fg and g^q are μ-integrable, ϕ and ψ blong, respectively, to $L^{1/p}_\mu(X)$ and $L^{-q/p}_\mu(X)$.

$1/p$ and $-q/p$ being conjugate exponents, we can apply Hölder's inequality to the pair (ϕ, ψ). We have

$$\int \phi\psi \, d\mu \leq \left(\phi^{1/p} \, d\mu\right)^{p} \left(\int \psi^{-q/p} \, d\mu\right)^{-p/q},$$

that is,

$$\int f^p \, d\mu \leq \left(\int fg \, d\mu\right)^{p} \left(\int g^q \, d\mu\right)^{-p/q}.$$

If $\int g^q \, d\mu \neq 0$, we may multiply both sides by $\left(\int g^q \, d\mu\right)^{p/q}$, and relation (1) follows. If $\int g^q \, d\mu = 0$, relation (1) is satisfied.

(2) If we write

$$(f+g)^p = f(f+g)^{p-1} + g(f+g)^{p-1},$$

relation (1) yields

$$\int f(f+g)^{p-1} \, d\mu \geq \left(\int f^p \, d\mu\right)^{1/p} \left(\int (f+g)^{(p-1)q}\right)^{1/q},$$
$$\int g(f+g)^{p-1} \, d\mu \geq \left(\int g^p \, d\mu\right)^{1/p} \left(\int (f+g)^{(p-1)q}\right)^{1/q},$$

and taking into account that $(p-1)q = p$, we have

$$\int (f+g)^p \, d\mu \geq \left(\left(\int f^p \, d\mu\right)^{1/p} + \left(\int g^p \, d\mu\right)^{1/p}\right) \left(\int (f+g)^{(p-1)q}\right)^{1/q}.$$

If $\int (f+g)^p \, d\mu \neq 0$, we may divide both sides by $\left(\int (f+g)^p \, d\mu\right)^{1/q}$, and relation (2) follows. If $\int (f+g)^p \, d\mu = 0$, then $f+g=0$ a.e., that is, f and g are both equal to zero a.e. and relation (2) is satisfied.

4. Let (f_k) be a sequence of integrable functions that converges to f in $L^1_m(\mathbf{R})$. If the Riemann–Lebesgue lemma holds for all the functions of the sequence (f_k), then it holds for f, since

$$\begin{aligned} \left| \int_{-\infty}^{\infty} f(x) \cos nx \, dx \right| &\leq \left| \int_{-\infty}^{\infty} (f(x) - f_k(x)) \cos nx \, dx \right| \\ &\quad + \left| \int_{-\infty}^{\infty} f_k(x) \cos nx \, dx \right| \\ &\leq \int_{-\infty}^{\infty} |f(x) - f_k(x)| \, dx + \left| \int_{-\infty}^{\infty} f_k(x) \cos nx \, dx \right| \end{aligned}$$

and similarly for the integral $\int_{\infty}^{\infty} f(x)\sin nx\,dx$.

Since the vector space $C_c^0(\mathbf{R})$ is dense in $L_m^1(\mathbf{R})$, it is sufficient to establish the lemma for functions in $C_c^0(\mathbf{R})$. These functions are Riemann integrable, and can be approximated by step functions. Step functions being linear combinations of characteristic functions of intervals, it is finally sufficient to prove the Riemann–Lebesgue lemma for the function $f = 1_{[a,b]}$, where $b > a$. In this case, we have

$$\begin{aligned}\int_{-\infty}^{\infty} 1_{[a,b]}(x)\cos nx\,dx &= \int_a^b \cos nx\,dx\\ &= \frac{\sin nb - \sin na}{n}\\ \int_{-\infty}^{\infty} 1_{[a,b]}(x)\sin nx\,dx &= \int_a^b \sin nx\,dx\\ &= \frac{\cos nb - \cos na}{n},\end{aligned}$$

and the result follows when n tends to ∞.

5. If the measure μ is such that one space is included in the other, then it is easy to conclude. If $L_\mu^p(X) \subset L_\mu^q(X)$, it is clear that $L_\mu^p(X) \cap L_\mu^q(X) = L_\mu^p(X)$ is complete for the L^p-norm but not for the L^q-norm.

If one space is not included in the other, from Theorem 7 it follows that their intersection is dense in both spaces. More precisely,

$$\begin{aligned}&\overline{L_\mu^p(X) \cap L_\mu^q(X)} = L_\mu^p(X) \quad \text{for the } L^p\text{-norm}\\ &\overline{L_\mu^p(X) \cap L_\mu^q(X)} = L_\mu^q(X) \quad \text{for the } L^q\text{-norm}.\end{aligned}$$

That is, $L_\mu^p(X) \cap L_\mu^q(X)$ is complete for neither the L^p-norm nor the L^q-norm.

6. (1) Hölder's inequality implies that

$$\int_{-\infty}^{\infty} \mathrm{e}^{-\epsilon|x|}|f(x)|\,dx \le \left(\int_{-\infty}^{\infty} |f(x)|\,dx\right)^{1/p} \left(\int_{-\infty}^{\infty} \mathrm{e}^{-\epsilon q|x|}\,dx\right)^{1/q},$$

which shows that $f_\epsilon \in L_m^1(\mathbf{R})$ if $f \in L_m^p(\mathbf{R})$, since

$$\int_{-\infty}^{\infty} \mathrm{e}^{-\epsilon q|x|}\,dx = \frac{2}{\epsilon q} < \infty.$$

(2) The relation

$$\int_{-\infty}^{\infty} \mathrm{e}^{-\epsilon p|x|}|f(x)|\,dx \le \int_{-\infty}^{\infty} |f(x)|\,dx$$

shows that $f_\epsilon \in L^p_m(\mathbf{R})$ if $f \in L^p_m(\mathbf{R})$.

For all $x \in \mathbf{R}$, we have

$$\lim_{\epsilon \to 0} |f_\epsilon(x) - f(x)|^p = 0.$$

Since $|f_\epsilon(x) - f(x)|^p \leq 2^p |f(x)|^p$, Lebesgue's theorem applies, and we have

$$\int_{-\infty}^{\infty} \lim_{\epsilon \to 0} |f_\epsilon(x) - f(x)|^p = 0.$$

That is, f_ϵ tends to f in the mean of order p when ϵ tends to zero.

7. Let us first assume that $f \in C^0_c(\mathbf{R})$; then the function $x \mapsto \int_0^x f(t)\,dt$ is continuously differentiable, and its derivative is f. In this case, we have

$$xF'(x) + F(x) = f(x).$$

An integration by parts yields

$$\begin{aligned} \int_0^\infty F^p(x)\,dx &= -p \int_0^\infty F^{p-1}(x) x F'(x)\,dx \\ &= -p \int_0^\infty F^{p-1}(x) f(x)\,dx + p \int_0^\infty F^p(x)\,dx. \end{aligned}$$

Hence,

$$(p-1)\int_0^\infty F^p(x)\,dx = p \int_0^\infty F^{p-1}(x) f(x)\,dx,$$

and Hölder's inequality applied to the second term on the rigth-hand side gives

$$\int_0^\infty F^p(x)\,dx \leq \frac{p}{p-1} \|F^{p-1}\|_q \|f\|_p,$$

where $q = p/(p-1)$.

If f belongs to $L^p_m(\mathbf{R}_+)$ but not to $C^0_c(\mathbf{R}_+)$, it can be approximated by a Cauchy sequence (f_n) of functions in $C^0_c(\mathbf{R}_+)$. If, for all positive integers n,

$$F_n(x) = \frac{1}{x} \int_0^x f_n(t)\,dt,$$

then

$$\|F_r - F_s\| \leq \frac{p}{p-1} \|f_r - f_s\|,$$

since the mapping $f \mapsto F$ is linear. The sequence (F_n) is, therefore, a Cauchy sequence of functions belonging to $L^p_m(\mathbf{R}_+)$. Since $L^p_m(\mathbf{R}_+)$ is complete, this sequence converges to a function $F \in L^p_m(\mathbf{R}_+)$, and we have

$$\begin{aligned}\|F\|_p &= \lim_{n\to\infty} \|F_n\|_p \\ &\leq \lim_{n\to\infty} \frac{p}{p-1}\|f_n\|_p \\ &\leq \frac{p}{p-1}\|f\|_p,\end{aligned}$$

which is the required result. This relation, known as the *Hardy's inequality*, shows that the linear mapping $f \mapsto F$ from $L^p_m(\mathbf{R}_+)$ into itself is continuous.

8. We have

$$\begin{aligned}\widehat{f}(t) &= \int_{-\infty}^{\infty} |x|^\alpha \exp(-2\pi a|x| + 2i\pi tx)\,dx \\ &= \int_{-\infty}^{0} (-x)^\alpha \exp(2\pi ax + 2i\pi tx)\,dx \\ &\qquad + \int_0^\infty x^\alpha \exp(-2\pi ax + 2i\pi tx)\,dx. \\ &= \int_0^\infty x^\alpha \exp\left(-2\pi(a+it)x\right)\,dx + \int_0^\infty x^\alpha \exp\left(-2\pi(a-it)x\right)\,dx.\end{aligned}$$

If we substitute $2\pi x(a+it)$ by z in the first integral and $2\pi x(a+it)$ by z in the second one and apply Cauchy's theorem we finally obtain

$$\widehat{f}(t) = \frac{\Gamma(\alpha+1)}{(2\pi(a+it))^{\alpha+1}} + \frac{\Gamma(\alpha+1)}{(2\pi(a-it))^{\alpha+1}}.$$

9. If

$$\widehat{\phi}(\xi,t) = \int_{-\infty}^{\infty} \phi(x,t)\mathrm{e}^{2i\pi\xi x}\,dx,$$

the function $t \mapsto \widehat{\phi}(\xi,t)$ satisfies the differential equation

$$\frac{\partial^2\widehat{\phi}}{\partial t^2} + 4\pi\xi^2\widehat{\phi} = 0$$

and the initial conditions

$$\widehat{\phi}(\xi,0) = \widehat{f_1}(\xi) \quad \text{and} \quad \frac{\partial\widehat{\phi}}{\partial t}(\xi,0) = \widehat{f_2}(\xi),$$

where $\widehat{f_1}$ and $\widehat{f_2}$ are the Fourier transforms, respectively, of f_1 and f_2.

The solution of the differential equation is

$$\begin{aligned}\widehat{\phi}(\xi,t) &= \widehat{f_1}\cos 2\pi\xi t + \frac{1}{2\pi\xi}\widehat{f_2}\sin 2\pi\xi t \\ &= \frac{1}{2}\left(\widehat{f_1} + \frac{1}{2i\pi\xi}\widehat{f_2}\right)\mathrm{e}^{2i\pi\xi t} + \frac{1}{2}\left(\widehat{f_1} - \frac{1}{2i\pi\xi}\widehat{f_2}\right)\mathrm{e}^{-2i\pi\xi t}.\end{aligned}$$

The required result follows from Theorem 7(2), and the fact that $-\widehat{f_2}/2i\pi\xi$ is the Fourier inverse of a primitive of f_2. f_1 should be twice continuously differentiable, and f_2 continuously differentiable. See Problem 13, Chapter 4, for another method.

10. The Fourier transform of f is given by

$$\widehat{f}(t) = \int_{-1}^{1}(1-x^2)^{-1/2}\mathrm{e}^{2i\pi tx}\,dx.$$

Making the substitution $x = \sin\theta$, we find that

$$\begin{aligned}\widehat{f}(t) &= \int_{-\pi/2}^{\pi/2}\mathrm{e}^{2i\pi t\sin\theta}\,d\theta \\ &= 2\int_{0}^{\pi/2}\cos(2\pi t\sin\theta)\,d\theta \\ &= \pi J_0(2\pi t).\end{aligned}$$

11. (1) We have (Example 5.4)

$$L\left(z, \theta(x)\frac{\sin x}{x}\right) = \frac{\pi}{2} - \tan^{-1} z.$$

Therefore, by Theorem 11(2), it follows that

$$L\left(z, \operatorname{Si} x\right) = \frac{1}{z}\left(\frac{\pi}{2} - \tan^{-1} z\right).$$

(2) If the integral

$$\int_0^\infty \mathrm{e}^{-zx}\,dx\int_x^\infty \frac{f(u)}{u}\,du$$

is convergent, we can interchange the order of the integrations (see Example 5.4). We find

$$\int_0^\infty \frac{f(u)}{u}\,du\int_0^u \mathrm{e}^{-zx}\,dx = \frac{1}{z}\int_0^\infty \frac{1-\mathrm{e}^{-zu}}{u}f(u)\,du.$$

Similarly, we have

$$\int_0^z L(\zeta, f)\, d\zeta = \int_0^z d\zeta \int_0^\infty \mathrm{e}^{-\zeta x}$$
$$= \int_0^\infty \frac{1 - \mathrm{e}^{-zu}}{u} f(u)\, du.$$

Hence,

$$L\left(z, \int_x^\infty \frac{f(u)}{u}\, du\right) = \frac{1}{z} \int_0^z L(\zeta, f)\, d\zeta.$$

If we replace $f(u)$ by $\cos u$, we obtain

$$L\,(z, \operatorname{Ci} x) = -\frac{1}{2z} \log(1 + z^2).$$

12. We have

$$L\left(z, \theta(x)\exp(-ax^2)\right) = \int_0^\infty \exp(-zx - ax^2)\, dx$$
$$= \exp(z^2/4a) \int_0^\infty \exp\left(-a\left(x + \frac{z}{2a}\right)^2\right) dx$$
$$= \frac{\exp(z^2/4a)}{\sqrt{a}} \int_{z/2\sqrt{a}}^\infty \exp(-u^2)\, du$$
$$= \frac{1}{2}\sqrt{\frac{\pi}{a}} \exp(z^2/4a) \operatorname{Erfc}(z/2\sqrt{a}),$$

where Erfc is the *complement of the error function* Erf such that

$$\operatorname{Erf}(z) + \operatorname{Erfc}(z) = 1.$$

Note that when we substitute a complex variable for a real variable, we have to apply Cauchy's theorem to justify the result.

13. We have

$$L(z, \operatorname{Erf}) = \int_0^\infty \mathrm{e}^{-zx}\, dx \frac{2}{\sqrt{\pi}} \int_0^x \mathrm{e}^{-u^2}\, du.$$

By Fubini's theorem (we can verify that its application is legitimate), we find

$$L(z, \operatorname{Erf}) = \frac{2}{\sqrt{\pi}} \int_0^\infty \mathrm{e}^{-u^2}\, du \int_u^\infty \mathrm{e}^{-zx}\, dx$$
$$= \frac{2}{z\sqrt{\pi}} \int_0^\infty \mathrm{e}^{-u^2 - zu}\, du$$
$$= \frac{2}{z\sqrt{\pi}} \mathrm{e}^{z^2/4} \int_0^\infty \exp\left(-\left(u - \frac{1}{2}z\right)^2\right) du$$
$$= \frac{\mathrm{e}^{z^2/4}}{z} \operatorname{Erfc}(z/2).$$

As in the preceding problem, when we substitute a complex variable for a real variable, we have to apply Cauchy's theorem to justify the result.

14. Since

$$L\left(z, \theta(x)x^n\right) = \frac{n!}{z^{n+1}},$$

we have

$$L\left(z, \theta(x)x^n \mathrm{e}^{-x}\right) = \frac{n!}{(z+1)^{n+1}}.$$

Hence,

$$L(z, \theta L_n) = n! \frac{1}{z}\left(1 - \frac{1}{z}\right)^n.$$

Therefore,

$$L\left(z, \sum_{n=0}^{\infty} \theta(x) L_n(x) \frac{t^n}{n!}\right) = \sum_{n=0}^{\infty} \frac{1}{z}\left(1 - \frac{1}{z}\right)^n t^n.$$

This relation makes sense if the series is convergent, that is, if $|t - t/z| < 1$. In this case

$$L\left(z, G(x,t)\right) = \frac{1}{1-t} \frac{1}{z + \frac{t}{1-t}}.$$

So

$$G(x,t) = \frac{1}{1-t} \mathrm{e}^{xt/(1-t)}.$$

15. This formula is a direct consequence of Theorem 12. The integral on the left-hand side is the value at x of the convolution of θJ_0 by itself. Since the Laplace transform of θJ_0 is $(z^2+1)^{-1/2}$, the Laplace transform of $(\theta J_0)^{*2}$ is $(z^2+1)^{-1}$, which is the Laplace transform of $x \mapsto \theta(x)\sin x$.

16. Making use of the Laplace transform with respect to t, the equation becomes

$$\frac{\partial^2 L\left(z, \phi(x,t)\right)}{\partial x^2} - z^2 L\left(z, \phi(x,t)\right) = 0.$$

Its solution is

$$L\left(z, \phi(x,t)\right) = A\mathrm{e}^{zx} + B\mathrm{e}^{-zx},$$

where A and B are two constants. Since the transform should exist in the right half-plane, $\lim_{x\to\infty} L\left(z, \phi(x,t)\right) = 0$ when $\Re z$ is large enough. This implies that $A = 0$. Taking into account the boundary condition $\phi(0,t) = g(t)$, we find that $B = L(z,g)$, that is,

$$L\left(z, \phi(x,t)\right) = L\left(z, g(t)\right)\mathrm{e}^{-zx}.$$

Hence,

$$\phi(x,t) = \theta(t - x)g(t - x).$$

It is easy to verify that this is the right solution to the problem provided g is twice differentiable.

Chapter 3

Hilbert Spaces

Hilbert spaces are Banach spaces on which a scalar product has been defined. They are the most natural generalization of finite dimensional Euclidean spaces. In this chapter we particularly insist on orthogonal systems and Hilbert bases that play an essential role in quantum mechanics and numerical analysis.

1. Scalar Product

Definition 1. A *scalar product* or *inner product* on a real (resp. complex) vector space E is a mapping f of $E \times E$ into $\mathbf{R}$ (resp. $\mathbf{C}$), which has the following properties

$$
\begin{aligned}
&(1) \quad f(x + x', y) = f(x, y) + f(x', y),\\
&(2) \quad f(\lambda x, y) = \overline{\lambda} f(x, y),\\
&(3) \quad f(y, x) = \overline{f(x, y)},\\
&(4) \quad f(x, x) \geq 0,\\
&(5) \quad f(x, x) = 0 \iff x = 0,
\end{aligned}
$$

where x, x', and y belong to E and λ belongs to $\mathbf{R}$ (resp. $\mathbf{C}$).

Many authors adopt a different definition and put $f(\lambda x, y) = \lambda f(x, y)$. Our convention is the most widely used by physicists.

From this definition, it immediately follows that

$$
\begin{aligned}
&(1) \quad f(x, y + y') = f(x, y) + f(x, y'),\\
&(2) \quad f(x, \lambda y) = \lambda f(x, y).
\end{aligned}
$$

When E is a real vector space, the mapping f is a symmetric bilinear form. When E is complex, f is semilinear in x and linear in y. In this case

f is said to be *sesquilinear*. f is completely determined by the values taken on the *diagonal* of $E \times E$,[1] since

$$f(x,y) = \frac{1}{4}\big(f(x+y,x+y) - if(x+iy,x+iy) \\ - f(x-y,x-y) + if(x-iy,x-iy)\big).$$

This relation is easy to verify if one expands all the terms of the right-hand side.

Remark. In this text, $f(x,y)$, which is called the *scalar product* of x and y, will be denoted $\langle x \mid y\rangle$. This notation, adopted by the majority of physicists, is not the only one. The scalar product is also often denoted $\langle x, y\rangle$ or $(x \mid y)$. The notation (x,y), used by some authors, is to be avoided since the scalar product could be mistaken for the ordered pair of vectors x, y.

The mapping $x \mapsto \sqrt{\langle x \mid x\rangle}$ is a *norm* on E, and we shall put

$$\|x\| = \sqrt{\langle x \mid x\rangle}.$$

A vector space on which a scalar product has been defined is called a *prehilbert space* or an *inner product space*.

Definition 2. Two vectors x and y of a prehilbert space E are *orthogonal* if $\langle x \mid y\rangle = 0$. If A is a nonempty subset of E, the set of all elements of E that are orthogonal to all the elements of A is called the *orthogonal supplement* of A in E and is denoted $A^{\perp}$.

Theorem 1. Cauchy–Schwarz inequality. *For any pair of vectors x, y in a prehilbert space E, we have*

$$|\langle x \mid y\rangle| \leq \|x\|\,\|y\|.$$

For any scalar λ, we have

$$\langle x + \lambda y \mid x + \lambda y\rangle \geq 0,$$

i.e.,

$$\|x\|^2 + \lambda\langle x \mid y\rangle + \overline{\lambda\langle x \mid y\rangle} + \lambda\overline{\lambda}\|y\|^2 \geq 0.$$

[1] The diagonal of $E \times E$ is the set of all pairs of the form (x,x).

If $\|y\| \neq 0$, substituting

$$\lambda = -\frac{\overline{\langle x \mid y \rangle}}{\|y\|^2}$$

yields the inequality. If $\|y\| = 0$ and $\|x\| \neq 0$, consider $\langle \lambda x + y \mid \lambda x + y \rangle \geq 0$, and substituting

$$\lambda = -\frac{\langle x \mid y \rangle}{\|x\|^2}$$

yields $|\langle x \mid y \rangle|^2 = 0$. Finally, if $\|x\| = 0$ and $\|y\| = 0$, consider $\langle x + \lambda y \mid x + \lambda y \rangle \geq 0$, and substituting $\lambda = -\overline{\langle x \mid y \rangle}$ yields $|\langle x \mid y \rangle|^2 = 0$. This proves the inequality in all cases.

Remark. The Cauchy–Schwarz inequality reduces to an equality if, and only if, the two vectors x and y are linearly dependent, since

$$|\langle x \mid y \rangle|^2 = \|x\|^2 \, \|y\|^2$$

implies

$$\langle x - \frac{\langle y \mid x \rangle}{\|y\|^2} y \mid x - \frac{\langle y \mid x \rangle}{\|y\|^2} y \rangle = 0,$$

and therefore

$$x = \frac{\langle y \mid x \rangle}{\|y\|^2} y.$$

Moreover, if $x = \lambda y$, where λ is a scalar, we have $|\langle x \mid y \rangle| = \|x\| \, \|y\|$.

Theorem 2. *For any pair of vectors x, y in a prehilbert space E we have*

$$(1) \qquad \|x + y\|^2 = \|x\|^2 + \|y\|^2 + 2\Re\langle x \mid y \rangle$$

$$(2) \qquad \|x + y\|^2 + \|x - y\|^2 = 2\|x\|^2 + 2\|y\|^2$$

$$(3) \qquad \|x\|^2 + \|y\|^2 = 2\|\frac{(x + y)}{2}\|^2 + \frac{1}{2}\|x - y\|^2.$$

Expanding $\langle x + y \mid x + y \rangle = \|x + y\|^2$, we obtain the first relation. Relations (2) and (3) follow at once from (1).

Similar relations are well-known in elementary Euclidean geometry. If x and y are orthogonal, $\langle x \mid y \rangle = 0$, and (1) is *Pythagoras theorem.* (2) and (3) are, respectively, the *parallelogram* and *median* laws.

Remark 1. If two vectors x and y in a prehilbert space satisfy the relation $\|x + y\|^2 = \|x\|^2 + \|y\|^2$, they are not necessarily orthogonal. Pythagoras relation is a characterization of orthogonality only in the case of real prehilbert spaces.

Remark 2. Taking into account Cauchy–Schwarz inequality, $\Re\langle x \mid y\rangle \leq |\langle x \mid y\rangle|$ implies that

$$\|x+y\|^2 \leq \|x\|^2 + \|y\|^2 + 2\|x\|\,\|y\|,$$

i.e.,

$$\|x+y\| \leq \|x\| + \|y\|,$$

which is Minkowski's inequality (corollary to Theorem 2, Chapter 2).

2. Hilbert Spaces

Definition 3. A complete prehilbert space is called a *Hilbert space.*

Example 1. The Lebesgue space $L^2_\mu(X)$ is a Hilbert space for the scalar product defined by

$$(\forall f \in L^2_\mu(X)) \quad (\forall g \in L^2_\mu(X)) \qquad \langle f \mid g\rangle = \int \overline{f} g \, d\mu.$$

There are as many Hilbert spaces as spaces X and measures μ. We shall frequently consider the case of functions defined on an interval $]a,b[$ of $\mathbf{R}$ ($X =]a,b[$) and measures μ such that $d\mu(x) = w(x)\,dx$, where w is a positive real-valued function defined on $]a,b[$. In this case the scalar product is said to be associated to the *weight function* w (see Examples 3.3 and 3.4).

Example 2. The space l^2 of all complex sequences (x_n) such that

$$\sum_{n=1}^{\infty} |x_n|^2 \leq \infty$$

is a Hilbert space if the scalar product of two sequences x and y is defined by

$$\langle x \mid y\rangle = \sum_{n=1}^{\infty} \bar{x}_n y_n.$$

l^2 is a particular $L^2_\mu(X)$ space, X is the set $\mathbf{N}$ of all positive integers, and μ is the counting measure.

l^2 is a very important Hilbert space since any separable Hilbert space is isometric to l^2 (see appendix Theorem A2).

Theorem 3. Orthogonal projection theorem. *Let E be a prehilbert space and F a complete vector subspace of E. For any $x \in E$, there is one and only one $y \in F$ such that*

$$\|x - y\| = d(x, F).$$

The vector y, called the orthogonal projection *of x on F and denoted $P_F(x)$, is the only vector of F such that $x - y$ is orthogonal to F.*

Consider a sequence (y_n) of vectors of F such that

$$\lim_{n \to \infty} \|x - y_n\| = d(x, F).$$

We shall first prove that (y_n) is a Cauchy sequence. It follows from Theorem 2(2) that

$$\|y_p - y_q\|^2 = 2(\|x - y_p\|^2 + \|x - y_q\|^2) - 4\|x - (y_p + y_q)\|^2.$$

Since $\frac{1}{2}(y_p + y_q) \in F$,

$$\|x - (y_p + y_q)\|^2 \geq d^2(x, F),$$

and, for any $\epsilon > 0$, there exists $n_0(\epsilon)$ such that, for $n > n_0(\epsilon)$, we have

$$\|x - y_n\|^2 \leq d^2(x, F) + \epsilon.$$

Therefore, for $p > n_0(\epsilon)$ and $q > n_0(\epsilon)$, we have

$$\|y_p - y_q\|^2 \leq 4\epsilon,$$

which proves that (y_n) is a Cauchy sequence. Since F is complete, y_n tends to a limit $y \in F$ such that $\|x - y\| = d(x, F)$.

To prove that y is the only vector such that $\|x - y\| = d(x, F)$, suppose that $y' \in F$ is also such that $\|x - y'\| = d(x, F)$; then, from Theorem 2(2), it follows that

$$\|y - y'\|^2 = 2(\|x - y\|^2 + \|x - y'\|^2) - 4\|x - \frac{1}{2}(y + y')\|^2.$$

Since

$$\|x - \frac{1}{2}(y + y')\|^2 \geq d^2(x, F),$$

$$\|y - y'\|^2 \leq 2(\|x - y\|^2 + \|x - y'\|^2) - 4d^2(x, F),$$

which implies $\|y - y'\| = 0$, i.e. $y = y'$.

To prove finally that y is the only vector of F such that $x - y$ is orthogonal to all vectors of F, let $z \neq 0$ be a vector of F and write

$$\|x - (y + \lambda z)\|^2 > d^2(x, F)$$

for any real scalar $\lambda \neq 0$. From Theorem 2(1) we have

$$\|x - y\|^2 - 2\lambda \Re\langle x - y \mid z\rangle + \lambda^2 \|z\|^2 > d^2(x, F),$$

i.e.,

$$-2\lambda \Re\langle x - y \mid z\rangle + \lambda^2 \|z\|^2 > 0.$$

A suitable choice of λ would yield a contradiction if $\Re\langle x - y \mid z\rangle \neq 0$. Hence, $\Re\langle x - y \mid z\rangle = 0$. If E is a complex prehilbert space, replacing z by iz shows that $\Im\langle x - y \mid z\rangle = 0$. Hence, $\langle x - y \mid z\rangle = 0$.

From the Pythagoras theorem,

$$\|x\|^2 = \|P_F(x)\|^2 + \|x - P_F(x)\|^2,$$

hence,

$$\|P_F(x)\| \leq \|x\|.$$

The equal sign corresponds to the case $x \in F$.

Since $x \mapsto P_F(x)$ is a linear mapping,[2] the last result shows that it is continuous and has a norm equal to one.

The kernel of P_F consists of the vectors orthogonal to F. For any x in E we have

$$x = P_F(x) + (x - P_F(x)),$$

where $P_F(x)$ is in F and $x - P_F(x)$ belongs to the kernel of P_F. E is therefore the direct sum of F and of its orthogonal supplement $F^\perp$. This is denoted

$$E = F \oplus F^\perp.$$

---

[2] It is indeed easy to verify that

$$(\forall x \in E) \quad (\forall \lambda \in \mathbf{C}) \quad P_F(\lambda x) = \lambda P_F(x)$$
$$(\forall x \in E) \quad (\forall x' \in E) \quad P_F(x + x') = P_F(x) + P_F(x').$$

If F_1 is a complete subspace of a prehilbert space E and F_2 a complete subspace of F_1, then

$$P_{F_2}(P_{F_1}(x)) = P_{F_2}(x).$$

This result is the generalization of a well known theorem of elementary Euclidean geometry. To prove it, write

$$\langle P_{F_2}(P_{F_1}(x)) - x \mid z\rangle = \langle P_{F_2}(P_{F_1}(x)) - P_{F_1}(x) \mid z\rangle + \langle P_{F_1}(x) - x \mid z\rangle = 0,$$

where $z \in F_2$. Since z also belongs to F_1, the right-hand side is equal to zero, and, therefore, $P_{F_2}(P_{F_1}(x))$ is the orthogonal projection of x on F_2 i.e., $P_{F_2}(P_{F_1}(x)) = P_{F_2}(x)$.

If $F_1 = F_2$, this last result shows that the orthogonal projection is an idempotent mapping.

Theorem 4. Riesz's theorem. *Let E be a Hilbert space and u a continuous linear form on E. There exists one and only one vector $a \in E$ such that, for any $x \in E$, $u(x) = \langle a \mid x\rangle$.*

This theorem, which determines the dual space E'[3] of a Hilbert space E, shows that any continuous linear form on a Hilbert space is a scalar product.

We shall first prove the following lemma

Lemma. *Let E be a prehilbert space; then for any $a \in E$ the mapping $x \mapsto \langle a \mid x\rangle$ is a continuous linear form on E of norm $\|a\|$.*

This follows from the Cauchy–Schwarz inequality. Indeed,

$$|\langle a \mid x\rangle| \leq \|a\|\,\|x\|$$

shows that $x \mapsto \langle a \mid x\rangle$ is continuous and has a norm less than or equal to $\|a\|$; on the other hand, if $a \neq 0$,[4] then, for $x_0 = a/\|a\|$, $\langle a \mid x_0\rangle = \|a\|$.

We can now prove Riesz's theorem. If $u = 0$, then $a = 0$. If $u \neq 0$, $\ker u = u^{-1}(0)$[5] is a closed hyperplane in E,[6] and its orthogonal supplement is a one-dimensional subspace. Let b be a vector of this subspace;

---

[3] The dual space of a topological vector space E is the vector space E' whose elements are the continuous linear forms on E.

[4] If $a = 0$, $\langle a \mid x\rangle = 0$, and the result is obvious in this case.

[5] $\ker u$ denotes the kernel of u.

[6] Since u is continuous, the inverse image of any closed set is closed.

then from Theorem 3, for any $x \in \ker u$, $\langle b \mid x\rangle = 0$. The two continuous linear forms $x \mapsto u(x)$ and $x \mapsto \langle b \mid x\rangle$ have then the same kernel, and, therefore, there is a scalar λ such that $u(x) = \lambda\langle b \mid x\rangle$,[7] and $a = \overline{\lambda} b$. The uniqueness of a follows from the fact that if $\langle a \mid x\rangle = \langle a' \mid x\rangle$, for any $x \in E$, then $\langle a - a' \mid x\rangle = 0$, i.e., $a = a'$.

Theorem 5. *Let (E_n) be a sequence of Hilbert spaces; the set E of all sequences $x = (x_1, x_2, \ldots, x_n, \ldots)$, such that $x_n \in E_n$ for each n and the series $\sum_{n=1}^{\infty} \|x_n\|^2$ is convergent, is a Hilbert space if we define on E a structure of vector space by*

$$x + y = (x_1 + y_1, x_2 + y_2, \ldots, x_n + y_n, \ldots)$$
$$\lambda x = (\lambda x_1, \lambda x_2, \ldots, \lambda x_n, \ldots)$$

and a scalar product by

$$\langle x \mid y\rangle = \sum_{n=1}^{\infty} \langle x_n \mid y_n\rangle.$$

The Hilbert space E is called the Hilbert sum *of the sequence of Hilbert spaces (E_n).*

The verification of the axioms of vector spaces is simple. If x belongs to E, λx belongs to E since the series $\sum_{n=1}^{\infty} |\lambda|^2 \|x\|^2 = |\lambda|^2 \sum_{n=1}^{\infty} \|x\|^2$ is convergent. On the other hand,

$$\|x_n + y_n\|^2 = \|x_n\|^2 + \|y_n\|^2 + 2\Re\langle x_n \mid y_n\rangle,$$

and taking into account that

$$2\Re\langle x_n \mid y_n\rangle \leq \|x_n\|^2 + \|y_n\|^2$$

(since $\|x_n - y_n\|^2 \geq 0$), we have

$$\|x_n + y_n\|^2 \leq 2(\|x_n\|^2 + \|y_n\|^2),$$

hence the series $\sum_{n=1}^{\infty} \|x_n + y_n\|^2$ is convergent.

From the Cauchy–Schwarz inequality it follows that

$$|\langle x_n \mid y_n\rangle| \leq \|x_n\| \, \|y_n\| \leq \frac{1}{2}(\|x_n\|^2 + \|y_n\|^2),$$

---

[7] $u(x) = 0$ and $\langle b \mid x\rangle = 0$ are two equations of the same hyperplane $\ker u$; therefore, $u(x)$ and $\langle b \mid x\rangle$ must be proportional.

which shows that the series $\sum_{n=1}^{\infty} \langle x_n \mid y_n \rangle$ is absolutely convergent, and it is easy to verify that the mapping $(x, y) \mapsto \langle x \mid y \rangle$ is a scalar product, for

$$\langle x \mid x \rangle = \sum_{n=1}^{\infty} \|x_n\|^2 \geq 0$$

and

$$\langle x \mid x \rangle = 0 \iff x = 0.$$

To prove that E is a Hilbert space, we have to prove that it is complete. Let the sequence (x^m), where $x^m = (x_1^m, x_2^m, \ldots, x_n^m, \ldots)$, be a Cauchy sequence in E; then, for any $\epsilon > 0$, there is an $m_0(\epsilon)$ such that for $p \geq m_0(\epsilon)$ and $q \geq m_0(\epsilon)$, we have $\|x^p - x^q\| \leq \epsilon$, i.e.,

$$\sum_{n=1}^{\infty} \|x_n^p - x_n^q\|^2 \leq \epsilon^2.$$

For each fixed n, this implies $\|x_n^p - x_n^q\| \leq \epsilon$, hence (x_n^m) is a Cauchy sequence in E_n and therefore converges to $y_n \in E_n$. On the other hand, since, for any fixed N, we have

$$\sum_{n=1}^{N} \|x_n^p - x_n^q\|^2 \leq \epsilon^2$$

as soon as $p \geq m_0(\epsilon)$ and $q \geq m_0(\epsilon)$, the continuity of the norm implies that

$$\sum_{n=1}^{N} \|x_n^p - y_n\|^2 \leq \epsilon^2$$

for $p \geq m_0(\epsilon)$. This result is true for any integer N, hence,

$$\sum_{n=1}^{\infty} \|x_n^p - y_n\|^2 \leq \epsilon^2.$$

Therefore the two sequences $(x_n^p - y_n)$ and (y_n) belong to E and, for $p \geq m_0(\epsilon)$, $\|x^p - y\| \leq \epsilon$ where $y = (y_n)$. This proves that the sequence (x^m) converges to $y \in E$, i.e., E is complete.

Remark 1. We can define an isomorphism j_n of E_n onto a closed subspace of E by

$$j_n(x_n) = (0, 0, \ldots, 0, x_n, 0, \ldots),$$

and we have

$$x = \sum_{n=1}^{\infty} j_n(x_n).$$

Remark 2. If, for each n, $E_n = \mathbf{R}$ (or $\mathbf{C}$), then the Hilbert sum E of the Hilbert spaces E_n is l^2 (Example 2.2).

Remark 3. In order to simplify, we assumed that the set of indices was the set $\mathbf{N}$ of natural integers. In this chapter we shall always assume that this is the case although many results remain valid for more general sets of indices.

3. Orthogonal Systems

Definition 4. Let E be a prehilbert space; a sequence (e_n) in E is an *orthonormal system* if, for any pair of indices m and n, $\langle e_m \mid e_n \rangle = \delta_{mn}$. If x belongs to E, the number $c_n(x) = \langle e_n \mid x \rangle$ is the nth *coefficient* (or n^{th} *coordinate*) of x relative to the system (e_n).

Example 1. In l^2, denote by e_n the sequence $(0, 0, \ldots, 0, 1, 0, \ldots)$ having all its terms equal to zero except the nth term equal to 1, i.e., e_n is the sequence (δ_{nm}) where n is fixed and m is the index of the sequence; then $\langle e_m \mid e_n \rangle = \delta_{mn}$, which means that the sequence (e_n) is an orthonormal system.

Example 2. In the vector space of all complex-valued functions defined on the interval [-1,1], define the scalar product of two functions x and y by

$$\langle x \mid y \rangle = \int_{-1}^{1} \overline{x(t)} y(t)\, dt.$$

If, for any integer n (positive or negative),

$$e_n(t) = \frac{1}{\sqrt{2}} \exp(in\pi t),$$

then the sequence (e_n) is an orthonormal system called the *trigonometric system*. The coefficients $c_n(x)$ are, in this case, called the *Fourier coefficients* of the function x (see Example 8).

Example 3. *Legendre polynomials.* Let (P_n) be a sequence of polynomials defined on the interval [-1,1] by the formula[8]

$$P_n(t) = \frac{1}{2^n\, n!} \frac{d^n}{dt^n} (t^2 - 1)^n.$$

---

[8] A formula of this type is called a *Rodrigues formula.*

If the scalar product of two polynomials is defined by

$$\langle P_m \mid P_n \rangle = \int_{-1}^{1} P_m(t) P_n(t)\, dt,$$

we have

$$\langle P_m \mid P_n \rangle = \frac{1}{2^m\, m!} \frac{1}{2^n\, n!} \int_{-1}^{1} \frac{d^m}{dt^m}(t^2-1)^m \frac{d^n}{dt^n}(t^2-1)^n\, dt,$$

and integrating m times by parts yields

$$\langle P_m \mid P_n \rangle = \frac{(-1)^m}{2^{m+n}\, m!\, n!} \int_{-1}^{1} (t^2-1)^m \frac{d^{m+n}}{dt^{m+n}}(t^2-1)^n\, dt,$$

since

$$\frac{d^{m-p}}{dt^{m-p}}(t^2-1)^m = 0 \quad \text{for} \quad t = \pm 1 \quad \text{and} \quad p = 1, 2, \ldots, m-1.$$

If $m > n$, $d^{m+n}(t^2-1)^n/dt^{m+n} = 0$
If $m = n$, $d^{m+n}(t^2-1)^n/dt^{m+n} = (2n)!$, and

$$\int_{-1}^{1} (t^2-1)^n\, dt = 2(-1)^n \int_0^{\pi/2} \cos^{2n+1}\theta\, d\theta = 2(-1)^n \frac{(2^n\, n!)^2}{(2n+1)!}.$$

Finally,

$$\langle P_m \mid P_n \rangle = \frac{2}{2n+1} \delta_{mn}.$$

If $m < n$, we obtain the same result by interchanging the roles of m and n.

The sequence (P_n) of Legendre polynomials is an orthogonal system. The sequence $(\sqrt{(2n+1)/2}P_n)$ obtained by normalizing the sequence of Legendre polynomials is an orthonormal system.

Since

$$(t^2-1)^n = \sum_{r=0}^{n} \frac{(-1)^r\, n!}{r!\,(n-r)!} t^{2n-2r},$$

we have

$$P_n(t) = \sum_{r=0}^{[n/2]} \frac{(-1)^r\,(2n-2r)!}{2^n\, r!\,(n-r)!\,(n-2r)!} t^{n-2r},$$

where $[n/2]$ is the integral part of $n/2$,i.e., the integer equal to $n/2$ if n is even and to $(n-1)/2$ if n is odd.

The first Legendre polynomials are

$$\begin{array}{ll} P_0(t) = 1 & P_3(t) = \frac{1}{2}(5t^3 - 3t) \\ P_1(t) = t & P_4(t) = \frac{1}{8}(35t^4 - 30t^2 + 3) \\ P_2(t) = \frac{1}{2}(3t^2 - 1) & P_5(t) = \frac{1}{8}(63t^5 - 70t^3 + 15t) \end{array}$$

Note that the polynomial P_n and its degree n have the same parity.

Example 4. *Orthogonal polynomials.* According to the definition of the scalar product one can construct various sequences of orthogonal polynomials. If w is a positive real-valued function defined on the interval $]a, b[$ such that the integral $\int_a^b w(t)\,dt$ exists, then the sequence of polynomials (p_n) is called an orthonormal system *with respect to the weight function* w if, for any pair of indices m and n,

$$\int_a^b w(t) p_m(t) p_n(t)\,dt = h_n \delta_{mn}.$$

h_n is the square of the norm of the polynomial p_n. As for Legendre polynomials, classical polynomials can be defined by a Rodrigues formula of the form[9]

$$p_n(t) = \frac{1}{e_n w(t)} \frac{d^n}{dt^n}\big(w(t) g^n(t)\big),$$

where e_n is a real and $g(t)$ a polynomial. The definition of three classical systems of orthogonal polynomials is given in the following table:

Name	Symbol	e_n	$w(t)$	$g(t)$	a	b
Jacobi	$P_n^{(\alpha,\beta)}$	$(-1)^n\, 2^n\, n!$	$(1-t)^\alpha (1+t)^\beta$	$1-t^2$	-1	1
Laguerre	$L_n^{(\alpha)}$	$n!$	$\mathrm{e}^{-t} t^\alpha$	t	0	∞
Hermite	H_n	$(-1)^n$	e^{-t^2}	1	$-\infty$	∞

Orthogonality relations for these polynomials can be established, as we did for Legendre polynomials, by performing m successive integrations by parts. This yields for h_n the following values

[9] In general there is no reason whatsoever for a sequence of orthogonal polynomials to satisfy such a relation.

Polynomials	h_n
Jacobi	$\dfrac{2^{\alpha+\beta+1}\,\Gamma(n+\alpha+1)\,\Gamma(n+\beta+1)}{(2n+\alpha+\beta+1)\,n!\,\Gamma(n+\alpha+\beta+1)}$
Laguerre	$\dfrac{\Gamma(n+\alpha+1)}{n!}$
Hermite	$\sqrt{\pi}\,2^n\,n!$

It is easy to verify that the degree of $P_n^{(\alpha,\beta)}$, $L_n^{(\alpha)}$ and H_n is equal to n.

The following particular Jacobi polynomials are frequently used

(i) Gegenbauer polynomials $G_n^{(p)}$, corresponding to $\alpha = \beta = p - 1/2$;

(ii) Chebyshev polynomials of the first kind T_n, corresponding to $\alpha = \beta = -1/2$;

(iii) Chebyshev polynomials of the second kind U_n, corresponding to $\alpha = \beta = 1/2$ (or $p = 1$).

The values of the parameters e_n and h_n for these polynomials are different from the values deduced from the table given above. For information only, here are the values of these parameters:

Polynomial	e_n	h_n
$G_n^{(p)}$	$\dfrac{(-1)^n\,2^n\,n!\Gamma(2p)\Gamma(n+p+1/2)}{\Gamma(p+1/2)\Gamma(n+2p)}$	$\dfrac{\pi 2^{1-2p}\Gamma(n+2p)}{n!\,(n+p)!\,(\Gamma(p))^2}$ if $p \neq 0$
		$\frac{2\pi}{n^2}$ if $p = 0$
T_n	$\dfrac{(-1)^n 2^{n+1}\Gamma(n+1/2)}{\sqrt{\pi}}$	$\frac{\pi}{2}$ if $n \neq 0$
		π if $n = 0$
U_n	$\dfrac{(-1)^n 2^{n+1}\Gamma(n+3/2)}{(n+1)\sqrt{\pi}}$	$\frac{\pi}{2}$

Note that Legendre polynomials are particular Jacobi polynomials corresponding to $\alpha = \beta = 0$.

Remark 1. The sequence $(\sqrt{w}p_n)$ is an orthogonal system associated to a unit weight function.

Remark 2. The vectors of an orthogonal system are linearly independent; since, if we had a relation of the form $\sum_{i\in I} \lambda_i e_i = 0$, where I is a finite subset of the set of indices, multiplying both sides of this relation by e_j, where $j \in I$, yields $\lambda_j = 0$.

Theorem 6. Bessel's inequality. *In a prehilbert space E, let (e_n) be an orthonormal system; then, for any x in E*

$$\sum_{n=1}^{\infty} |\langle e_n \mid x\rangle|^2 \leq \|x\|^2.$$

Let $(e_i)_{i\in I}$ be a finite subsystem of (e_n) (I is a finite subset of the set of indices). This system generates a finite dimensional (hence complete) subspace V of E. Then for any vector $y = \sum_{i\in I} \lambda_i e_i$ of V, we have

$$\|x-y\|^2 = \langle x - \sum_{i\in I} \lambda_i e_i \mid x - \sum_{i\in I} \lambda_i e_i\rangle$$

or

$$\|x-y\|^2 = \|x\|^2 - \sum_{i\in I} \overline{\lambda}_i \langle e_i \mid x\rangle \sum_{i\in I} \lambda_i \langle x \mid e_i\rangle + \sum_{i\in I} |\lambda_i|^2,$$

which may be written

$$\|x-y\|^2 = \|x\|^2 - \sum_{i\in I} |\langle e_i \mid x\rangle|^2 + \sum_{i\in I} |(\langle e_i \mid x\rangle - \lambda_i)|^2.$$

Hence, $\|x-y\|$ reaches its minimum value when, for any $i \in I$, $\lambda_i = \langle e_i \mid x\rangle$. In this case,

$$\sum_{i\in I} \langle e_i \mid x\rangle e_i = P_V(x),$$

and

$$d^2(x,V) = \|x\|^2 - \sum_{i\in I} |\langle e_i \mid x\rangle|^2,$$

which implies

$$\|x\|^2 \geq \sum_{i\in I} |\langle e_i \mid x\rangle|^2.$$

The upper bound of the right-hand side being equal to $\sum_{n=1}^{\infty} |\langle e_n \mid x\rangle|^2$, Bessel's inequality follows directly from this relation.

Remark. In the particular case of a finite dimensional subspace V, we proved the existence and the unicity of the orthogonal projection of a vector x of E on V (Theorem 3).

Definition 5. An orthonormal system (e_n) in a prehilbert space E is said to be *complete* or *total* if the set of all finite linear combinations of vectors of (e_n) is everywhere dense in E. A complete orthonormal system is also called a *Hilbert basis*.

If (e_n) is a complete orthonormal system, this means that, for any x in E, there exists a finite linear combination $\sum_{i \in I} \lambda_i e_i$ such that

$$\|x - \sum_{i \in I} \lambda_i e_i\| \leq \epsilon$$

for any $\epsilon > 0$.

In l^2 it is readily verified that the system (e_n), where $e_n = (\delta_{nm})_{m=1}^{\infty}$, is complete (see Example 3.1).

Remark. In the case of Hilbert spaces, dimensionality is a rather tricky concept. In the Euclidean space $\mathbf{R}^n$, any set of linearly independent vectors has at most n elements,[10] and any complete orthonormal basis also consists of n vectors. In the Hilbert space l^2, for example, any complete orthonormal system is countable, but it is easy to find in l^2 a set of linearly independent vectors of a greater cardinality. Consider, for example, the set of vectors

$$\{\, x_\alpha \mid x_\alpha = (\alpha, \alpha^2, \alpha^3, \ldots) \; \forall \alpha \in]0,1[\, \}.$$

Since, for any $\alpha \in]0,1[$, the series $\sum_{n=1}^{\infty} \alpha^{2n}$ converges,[11] all these vectors belong to l^2. This set is not countable, it is equipotent to the set of all reals belonging to the open interval]0,1[. On the other hand, the vectors in this set are linearly independent. Indeed, if I is a finite subset of indices, the relation

$$\sum_{i \in I} \lambda_i x_{\alpha_i} = 0, \quad \text{where} \quad (\forall i \in I) \quad \alpha_i \in]0,1[,$$

implies, for any $i \in I$, $\lambda_i = 0$.

The cardinality of a Hilbert basis is called the *orthogonal dimension* of the Hilbert space. For this concept to have any meaning we have to prove that all Hilbert bases of a Hilbert space E have the same cardinality. This is actually the case.

Note that what we are saying is valid only for Hilbert spaces since, in any Hilbert space, it is always possible to find a Hilbert basis (see appendix Theorem A1). In a prehilbert space this is not always the case.

---

[10] This result follows from the fact that $\mathbf{R}^n$ is an n-dimensional vector space and does not take into account the existence of a scalar product.

[11] Its sum is $\alpha^2/(1-\alpha^2)$.

Theorem 7. Parseval's identity. *Let (e_n) be an orthonormal system in a prehilbert space E; a necessary and sufficient condition for this system to be complete is that, for any $x \in E$, we have*

$$\|x\|^2 = \sum_{n=1}^{\infty} |\langle e_n \mid x\rangle|^2.$$

If (e_n) is complete, then for any $\epsilon > 0$, there exists a finite linear combination $\sum_{i \in I} \lambda_i e_i$ such that

$$\|x - \sum_{i \in I} \lambda_i e_i\|^2.$$

But (see the proof of Theorem 6)

$$\|x\|^2 - \sum_{i \in I} |\langle e_i \mid x\rangle|^2 \leq \|x - \sum_{i \in I} \lambda_i e_i\|^2,$$

hence,

$$\|x\|^2 \leq \sum_{i \in I} |\langle e_i \mid x\rangle|^2 + \epsilon.$$

Since ϵ may be as small as we want,

$$\|x\|^2 \leq \sum_{i \in I} |\langle e_i \mid x\rangle|^2,$$

and taking into account Bessel's inequality, this yields Parseval's equality.

If, conversely, any $x \in E$ satisfies Parseval's equality, then for any $\epsilon > 0$, there exists a finite set of indices I such that

$$\|x\|^2 - \sum_{i \in I} |\langle e_i \mid x\rangle|^2 \leq \epsilon.$$

But (see the proof of Theorem 6)

$$\|x\|^2 - \sum_{i \in I} |\langle e_i \mid x\rangle|^2 \leq \|x - \sum_{i \in I} \lambda_i e_i\|^2,$$

hence,

$$\|x - \sum_{i \in I} \langle e_i \mid x\rangle e_i\|^2 \leq \epsilon,$$

which means that the system is complete.

Corollary 1. *Let (e_n) be a complete orthonormal system in a prehilbert space E; for any pair x and y of vectors of E,*

$$\langle x \mid y\rangle = \sum_{n=1}^{\infty} \overline{\langle e_n \mid x\rangle}\langle e_n \mid y\rangle$$
$$= \sum_{n=1}^{\infty} \langle x \mid e_n\rangle\langle e_n \mid y\rangle.$$

This result follows directly from the fact that both sides of this relation are equal on the diagonal of $E \times E$. This relation is very frequently used in quantum mechanics.

Corollary 2. *Let (e_n) be a complete orthonormal system in a prehilbert space E; then, for any $x \in E$, the series $\sum_{n=1}^{\infty} \langle e_n \mid x\rangle e_n$ is convergent and its sum is equal to x.*

For any $\epsilon > 0$, there exists a finite subset I_0 of the set of indices such that

$$\|x\|^2 - \sum_{i \in I_0} |\langle e_i \mid x\rangle|^2 \leq \epsilon.$$

Then, for any finite subset $I \supset I_0$, we have

$$\|x - \sum_{i \in I} \langle e_i \mid x\rangle e_i\|^2 = \|x\|^2 - \sum_{i \in I} |\langle e_i \mid x\rangle|^2 \leq \epsilon,$$

which proves that the series is convergent and that its sum is equal to x. It is therefore entirely justified to write

$$x = \sum_{n=1}^{\infty} \langle e_n \mid x\rangle e_n.$$

Remark 1. This relation must be correctly understood. For example, let (f_n) be a complete orthonormal sequence of functions in the Hilbert space $L_m^2([a,b])$. If, for $g \in L_m^2([a,b])$, we write

$$g = \sum_{n=1}^{\infty} c_n f_n,$$

this means

$$\lim_{n \to \infty} \int_a^b \left|g(t) - \sum_{k=1}^{n} c_k f_k(t)\right|^2 dt = 0$$

but *does not mean*

$$(\forall t \in [a,b]) \qquad g(t) = \sum_{n=1}^{\infty} c_n f_n(t).$$

The series converges in the norm topology of the Hilbert space $L_m^2([a,b])$, but not necessarily pointwise.

Remark 2. Let (e_n) be an orthonormal system in a Hilbert space E. The one-dimensional subspace E_n of E generated by e_n is complete; it is a Hilbert space, and E is the Hilbert sum of the sequence of Hilbert spaces (E_n) (Theorem 5). The orthogonal projection of any $x \in E$ on E_n is

$$P_{E_n}(x) = \langle e_n \mid x \rangle e_n,$$

and

$$x = \sum_{n=1}^{\infty} P_{E_n}(x).$$

Corollary 3. *A necessary and sufficient condition for an orthonormal system (e_n) in a prehilbert space E to be complete is that the relation $\langle e_n \mid x \rangle = 0$, for every n, implies $x = 0$.*

This result readily follows from Parseval's equality.

A sequence (a_n) of vectors in a prehilbert space is said to be *dense* in E if the space generated by all finite linear combinations $\sum_{i \in I} \lambda_i a_i$ (I is a finite subset of the set of indices) is dense in E. If the vectors of a sequence (a_n) dense in E are not linearly independent, it is always possible to find a subsequence of linearly independent vectors that, by construction, will obviously be dense in E. As the first vector of the subsequence, choose a_{k_1} first vector of (a_n) different from 0, and as the nth vector a_{k_n} first vector of (a_n) that is not a linear combination of the $n-1$ vectors $a_{k_1}, a_{k_2}, \ldots, a_{k_{n-1}}$. A subsequence of linearly independent vectors dense in E is said to be *total*.

Given a sequence of linearly independent vectors in a prehilbert space E, the *Schmidt orthogonalization process*, as in the case of finite-dimensional Euclidean spaces, can be used to obtain an orthogonal system.

Theorem 8. Schmidt orthogonalization process. *Let (a_n) be a sequence of linearly independent vectors in a prehilbert space E, and V_p the p-dimensional subspace of E generated by the vectors $a_1, a_2, \ldots, a_p$; the sequence (e_n) defined by $e_1 = a_1$, and, for $p > 1$, by $e_p = a_p - P_{V_{p-1}}(a_p)$, is an orthogonal system such that $e_1, e_2, \ldots, e_p$ generate V_p.*

Use induction on p. $e_1 = a_1$ and $e_2 = a_2 - P_{V_1}(a_2)$ generate V_2, and, as a consequence of the orthogonal projection theorem (Theorem 3), we have

$\langle e_1 \mid e_2 \rangle = 0$. Assume then that $e_1, e_2, \dots, e_{p-1}$ constitute an orthogonal system generating V_{p-1}. Since a_p is linearly independent of $a_1, a_2, \dots, a_{p-1}$, it does not belong to V_{p-1}, and therefore $e_p \neq 0$. On the other hand it follows from the orthogonal projection theorem that e_p is orthogonal to V_{p-1} and is, consequently, orthogonal to $e_1, e_2, \dots, e_{p-1}$. The p vectors $e_1, e_2, \dots, e_{p-1}, e_p$ constitute, therefore, an orthogonal system generating the subspace V_p. Since all the vectors of the sequence (e_n) are different from zero, the sequence $(e_n/\|e_n\|)$ is an orthonormal system.

Example 5. *Orthogonalization of the sequence* (t^n). In the prehilbert space of real-valued continuous functions defined on the interval $[-1, 1]$, in which the scalar product of two functions x and y is defined by

$$\langle x \mid y \rangle = \int_{-1}^{1} x(t)y(t)\,dt,$$

the sequence (t^n) is a sequence of linearly independent functions. A sequence of orthogonal polynomials can be deduced from (t^n) by the Schmidt orthogonalization process. One readily verifies that 1 and t are orthogonal. t and t^2 are also orthogonal, but 1 and t^2 are not. Since $\int_{-1}^{1} 1.t^2\,dt = 2/3$, it follows that the first three orthogonal polynomials of the sequence are $1, t, t^2 - \frac{1}{3}$. Except for a factor $3/2$ for the last one, these polynomials are the first three Legendre polynomials P_1, P_2, P_3 (Example 3.3). This result is actually valid for all values of the index n. Up to a constant factor the Legendre polynomials can be deduced from the sequence (t^n) by the Schmidt orthogonalization process. To prove this result it is sufficient to verify that the Legendre polynomial P_n is orthogonal to the subspace generated by $1, t, t^2, \dots, t^{n-1}$, i.e.,

$$(\forall p < n) \qquad \int_{-1}^{1} t^p \frac{d^n}{dt^n}(t^2 - 1)^n\,dt = 0.$$

This result readily follows after $n + 1$ integrations by parts.

If the scalar product is associated to a weight function, the orthogonalization of the sequence (t_n) yields, according to the weight function w and the limits of integration a and b (Example 3.4), all the classical orthogonal polynomials.

All these polynomials have in common the following property: for any integer n, $p_n(t)$ has n distinct real roots all located between a and b.

If $n > 0$, $p_n(t)$ is orthogonal to 1, i.e.,

$$\int_a^b w(t)p_n(t)\,dt = 0,$$

which implies that $p_n(t)$ has, at least, one real root t_1 between a and b. If $n > 1$, let $p_n(t) = (t - t_1)a_{n-1}(t)$ where $a_{n-1}(t)$ is a polynomial of degree $n-1$. Since $p_n(t)$ is orthogonal to $t - t_1$, we have

$$\int_a^b w(t)(t-t_1)^2 a_{n-1}(t)\,dt = 0,$$

which implies that $a_{n_1}(t)$ has a real root t_2 between a and b. We can then write $p_n(t) = (t-t_1)(t-t_2)a_{n-2}(t)$, where $a_{n-2}(t)$ is a polynomial of degree $n-2$. If $n > 2$ we carry on the proof expressing that $p_n(t)$ is orthogonal to $(t-t_1)(t-t_2)$. This procedure shows that $p_n(t)$ has n real roots between a and b but does not prove that the roots are necessarily distinct. To prove this, assume, for example, that $t_1 = t_2$, then put $p_n(t) = (t-t_1)^2 b_{n-2}(t)$; but since $p_n(t)$ is orthogonal to $b_{n-2}(t)$, we would have

$$\int_a^b w(t)(t-t_1)^2 b_{n-2}^2(t)\,dt = 0,$$

which is impossible, the integrand being positive.

Given a sequence of orthogonal polynomials obtained from (t^n) by the orthogonalization process, it is important to know whether the sequence is complete or not. The following theorem is helpful in this respect.

Theorem 9. *Let $]a,b[$ be a finite or infinite interval of $\mathbf{R}$ and w be a positive measurable function defined on $]a,b[$ such that*

$$\int_a^b e^{r|t|} w(t)\,dt < \infty,$$

where r is a positive real; if (p_n) is a sequence of orthogonal polynomials on $]a,b[$ associated to the weight function w, then this sequence is complete in the Hilbert space $L^2_\mu(]a,b[)$, where μ is the measure associated to w.

For any $n \in \mathbf{N}$, the function $t \mapsto t^n$ belongs to $L^2_\mu(]a,b[)$ since, if t is sufficiently large, one can always find $A > 0$ such that $|t|^{2n} < Ae^{r|t|}$.

If there exists $f \in L^2_\mu(]a,b[)$ different from zero and orthogonal to all the polynomials of the sequence (p_n), then

$$(\forall n \in \mathbf{N}) \qquad \int_a^b w(t)p_n(t)f(t)\,dt = 0,$$

or

$$(\forall n \in \mathbf{N}) \qquad \int_a^b w(t)t^n f(t)\,dt = 0.$$

Let

$$F(z) = \int_a^b w(t)f(t)e^{izt}\,dt = 0,$$

where $z = x + iy$. If y is sufficiently small, the Cauchy–Schwarz inequality shows that

$$|F(z)|^2 \le \int_a^b w(t)|f(t)|^2\,dt \int_a^b w(t)e^{-2yt}\,dt < \infty$$

and

$$|F'(z)|^2 \le \int_a^b w(t)|f(t)|^2\,dt \int_a^b w(t)|t|^2 e^{-2yt}\,dt < \infty.$$

$z \mapsto F(z)$ is therefore analytic in the neighborhood of zero and its expansion in a power series of z follows from Fubini's theorem

$$\begin{aligned} F(z) &= \int_a^b w(t)f(t)\Big(\sum_{k=0}^{\infty} \frac{(izt)^k}{k!}\Big)\,dt \\ &= \sum_{k=0}^{\infty} \frac{z^k}{k!} \int_a^b w(t)(it)^k f(t)\,dt. \end{aligned}$$

This last result shows that all the coefficients of the expansion are equal to zero, hence $F(z) = 0$. Note that the application of Fubini's theorem is legitimate since

$$\begin{aligned} \int_a^b w(t)|f(t)|\Big(\sum_{k=0}^{\infty} \frac{|zt|^k}{k!}\Big)\,dt &= \int_a^b w(t)|f(t)|e^{|zt|}\,dt \\ &\le \Big(\int_a^b w(t)|f(t)|^2\,dt \int_a^b w(t)e^{2|yt|}\,dt\Big)^{1/2} \\ &< \infty \quad \text{if} \quad 2|z| < r. \end{aligned}$$

From $F(z) = 0$ it follows that, for any real x,

$$F(x) = \int_a^b w(t)f(t)e^{itx}\,dt = 0.$$

If $]a, b[$ coincides with $\mathbf{R}$, F is the Fourier transform of wf, and from $F = 0$ it follows that f is almost everywhere equal to zero. Note that the existence of the Fourier transform of wf follows from the Cauchy–Schwarz inequality for

$$\int_a^b w(t)|f(t)|\,dt \le \Big(\int_a^b w(t)|f(t)|^2\,dt \int_a^b w(t)\,dt\Big)^{1/2}.$$

If $]a,b[$ does not coincide with $\mathbf{R}$, the result remains valid if we consider the extension $\widehat{wf}$ of wf defined as follows

$$\widehat{wf} = \begin{cases} wf, & \text{if } a < t < b; \\ 0, & \text{otherwise.} \end{cases}$$

This completes the proof.

The weight functions considered in Example 3.4 satisfy the condition of Theorem 9; Jacobi, Laguerre, and Hermite polynomials are therefore complete sequences of orthogonal polynomials. Gegenbauer, Chebyshev and Legendre polynomials, which are particular Jacobi polynomials, are also complete orthogonal sequences.

Example 6. *Best approximation.* Let f be a function belonging to $L^2_m([-1,1])$; the *best approximation* of f by a polynomial of degree n is a polynomial π_n that minimizes the integral $\int_{-1}^{1} |f(t) - \pi(t)|^2\, dt$.[12] π_n is the projection of f on the subspace of all polynomials of degree not greater than n. From the orthogonal projection theorem (Theorem 3) it follows that π_n exists and is unique.

If we expand f in a series of Legendre polynomials, we obtain

$$f = \sum_{k=0}^{\infty} c_k(f) P_k \qquad \text{and} \qquad \pi_n = \sum_{k=0}^{n} c_k(f) P_k,$$

where

$$c_k(f) = \frac{2k+1}{2} \int_{-1}^{1} P_k(t) f(t)\, dt.$$

The vector $f - \pi_n$ is called the *error*.

For example, the best approximation of e^t on the interval $[-1,1]$ by a polynomial of degree 2 is

$$\pi_2(t) = \frac{15}{4}(e - \frac{7}{e})t^2 + \frac{3}{e}t + \frac{33}{4e} - \frac{3e}{4}.$$

Example 7. *Generating functions of Laguerre and Hermite polynomials.* The integral $\int_0^{\infty} e^{(2a-1)t} t^{\alpha}\, dt$ is convergent if $a < 1/2$ and $\alpha > -1$; this

[12] This definition of the best approximation is the *mean square approximation* There exist many other "best approximations." If E is a normed vector space and F a subspace of E, the best approximation of $x \in E$ by an element of F is the element $y \in F$ such that $\|x - y\|$ is minimum. If E is not a Hilbert space, y does not necessarily exist, and if it exists it is not necessarily unique.

implies that the function $t \mapsto e^{at}$ belongs to the Hilbert space $L^2_\mu(]0,\infty[)$ where μ is the measure associated to the weight function $w\colon t \mapsto t^\alpha e^{-t}$. We can therefore expand it in a series of Laguerre polynomials. We have

$$e^{at} = \sum_{n=0}^{\infty} c_n L_n^{(\alpha)}(t),$$

with

$$c_n = \frac{n!}{\Gamma(n+\alpha+1)} \int_0^\infty e^{(a-1)t} t^\alpha L_n^{(\alpha)}(t)\, dt.$$

The Rodrigues formula for $L_n^{(\alpha)}$ yields

$$c_n = \frac{n!}{\Gamma(n+\alpha+1)} \int_0^\infty e^{at} \frac{d^n}{dt^n} t^{\alpha+n)}\, dt,$$

and after n integrations by parts,

$$\begin{aligned} c_n &= \frac{(-1)^n a^n}{\Gamma(n+\alpha+1)} \int_0^\infty e^{(a-1)t} t^{(\alpha+n)}\, dt \\ &= \frac{(-1)^n a^n}{(1-a)^{n+\alpha+1}}, \end{aligned}$$

i.e.,

$$e^{at} = \frac{1}{(1-a)^{\alpha+1}} \sum_{n=0}^{\infty} \Big(\frac{a}{a-1}\Big)^n L_n^{(\alpha)}(t).$$

Replacing $a/(a-1)$ by z, the preceding result can be written

$$\frac{e^{-tz/(1-z)}}{(1-z)^{(\alpha+1)}} = \sum_{n=0}^{\infty} L_n^{(\alpha)}(t) z^n,$$

which shows that

$$(t,z) \mapsto \frac{e^{-tz/(1-z)}}{(1-z)^{(\alpha+1)}}$$

is the generating function of the Laguerre polynomials.

The integral $\int_{-\infty}^{\infty} e^{2at-t^2}\, dt$ is convergent for any a (real or complex); this implies that the function $t \mapsto e^{at}$ belongs to the Hilbert space $L^2_\mu(\mathbf{R})$ where μ is the measure associated to the weight function $w\colon t \mapsto e^{-t^2}$. We can therefore expand it in a series of Hermite polynomials. We have

$$e^{at} = \sum_{n=0}^{\infty} c_n H_n(t),$$

with

$$c_n = \frac{1}{2^n\, n!\, \sqrt{\pi}} \int_{-\infty}^{\infty} e^{at-t^2} H_n(t)\, dt.$$

The Rodrigues formula for $H_n(t)$ yields

$$c_n = \frac{(-1)^n}{2^n\, n!\, \sqrt{\pi}} \int_{-\infty}^{\infty} e^{at} \frac{d^n}{dt^n} e^{-t^2}\, dt,$$

and after n integrations by parts,

$$\begin{aligned} c_n &= \frac{a^n}{2^n\, n!\, \sqrt{\pi}} \int_{-\infty}^{\infty} e^{at-t^2}\, dt, \\ &= \frac{a^n}{2^n\, n!} e^{a^2/4}, \end{aligned}$$

i.e.,

$$e^{at} = e^{a^2/4} \sum_{n=0}^{\infty} \frac{a^n}{2^n\, n!} H_n(t).$$

Replacing $a/2$ by z, the preceding result can be written

$$e^{2zt-t^2} = \sum_{n=0}^{\infty} \frac{1}{n!} H_n(t) z^n,$$

which shows that

$$(t,z) \mapsto e^{2zt-z^2}$$

is the generating function of Hermite polynomials.

There exist many complete sequences of functions that are not polynomials. In what follows we shall establish two useful criteria to prove whether or not a sequence of functions is complete. The first one is a consequence of the following lemma

Lemma. *Let $[a,b]$ be a finite interval of* $\mathbf{R}$; *the family of characteristic functions* $(1_{[a,u]})_{u\in[a,b]}$ *is total in* $L^2_m([a,b])$.[13]

To see this, suppose that f is a function in $L^2_m([a,b])$ such that

$$(\forall u \in [a,b]) \qquad \int_a^b 1_{[a,u]} f\, dm = \int_a^b f(t)\, dt = 0.$$

[13] The result remains valid if u is required only to range over a dense subset of $[a,b]$.

To prove the lemma we have to prove that $f = 0$. This is indeed the case, since f is integrable because $L^2_m([a,b]) \subset L^1_m([a,b])$ and the function $u \mapsto \int_a^u f(t)\,dt$ being equal to zero its derivative, which is, almost everywhere, equal to f, is null.

Theorem 10. Vitali's criterion. *Let (f_n) be an orthonormal sequence of functions in the Hilbert space $L^2_m([a,b])$, where $[a,b]$ is a finite interval of* **R**; *this sequence is complete if, and only if,*

$$\sum_{n=1}^{\infty} \Big| \int_a^u f_n(t)\,dt \Big|^2 = u - a$$

for every $u \in [a,b]$.

If (f_n) is complete, for any $u \in [a,b]$, we have

$$1_{[a,u]} = \sum_{n=1}^{\infty} c_n f_n,$$

where

$$c_n = \int_a^b 1_{[a,u]} f_n\,dm = \int_a^u f_n(t)\,dt,$$

and the result follows from Parseval equality.

The verification of Vitali's relation requires the summation of a series for at least countably many values of u; however, the sequence being complete only if

$$(\forall u \in [a,b]) \qquad \sum_{n=1}^{\infty} \Big| \int_a^u f_n(t)\,dt \Big|^2 = u - a,$$

we can integrate both sides of this relation. We obtain

$$\frac{2}{(b-a)^2} \sum_{n=1}^{\infty} \int_a^b \Big| \int_a^u f_n(t)\,dt \Big|^2 du = 1$$

if we interchange the order of integration and summation, which is legitimate in accordance with Fubini's theorem. This result is a necessary condition for the sequence (f_n) to be complete. It is also sufficient. Indeed, from Bessel's inequality, the function

$$F \colon u \mapsto a - u - \sum_{n=1}^{\infty} \Big| \int_a^u f_n(t)\,dt \Big|^2$$

is nonnegative on $[a, b]$, and since

$$\int_a^b F(u)\, du = 0,$$

it follows that F is almost everywhere null. Hence, for every u in a dense subset of $[a, b]$, $F(u) = 0$, and we have

Theorem 11. Dalzell's criterion. *Let (f_n) be an orthonormal sequence in the Hilbert space $L^2_m([a, b])$, where $[a, b]$ is a finite interval of $\mathbf{R}$; this sequence is complete if, and only if,*

$$\frac{2}{(b-a)^2} \sum_{n=1}^{\infty} \int_a^b \Big| \int_a^u f_n(t)\, dt \Big|^2 du = 1.$$

Example 8. *Fourier series. Convergence problems.* The trigonometric sequence $(\frac{1}{\sqrt{2}} \exp i\pi nt)_{n \in \mathbf{Z}}$ is complete in $L^2_m([-1, 1])$ since

$$\frac{1}{2} \sum_{n=-\infty}^{\infty} \int_{-1}^{1} \Big| \int_{-1}^{u} \frac{e^{i\pi nt}}{\sqrt{2}}, dt \Big|^2, du = 1,$$

taking into account the result (see Problem 7)

$$\sum_{n=1}^{\infty} \frac{1}{n^2} = \frac{\pi^2}{6}.$$

If f is a function belonging to the Hilbert space $L^2_m([-1, 1])$, the Fourier series of f converges to f in the norm topology of the Hilbert space but has no reason to converge pointwise.

Since Fourier series play an important role in many applications, we shall derive some useful results regarding the pointwise convergence.

If f is continuous and if its Fourier series is uniformly convergent, then the convergence to f is pointwise.

Denoting by $g(t)$ the sum of the Fourier series of f, we have

$$\frac{1}{\sqrt{2}} \sum_{n=-\infty}^{\infty} c_n(f) e^{i\pi nt} = g(t).$$

g is continuous on $[-1, 1]$, and the series obtained by multiplying the left-hand side by $\exp(-i\pi kt)$ is uniformly convergent. It is then legitimate to integrate term by term, which yields

$$(\forall k \in \mathbf{Z}) \qquad c_k(g) = c_k(f).$$

This implies that f and g are equal almost everywhere. Since they are continuous, they are equal.

If $\sum_{n=1}^{\infty} |c_n(f)| < \infty$, the Fourier series of f is normally convergent, hence uniformly convergent. Therefore, if f is continuous, we have

$$(\forall t \in [-1,1]) \qquad \frac{1}{\sqrt{2}} \sum_{-\infty}^{\infty} c_n(f) \mathrm{e}^{i\pi n t} = f(t).$$

For example, for any t in [-1,1], we have

$$\frac{\pi^2}{12}(1-3t^2) = \sum_{n=1}^{\infty} (-1)^{n+1} \frac{\cos \pi n t}{n^2},$$

since $f: t \mapsto (\pi^2/12)(1-3t^2)$, defined on [-1,1], is continuous and its Fourier series is uniformly convergent.

The sum of an uniformly convergent Fourier series being continuous, the convergence condition we have just given does not give any information regarding the convergence of Fourier series of discontinuous functions. Since discontinuous functions are widely used in applications, we shall establish a sufficient condition of convergence valid for a rather large class of discontinuous functions.

If f is in $L_m^1([-1,1])$,[14] its Fourier coefficients are defined, and the partial sum

$$S_n(t) = \frac{1}{\sqrt{2}} \sum_{|k| \le n} c_k(f) \mathrm{e}^{i\pi k t}$$

can be written

$$\begin{aligned} S_n(t) &= \frac{1}{2} \sum_{|k| \le n} \mathrm{e}^{i\pi k t} \int_{-1}^{1} \mathrm{e}^{-i\pi k u} f(u)\, du \\ &= \frac{1}{2} \int_{-1}^{1} \sum_{|k| \le n} \mathrm{e}^{i\pi k (t-u)} f(u)\, du \\ &= \int_{-1}^{1} D_n(t-u) f(u)\, du, \end{aligned}$$

---

[14] Since the Lebesgue measure of [-1,1] is bounded, if f is in $L_m^2([-1,1])$, it is also in $L_m^1([-1,1])$.

where

$$D_n(t-u) = \frac{1}{2} \sum_{|k|\le n} e^{i\pi k(t-u)}$$
$$= \frac{\sin(2n+1)\pi(t-u)/2}{2\sin \pi(t-u)/2}.$$

The function D_n is called the *Dirichlet kernel*. In the sum that defines D_n, the only term that is not periodic is the term corresponding to $k = 0$. Hence,

$$\int_{-1}^{1} D_n(u)\,du = 1,$$

which, after a change of variable, yields

$$S_n(t) - f(t) = \frac{1}{2}\int_{-1}^{1} \big(f(t+u) - f(t)\big)\frac{\sin\frac{(2n+1)u}{2}}{\sin\frac{u}{2}}\,du.$$

If, t being fixed, there exists $\epsilon > 0$ such that the integral

$$\int_{-\epsilon}^{\epsilon} \left|\frac{f(t+u)-f(u)}{u}\right| du$$

is convergent, then $S_n(t)$ converges to $f(t)$.

This follows directly from the Riemann–Lebesgue lemma[15] applied to the integrable function

$$u \mapsto \frac{f(t+u)-f(t)}{u}\,\frac{u}{2\sin\frac{u}{2}}.$$

If the function f has at t a *discontinuity of the first kind*, and if $f(t-0)$ and $f(t+0)$ denote, respectively, the left and right limit of f at t, the preceding convergence condition is easy to extend.

If there exists $\epsilon > 0$ such that the integrals

$$\int_{-\epsilon}^{0} \left|\frac{f(t+u)-f(t-0)}{u}\right| du \qquad \text{and} \qquad \int_{0}^{\epsilon} \left|\frac{f(t+u)-f(t+0)}{u}\right| du$$

are convergent, then $S_n(t)$ converges to $\frac{1}{2}\big(f(t-0)+f(t+0)\big)$.

---

15 This lemma states that if g is an integrable function on $[a,b]$, then $\lim_{k\to\infty} \int_a^b g(u)\sin ku\,du = 0$ (see Chapter 2, Problem 4).

In this case the result follows from the Riemann–Lebesgue lemma applied to the functions

$$u \mapsto \frac{f(t+u)-f(t-0)}{u}\frac{u}{2\sin u/2}$$

and

$$u \mapsto \frac{f(t+u)-f(t+0)}{u}\frac{u}{2\sin u/2},$$

integrable, respectively on [-1,0] and [0,1].

This result applies, for example, to the function f defined by

$$f(t) = \begin{cases} -1, & \text{if } -1 < t < 0, \\ 1, & \text{if } 0 < t < 1, \end{cases}$$

whose Fourier series is

$$\frac{4}{\pi}\sum_{k=0}^{\infty}\frac{\sin(2k+1)\pi t}{2k+1}.$$

Remark 1. If f and g are two continuous functions on [-1,1] such that, for any integer k, $c_k(f) = c_k(g)$, the two functions are equal. Therefore if f is a continuous function, although its Fourier series has no reason to be convergent, the set of its Fourier coefficients characterizes f, and it is natural to address the following problem

Given all the Fourier coefficients of a continuous function on [-1,1], *how can we determine this function?*

The following theorem, due to Fejér, solves this problem

Let f be a continuous function on [-1,1]; *the sequence of its Fejér sums defined by*

$$\sigma_n(t) = \frac{1}{2}\big(S_0(t) + S_1(t) + \cdots + S_{n-1}(t)\big),$$

where $S_j(t)$ is the partial sum defined above, converges uniformly to $f(t)$ for every t in]-1,1[.

Replacing the partial sums by their expressions yields

$$\sigma_n(t) = \frac{1}{2n}\int_{-1}^{1}\left(\sum_{j=0}^{n-1}\frac{\sin\frac{2j+1}{2}\pi(t-u)}{\sin\frac{1}{2}\pi(t-u)}\right) f(u)\,du$$

or

$$\sigma_n(t) = \int_{-1}^{1} F_n(t-u)f(u)\,du,$$

where F_n denotes the *Fejér kernel* given by

$$F_n(u) = \frac{1}{2n} \sum_{j=0}^{n-1} \frac{\sin \frac{2j+1}{2} \pi u}{\sin \frac{1}{2} \pi u} = \frac{1}{2n} \frac{\sin^2 n \frac{\pi u}{2}}{\sin^2 \frac{\pi u}{2}}.$$

Here again, we have

$$\int_{-1}^{1} F_n(u)\, du = 1,$$

which follows from $F_n = \frac{1}{n} \sum_{j=0}^{n-1} D_j$, and $(\forall j > 0)$ $\int_{-1}^{1} D_j(u)\, du = 1$.

Hence,

$$\sigma_n(t) - f(t) = \int_{-1}^{1} (f(t+u) - f(t))\, F_n(u)\, du.$$

Since f is continuous on the compact interval [-1,1], it is uniformly continuous, and therefore, for any $\epsilon > 0$, we can find $\delta > 0$ such that for $|u| < \epsilon$ we have $|f(t+u) - f(t)| < \delta$ and this for any t. It follows that

$$\int_{|u|<\delta} (f(t+u) - f(t))\, F_n(u)\, du < \epsilon \int_{|u|<\delta} F_n(u)\, du < \epsilon,$$

which shows that the integral of $\big(f(t+u) - f(t)\big)F_n(u)$ over $]-\delta, \delta[$ tends to zero.

Furthermore, since f is continuous on [-1,1] it is bounded, and there exists $M > 0$ such that, for any t in [-1,1], $|f(t)| \leq M$. We have, then,

$$\int_{\delta \leq |u| \leq 1} |f(t+u) - f(t)| F_n(u)\, du \leq 2M \int_{\delta \leq |u| \leq 1} F_n(u)\, du.$$

But for $\delta \leq |u| \leq 1$, we have

$$\frac{\sin^2 n \frac{\pi u}{2}}{\sin^2 \frac{\pi u}{2}} \leq \frac{1}{\delta^2},$$

since $\sin^2 n\pi u/2 \leq 1$, and for $0 \leq x \leq \pi/2$, $\sin x \geq 2x/\pi$ or, after a change of variable, $\sin \pi u/2 \geq \delta$ if $\delta < u \leq 1$.

Finally

$$\int_{\delta \leq |u| \leq 1} |f(t+u) - f(t)| F_n(u)\, du \leq \frac{M}{n\delta^2},$$

which shows that the integral tends to 0 when n tends to ∞.

This completes the proof of the Fejér theorem.

Remark 2. If F is a complex analytic function in an annulus containing the closed path $t \mapsto \exp(i\pi t)$ where t belongs to [-1,1] then, from the Laurent theorem, it follows that $F(z)$ is the sum of a uniformly convergent series of the form $\sum_{-\infty}^{\infty} c_n z^n$, which is uniformly convergent in the annulus and in particular for $|z| = 1$. Therefore, if we put $z = \exp(i\pi t)$, we obtain the Fourier series expansion of the function $t \mapsto F(\mathrm{e}^{i\pi t})$, and this expansion converges uniformly to $F(\mathrm{e}^{i\pi t})$.

Similarly if F is a complex analytic function in an open disk containing the closed path $t \mapsto \exp(i\pi t)$, where t belongs to [-1,1], $F(z)$ is the sum of a series of the form $\sum_{n=0}^{\infty} c_n z^n$, which is uniformly convergent in the open disk and in particular for $|z| = 1$. Therefore, if we put $z = \exp(i\pi t)$, we obtain the following Fourier series expansions

$$\Re F(\mathrm{e}^{i\pi t}) = \sum_{n=0}^{\infty} \frac{1}{2}(c_n + \overline{c}_n)\cos n\pi t + \sum_{n=1}^{\infty} \frac{i}{2}(c_n - \overline{c}_n)\sin n\pi t,$$

$$\Im F(\mathrm{e}^{i\pi t}) = \sum_{n=0}^{\infty} \frac{1}{2i}(c_n - \overline{c}_n)\cos n\pi t + \sum_{n=1}^{\infty} \frac{1}{2}(c_n + \overline{c}_n)\sin n\pi t,$$

and, as in the previous case, the convergence is uniform.

For example, it is not difficult to sum the Fourier series

$$-2\sum_{n=1}^{\infty} \frac{a^n}{n} \cos n\pi t,$$

where $0 < a < 1$. The series is uniformly convergent, and can be written

$$-2\Re \sum_{n=1}^{\infty} \frac{z^n}{n},$$

with $z = a\exp(in\pi t)$. Since

$$-\sum_{n=1}^{\infty} \frac{z^n}{n} = \log(1 - z) \qquad (|z| < 1),$$

we have

$$-2\sum_{n=1}^{\infty} \frac{a^n}{n} \cos n\pi t = \log(1 + a^2 - 2a\cos \pi t).$$

Example 9. *The Gibbs phenomenon.* Let us study the convergence of a Fourier series of a function in the neighborhood of a point of discontinuity of the first kind. Consider, for example, the function f defined by

$$f(t) = \begin{cases} -1, & \text{if } t \in]\text{-}1,0[; \\ 1, & \text{if } t \in]0,1[. \end{cases}$$

The partial sum $S_n(t)$ of its Fourier series can be written

$$\begin{aligned} S_n(t) &= \int_{-1}^{1} D_n(u-t) f(u)\, du \\ &= \int_{-1}^{0} D_n(u-t)\, du + \int_{0}^{1} D_n(u-t)\, du \\ &= -\int_{-1-t}^{-t} D_n(u)\, du + \int_{-t}^{1-t} D_n(u)\, du. \end{aligned}$$

Changing u in $-u$ in the first integral yields

$$S_n(t) = -\int_{t}^{1+t} D_n(u)\, du + \int_{-t}^{1-t} D_n(u)\, du,$$

i.e.,

$$S_n(t) = \int_{-t}^{t} D_n(u)\, du - \int_{1-t}^{1+t} D_n(u)\, du.$$

Hence,

$$\left| S_n(t) - \int_{-t}^{t} D_n(u)\, du \right| = \left| \int_{1-t}^{1+t} D_n(u)\, du \right|.$$

But

$$\int_{1-t}^{1+t} D_n(u)\, du = \int_{1-t}^{1+t} \frac{\sin \frac{2n+1}{2}\pi u}{\sin \frac{1}{2}\pi u}\, du.$$

Integrating by part yields

$$\begin{aligned} \int_{1-t}^{1+t} D_n(u)\, du = &-\frac{2}{(2n+1)\pi} \left[\frac{\cos \frac{2n+1}{2}\pi u}{2 \sin \frac{1}{2}\pi u} \right]_{1-t}^{1+t} \\ &- \frac{1}{2n+1} \int_{1-t}^{1+t} \cos \frac{2n+1}{2}\pi u \frac{\cos \frac{1}{2}\pi u}{\sin^2 \frac{1}{2}\pi u}\, du \end{aligned}$$

which shows that

$$\left| \int_{1-t}^{1+t} D_n(u)\, du \right| \leq \frac{A}{n} \qquad (A > 0).$$

Therefore, for $-1 < t < 1$, we have

$$\lim_{n\to\infty} S_n(t) = \lim_{n\to\infty} \int_{-t}^{t} D_n(u)\,du.$$

As above, an integration by part shows that

$$\left| \int_{-t}^{t} \left(\frac{1}{2\sin\frac{1}{2}\pi u} - \frac{1}{u} \right) \sin\frac{2n+1}{2}\pi u\,du \right| \leq \frac{B}{n} \qquad (B > 0),$$

hence,

$$\begin{aligned}\lim_{n\to\infty} S_n(t) &= \lim_{n\to\infty} \int_{-t}^{t} \frac{\sin\frac{2n+1}{2}\pi u}{\pi u}\,du \\ &= \lim_{n\to\infty} \frac{2}{\pi} \int_0^{\frac{2n+1}{2}\pi t} \frac{\sin u}{u}\,du \\ &= f(t).\end{aligned}$$

But, if n is fixed, the function

$$t \mapsto \frac{2}{\pi} \int_0^{\frac{2n+1}{2}\pi t} \frac{\sin u}{u}\,du$$

oscillates around its limit values. If $t > 0$, its first maximum is reached for $t = 2/(2n+1)$, and at this point the value taken by the function is

$$\frac{2}{\pi} \int_0^{\pi} \frac{\sin u}{u}\,du = 1.179\ldots.$$

Therefore, even for a very large fixed n, there exist points in the neighborhood of the origin where the partial sum $S_n(t)$ differs from $f(t)$ by almost 18%. This behavior of the partial sum $S_n(t)$ in the neighborhood of the origin is not particular to the function f we have considered; any partial sum of the Fourier series of function that has a discontinuity of the first kind will exhibit such a behavior in the neighborhood of the point of discontinuity. This behavior is called the *Gibbs phenomenon.*

Appendix

When we discussed problems of dimensionality in the remark following Definition 5 we claimed that

Theorem A1. *Every Hilbert space contains a Hilbert basis.*

The proof of this theorem rests on Zorn's axiom. In order to state this axiom, we need the following definitions

Let R be a relation on a set E such that, for all x, y, and z in E,

(i) xRx;

(ii) xRy and yRx implies $x = y$;

(iii) xRy and yRz implies xRz.

Then E is said to be *ordered by* R.

If E is an ordered set by R and either xRy or yRx for all x and y in E, then E is said to be *totally ordered.*

Let E be an ordered set by R; an element z in E is called a *maximal element* if zRx implies $x = z$.

Let E be an ordered set by R and E' a subset of E that is totally ordered by R; if there exists z in E such that xRz for any x in E', then z is called an *upper bound* for E'.

Zorn's axiom asserts as follows:

Let E be a nonempty set ordered by R; if every totally ordered subset of E has an upper bound in E, then E has at least one maximal element.[16]

Consider the set of all orthonormal systems in a Hilbert space E. This set can be ordered by inclusion. From Zorn's axiom it follows that there exists a maximal orthonormal system that is necessarily complete, for if there exists a nonnull element in E orthogonal to all the elements of this system, then maximality is contradicted.

Theorem A2. *Any Hilbert space is isometric*[17] *to $l_2(I)$.*

The Hilbert space $l_2(I)$ is the space of all the families of complex numbers $(x_i)_{i\in I}$ such that the family $(|x_i|^2)_{i\in I}$ is summable.[18]. This sum

[16] Zorn's axiom is equivalent to the more intuitive *axiom of choice*, which says: If $(A_i)_{i\in I}$ is a family of nonempty subsets of a set E, then there exists a mapping f of I in E such that, for any index i in I, we have $f(i) = A_i$.

[17] See Note 2.

[18] A family of reals $(\lambda_i)_{i\in I}$ is *summable* if there exists a real λ such that, for

is denoted $\sum_{i\in I} |x_i|^2$, and the scalar product of two elements x and y of $l_2(I)$ is defined by $\langle x \mid y\rangle = \sum_{i\in I} \overline{x}_i y_i$.

The Hilbert space l_2 of example 2.2 corresponds to the case $I = \mathbf{N}$.

Let E be a Hilbert space and $(e_i)i \in I$ a Hilbert basis for E, which exists according to the previous theorem. The linear mapping u that, to any $x \in E$, associates the family $(x_i)_{i\in I}$ of coefficients of x relative to the basis $(e_i)_{i\in I}$ ($x_i = \langle e_i \mid x\rangle$) is an isometry of the Hilbert space E in the Hilbert space $l_2(I)$, which conserves the scalar product. $u(E)$ is dense in $l_2(I)$ because it contains all the vectors of the basis of $l_2(I;)$[19] but since $u(E)$ is a complete, i.e. closed, subset of $l_2(I)$, it is indeed identical to $l_2(I)$.

In physical applications Hilbert spaces are, in general, *separable*.[20]. Let E be such a Hilbert space; in this case it is not difficult to construct a total system (see the discussion following Corollary 3 of Theorem 7), and, using the Schmidt orthogonalization process, to obtain a denumerable Hilbert basis. E is therefore isometric to $l_2(\mathbf{N})$ (which has been denoted simply l_2). Conversely, if a Hilbert space E has a denumerable basis, the set of the vectors that have a finite number of rational nonzero coefficients is a dense denumerable subset of E, which is therefore separable.

---

any $\epsilon > 0$, there exists a finite subset J_0 of the set of indices I such that, for any finite subset J of I containing J_0, we have $|\sum_{i\in J} \lambda_i - \lambda| \leq \epsilon$.

[19] This basis $(e'_i)_{i\in I}$ is defined by $e'_i = (\delta_{ij})_{j\in I}$.

[20] A topological space is *separable* if it contains a dense denumerable subset. $\mathbf{R}$ is separable since it contains $\mathbf{Q}$ (the set of rationals) which is denumerable and dense in $\mathbf{R}$.

Notes

Note 1.
Let E and F be two normed vector spaces and u a linear mapping of E in F; for u to be continuous it is necessary and sufficient that there exists $a > 0$ such that, for any $x \in E$, $\|u(x)\| \leq a\|x\|$.

The condition is sufficient because

$$\|u(x) - u(x_0)\| = \|u(x - x_0)\| \leq a\|x - x_0\|.$$

To prove that the condition is necessary, assume that u is continuous at zero; then it exists a closed ball $B(0, r)$ of radius r centered at the origin such that, for any z belonging to this ball, we have $\|u(z)\| \leq 1$. Let $x \neq 0$ be a point in E; if we put $z = rx/\|x\|$, we have $\|z\| = r$ and therefore $\|u(z)\| \leq 1$; but $u(z) = (r/\|x\|)u(x)$, hence $\|u(x)\| \leq (1/r)\|x\|$. If $x = 0$, the condition $\|u(x)\| \leq a\|x\|$ is obviously satisfied. Finally, if u is continuous at the origin, it is continuous everywhere in E.

If we define the norm of a linear continuous mapping by

$$\|u\| = \sup_{\|x\| \leq 1} \|u(x)\|,$$

the set of all continuous linear mappings is a normed vector space.

Note 2.
Let (E, d) and (E', d') be two metric spaces; a bijection f of E onto E' such that, for any x and y in E,

$$d'\big(f(x), f(y)\big) = d(x, y)$$

is called an *isometry*.

Given (E, d) and a bijection f of E onto E', we can define d' by the preceding relation. d' is then called the *transported distance* by f of E onto E.

Problems

1. Expand the derivative P_n' of the Legendre polynomial P_n in a series of Legendre polynomials.

2. If P_n denotes the Legendre polynomial of degree n, show that

$$(1) \qquad (1-x)^n P_n\left(\frac{1+x}{1-x}\right) = \sum_{k=0}^{\infty} \binom{n}{k}^2 x^k,$$

$$(2) \qquad P_n(x) = \frac{1}{\pi}\int_0^{\pi} \left(x + \sqrt{x^2-1}\cos\phi\right)^n d\phi.$$

3. Let (P_n) be the sequence of Legendre polynomials; using the orthogonality relations, show that Legendre polynomials satisfy the following recursion relations

$$(n+1)P_{n+1}(x) - (2n+1)xP_n(x) + nP_{n-1}(x) = 0.$$

Show then that the expansion in series of powers of z of the function $z \mapsto (1-2xz+z^2)^{-1/2}$ is $\sum_{n=0}^{\infty} P_n(x)z^n$. Determine the values of $P_n(1)$ and $P_n(-1)$.

4. Let (p_n) be a sequence of orthogonal polynomials that has been obtained from the sequence (x_n) using the Schmidt orthogonalization process; show that these polynomials satisfy the following recursion relation

$$p_n(x) = (A_n x + B_n)p_{n-1}(x) + C_n p_{n-2}(x),$$

where $n > 2$, and A_n, B_n and C_n depend only on n.

5. Let (p_n) be a sequence of orthogonal polynomials on the interval $[a,b]$ associated to the weight function w, and ρ a nonnegative polynomial on $[a,b]$ all of whose zeros are real and distinct; if we denote by x_k $(k = 1,2,\ldots,l)$ the zeros of ρ, show that the sequence (q_n) defined by

$$\rho(x)q_n(x) = \begin{vmatrix} p_n(x) & p_{n+1}(x) & \cdots & p_{n+l}(x) \\ p_n(x_1) & p_{n+1}(x_1) & \cdots & p_{n+l}(x_1) \\ \vdots & \vdots & \ddots & \vdots \\ p_n(x_l) & p_{n+1}(x_l) & \cdots & p_{n+l}(x_l) \end{vmatrix}$$

is a sequence of orthogonal polynomials on $[a,b]$ associated to the weight function ρw.

6. Expand in Fourier series the following functions defined on the interval $]-\pi,\pi[$

$$\begin{aligned} &(1) \qquad f_1 : x \mapsto \frac{1}{2}x, \\ &(2) \qquad f_2 : x \mapsto \frac{1}{12}(\pi^2 x - x^3), \\ &(3) \qquad f_3 : x \mapsto \begin{cases} \frac{\pi}{8}(\pi x + x^2) & \text{if } x \in]-\pi, 0] \\ \frac{\pi}{2}(\pi x - x^2) & \text{if } x \in [0, \pi[. \end{cases} \end{aligned}$$

Are these Fourier series pointwise convergent?

7. Find the Fourier series of the function f defined on [-1,1] by

$$f(t) = \begin{cases} -1 - x, & \text{if } x \in [-1, 0[, \\ 1 - x, & \text{if } x \in]0, 1], \end{cases}$$

and show that

$$\sum_{n=0}^{\infty} \frac{(-1)^n}{2n+1} = \frac{\pi}{4} \quad \text{and} \quad \sum_{n=1}^{\infty} \frac{1}{n^2} = \frac{\pi^2}{6}.$$

8. Find the Fourier series of the function f defined on $[-\pi, \pi]$ by $f(x) = \cos px$ where p is not an integer, and show that

$$\begin{aligned} &(1) \qquad \pi \cot p\pi = \frac{1}{p} + 2\sum_{n=1}^{\infty} \frac{p}{p^2 - n^2}, \\ &(2) \qquad \frac{\pi}{\sin p\pi} = \frac{1}{p} + 2\sum_{n=1}^{\infty} \frac{(-1)^n p}{p^2 - n^2}. \end{aligned}$$

9. Let f and g be two functions belonging to $L^2_m([0,1])$; show that

$$c_n(fg) = \sum_{p=-\infty}^{\infty} c_{n-p}(f) c_p(g),$$

where $c_k(x)$ is the kth coefficient of x relative to the orthogonal system $(\exp(2i\pi kt))_{k \in \mathbf{Z}}$.

10. Determine the values of a, b, c, d, and e that minimize the integral

$$\int_{-1}^{1} (x^5 - ax^4 - bx^3 - cx^2 - dx - e)^2 \, dx.$$

11. Determine the values of a, b, and c that minimize the integral

$$\int_0^{\infty} (x^3 - ax^2 - bx - c)^2 e^{-x}\, dx.$$

12. Let (p_n) be a sequence of orthonormal polynomials on $[a, b]$, and f a continuous function on $[a, b]$; show that the polynomial $\sum_{k=0}^{n} \langle p_k \mid f \rangle p_k(x)$ coincides with $f(x)$ at, at least, $n + 1$ points on $[a, b]$.

13. Let (p_n) be a sequence of orthonormal polynomials and K_n the function of two variables defined by

$$K_n(x, y) = \sum_{k=0}^{n} p_k(x) p_k(y);$$

show that, for any polynomial P of degree not greater than n, we have

$$\langle P(x) \mid K(x, y) \rangle = P(y).$$

The notation adopted in the left-hand side means that to calculate the scalar product we have to integrate over x.

14. Show that there exists a unique polynomial $S_n(x)$ of degree n such that

$$\int_{-1}^{1} S_n(x) Q(x)\, dx = Q(1),$$

where $Q(x)$ is any polynomial whose degree is not greater than n.

Determine the expansion of S_n a in series of Legendre polynomials, and show that the sequence (S_n) satisfies the following orthogonality relations

$$\int_{-1}^{1} (1 - x) S_n(x) S_m(x)\, dx = \begin{cases} 0, & \text{if } m \neq n; \\ \frac{n+1}{2}, & \text{if } m = n. \end{cases}$$

15. Let (r_n) be the sequence of the so-called *Rademacher functions* defined on [0,1] by

$$r_n(x) = \operatorname{sgn}(\sin 2^n \pi x).$$

Is this sequence complete in $L^2_m([0, 1])$?

Solutions

1. The degree of P'_n is equal to $n-1$. Therefore, taking into account the parity of Legendre polynomials, we have

$$P'_n = \sum_{p=0}^{[n-1/2]} c_{n-2p-1} P_{n-2p-1},$$

where $[(n-1)/2]$ denotes the integral part of $(n-1)/2$, and

$$c_{n-2p-1} = \frac{2n-4p-1}{2} \int_{-1}^{1} P'_n(x) P_{n-2p-1}(x)\, dx.$$

An integration by parts yields

$$\int_{-1}^{1} P'_n(x) P_{n-2p-1}(x)\, dx = [P_n(x) P_{n-2p-1}(x)]_{-1}^{1} - \int_{-1}^{1} P_n(x) P'_{n-2p-1}(x)\, dx = 2,$$

because the last integral is null (P_n is orthogonal to all polynomials of degree less than n). Finally,

$$P'_n = \sum_{p=0}^{[n-1/2]} (2n-4p-1) P_{n-2p-1}.$$

2. From Leibniz's rule applied to Rodrigues's formula it follows that

$$P_n(x) = \frac{1}{2^n\, n!} \sum_{k=0}^{n} \binom{n}{k} \frac{n!}{(n-k)!} (x-1)^{n-k} \frac{n!}{k!} (x+1)^k.$$

Hence, (1) is obtained by changing x in $(1+x)/(1-x)$.

To obtain (2) we write

$$(x + \sqrt{x^2-1}\cos\phi)^n = \left(\sqrt{\frac{x+1}{2}} + \sqrt{\frac{x-1}{2}} e^{i\phi}\right)^n \times \left(\sqrt{\frac{x+1}{2}} + \sqrt{\frac{x-1}{2}} e^{-i\phi}\right)^n,$$

and apply the binomial formula to the right-hand side. This yields

$$\begin{aligned}\frac{1}{\pi}\int_0^{2\pi}&(x+\sqrt{x^2-1}\cos\phi)^n\,d\phi\\&=\frac{1}{2\pi}\int_0^{2\pi}\left(\sum_{j=0}^{n}\binom{n}{j}\left(\frac{x+1}{2}\right)^{\frac{n-j}{2}}\left(\frac{x-1}{2}\right)^{\frac{j}{2}}\mathrm{e}^{ij\phi}\right)\\&\qquad\times\left(\sum_{l=0}^{n}\binom{n}{l}\left(\frac{x+1}{2}\right)^{\frac{n-l}{2}}\left(\frac{x-1}{2}\right)^{\frac{l}{2}}\mathrm{e}^{-ij\phi}\right)d\phi.\end{aligned}$$

Expanding the product and taking into account

$$\frac{1}{2\pi}\int_0^{2\pi}\mathrm{e}^{i\lambda\phi}\,d\phi=\begin{cases}0, & \text{if } \lambda\neq 0,\\ 1, & \text{if } \lambda=0,\end{cases}$$

we finally obtain (2).

This result could also have obtained applying Cauchy's formula for the derivative of an analytic function to Rodrigues's formula.

3. Since $xP_n(x)$ is a polynomial of degree $n+1$, taking into account the parity of Legendre polynomials, we have

$$xP_n(x)=c_{n+1}P_{n+1}(x)+c_{n-1}P_{n-1}(x)+c_{n-3}P_{n-3}(x)+\cdots.$$

where the coefficients are given by

$$\begin{aligned}c_k&=\frac{2k+1}{2}\langle P_k\mid xP_n\rangle\\&=\frac{2k+1}{2}\langle xP_k\mid P_n\rangle.\end{aligned}$$

Because P_n is orthogonal to all polynomials of degree less than n, only c_{n-1} and c_{n+1} are different from zero. Hence,

$$xP_n(x)=c_{n+1}P_{n+1}(x)+c_{n-1}P_{n-1}(x).$$

Equating the coefficients of x^{n+1} in both sides of this relation gives c_{n+1}:

$$\frac{(2n)!}{2^n(n!)^2}=c_{n+1}\frac{(2n+2)!}{2^{n+1}\big((n+1)!\big)^2},$$

i.e.,

$$c_{n+1}=\frac{n+1}{2n+1}.$$

To determine c_{n-1} it is sufficient to change n in $n-1$ in the right-hand side of

$$\frac{c_{n+1}}{2n+3} = \frac{1}{2}\langle P_{n+1} \mid xP_n\rangle$$

for

$$\begin{aligned}\frac{c_{n-1}}{2n-1} &= \frac{1}{2}\langle P_{n-1} \mid xP_n\rangle \\ &= \frac{1}{2}\langle xP_{n-1} \mid P_n\rangle.\end{aligned}$$

Hence,

$$c_{n-1} = \frac{n}{2n+1},$$

and finally

$$(2n+1)xP_n(x) = (n+1)P_{n+1}(x) + nP_{n-1}(x).$$

For a fixed x, the function $z \mapsto (1-2xz+z^2)^{-1/2}$ can be expanded in powers of z if $|2xz - z^2| < 1$. If $|2xz| + |z^2| < 1$, we can rearrange the terms of the expansion and order them in increasing powers of z. Moreover, if a, b and ϵ are three positive numbers such that $2ab + a^2 < 1 - \epsilon$, the expansion will be uniformly convergent for $|z| \leq a$ and $|x| \leq b$.

Since the nth derivative of $(1-2xz+z^2)^{-1/2}$ with respect to z is of the form $(1-2xz+z^2)^{-n-1/2} f_n(x,z)$, where $f_n(x,z)$ is a polynomial of degree n in x and z, it follows that the coefficient of z^n in the expansion is a polynomial of degree n in x.

To show that the sequence of polynomials (Q_n) defined by

$$(1-2xz+z^2)^{-1/2} = \sum_{n=0}^{\infty} Q_n(x) z^n$$

is actually the sequence (P_n) of Legendre polynomials, it will be sufficient to verify that they satisfy the same recursion relation as the Legendre polynomials and that the first two polynomials of the sequence Q_0 and Q_1 are equal, respectively, to P_0 and P_1. If we denote by F the function $(x,z) \mapsto (1-2xz+z^2)^{-1/2}$, we have

$$(1-2xz+z^2)\frac{\partial F}{\partial z} = (x-z)F,$$

i.e.,

$$(1-2xz+z^2)\sum_{k=1}^{\infty} kQ_k(x)z^{k-1} = (x-z)\sum_{l=0}^{\infty} Q_l(x)z^l;$$

the power series in z whose sum is $F(x,z)$ being uniformly convergent, it was legitimate to take the derivative term by term.

If we identify the coefficients of z^n in both sides, we find that the sequence (Q_n) satisfies the same recursion relation as the sequence (P_n), and since

$$Q_0(x) = 1 = P_0(x) \quad \text{and} \quad Q_1(x) = x = P_n(x),$$

it follows that, for any positive integer n, $Q_n(x) = P_n(x)$.

If we replace z by 1 or -1 in $F(x,z)$, we obtain $P_n(1) = 1$, and $P_n(-1) = (-1)^n$.

4. If A_n is chosen such that $p_n(x) - A_n x p_{n-1}(x)$ is a polynomial of degree $n-1$, it can be represented by a linear combination of the form $\sum_{k=0}^{n-1} c_k p_k(x)$. By construction, p_n is orthogonal to all polynomials of degree less than n, and p_{n-1} to all polynomials of degree less than $n-1$. As a consequence the scalar product $\langle p_n - A_n x p_{n-1} \mid p_k\rangle$ is null if $k < n-2$. This implies that the coefficients c_k are equal to zero if $k < n-2$. The linear combination contains therefore only two terms, one in p_{n-1} and one in p_{n-2}, and the result follows.

Note that the degrees of the various polynomials must be nonnegative integers; therefore n must be greater than 2.

5. The determinant is a polynomial of degree $n+l$ divisible by $\rho(x)$; it is consequently of the form $\rho(x)q_n(x)$, where q_n is a polynomial of degree n. But ρq_n is a linear combination of of the polynomials $p_n, p_{n+1}, \ldots, p_{n+l}$; therefore, for any polynomial π of degree less than n, we have

$$\int_a^b \rho(x) q_n(x) \pi(x) w(x)\, dx = 0,$$

which means that q_n is orthogonal to any polynomial of degree less than n for the scalar product associated to the weight function ρw. To complete the proof we have to show that ρq_n is not identically zero. But if this were the case the coefficient of $p_{n+l}(x)$ should be zero, and we could find l real constants $\lambda_0, \lambda_1, \ldots, \lambda_{l-1}$, such that the polynomial $\sum_{k=0}^{l-1} \lambda_k p_{n+k}(x)$ of degree $n+l-1$ would have $x_1, x_2, \ldots, x_l$ as roots. This polynomial, which is not identically zero, would be of the form $\rho(x)f(x)$, where $f(x)$ would be a polynomial of degree $n-1$. Orthogonality relations imposing

$$\int_a^b \rho(x) f(x) f(x) w(x)\, dx = 0,$$

$f(x)$ would be identically equal to zero in contradiction with our assumption.

If x_k is a zero of order m of ρ, the definition of q_n must be modified as follows: in the m lines corresponding to this multiple root we write the derivatives of order $0, 1, \ldots, m-1$ of the polynomials $p_n(x), p_{n+1}(x), \ldots, p_{n+l}(x)$ at $x = x_k$.

6. All these functions are odd, and we can, consequently, expand them in series of sines. We can thus write

$$(\forall i = 1, 2, 3) \qquad f_i(x) = \sum_{n=1}^{\infty} c_n \sin nx,$$

where

$$c_n = \frac{2}{\pi} \int_0^{\pi} f_i(x) \sin nx \, dx.$$

After simple calculations we obtain

$$f_1(x) = -\sum_{n=1}^{\infty} \frac{(-1)^n}{n} \sin nx,$$

$$f_2(x) = -\sum_{n=1}^{\infty} \frac{(-1)^n}{n^3} \sin nx,$$

$$f_3(x) = \sum_{p=0}^{\infty} \frac{1}{(2p+1)^3} \sin(2p+1)x.$$

The Fourier series of f_2 and f_3 are uniformly convergent, and they therefore, converge pointwise, respectively, to $f_2(x)$ and $f_3(x)$.

The periodic function, of period 2π, defined in $\mathbf{R}$ and equal to f_1 on the interval $]-\pi, \pi[$, is discontinuous at the points $x = (2k+1)\pi$, where k is an integer. Its Fourier series is not, of course, uniformly convergent; however, the sufficient condition of convergence to

$$\frac{1}{2}\big(f((2k+1)\pi + 0)) - f((2k+1)\pi - 0)\big)$$

is satisfied.

7. This function is odd and can therefore can be expanded in series of sines. We have

$$f(x) = \sum_{n=1}^{\infty} c_n \sin n\pi x$$

with

$$\begin{aligned} c_n &= \int_{-1}^{1} f(x) \sin n\pi x \, dx \\ &= 2 \int_0^1 (1-x) \sin n\pi x \, dx \\ &= \frac{2}{n\pi}. \end{aligned}$$

If we replace x by $1/2$ in this expansion, we obtain

$$\frac{1}{2} = \sum_{n=1}^{\infty} \frac{2}{n\pi} \sin \frac{n\pi}{2} = \frac{2}{\pi} \sum_{n=1}^{\infty} \frac{(-1)^n}{2n+1},$$

which is the first formula. The second one follows from the Parseval equality, which reads

$$\int_{-1}^{1} |f(x)|^2 \, dx = \sum_{n=1}^{\infty} |c_n|^2$$

or

$$2 \int_0^1 (1-x)^2 \, dx = \frac{4}{\pi^2} \sum_{n=1}^{\infty} \frac{1}{n^2},$$

i.e.,

$$\frac{2}{3} = \frac{4}{\pi^2} \sum_{n=1}^{\infty} \frac{1}{n^2}.$$

8. The function is even and can therefore be expanded in series of cosines. We have

$$\cos px = \sum_{n=0}^{\infty} c_n \cos nx,$$

with

$$\begin{aligned} c_0 &= \frac{1}{\pi} \int_0^{\pi} \cos px \, dx \\ &= \frac{\sin p\pi}{p}, \end{aligned}$$

and

$$\begin{aligned} c_n &= \frac{2}{\pi} \int_0^{\pi} \cos px \cos nx \, dx \\ &= \frac{1}{\pi} \left[\frac{\sin(p+n)x}{p+n} + \frac{\sin(p-n)x}{p-n} \right]_0^{\pi} \\ &= \frac{2p}{\pi} \sin p\pi \frac{(-1)^n}{p^2 - n^2}. \end{aligned}$$

Formulas (1) and (2) are obtained by replacing, respectively, x by π and zero.

9. We have

$$\begin{aligned} c_n(fg) &= \int_0^1 \mathrm{e}^{-2i\pi nt} f(t)g(t)\,dt \\ &= \langle \mathrm{e}^{2i\pi nt}\bar{f} \mid g\rangle \\ &= \sum_{-\infty}^{\infty} \overline{c_p(\mathrm{e}^{2i\pi nt}\bar{f})}c_p(g), \end{aligned}$$

but

$$\begin{aligned} \overline{c_p(\mathrm{e}^{2i\pi nt}\bar{f})} &= \overline{\int_0^1 \mathrm{e}^{-2i\pi pt}\mathrm{e}^{2i\pi nt}\bar{f}(t)\,dt} \\ &= \overline{c_{p-n}(\bar{f})} \\ &= c_{p-n}(f), \end{aligned}$$

and the result follows.

10. The problem is to determine the best approximation of x^5 by a polynomial of degree 4 on the interval [-1,1]. But taking into account the expression of the Legendre polynomial $P_5(x)$ (see Example 3.3), we have

$$x^5 = \frac{8}{63}P_5(x) + \frac{70}{63}x^3 - \frac{15}{63}x,$$

and, consequently, the best approximation of x^5 by a polynomial of degree less than 5 is $(70/63)x^3 - (15/63)x$. Then $a = 0$, $b = 70/63$, $c = 0$, $d = -15/63$, $e = 0$.

11. This problem is similar to the preceding one. In the Hilbert space $L^2_\mu(\mathbf{R}_+)$, where $d\mu(x)\mathrm{e}^{-x}\,dx$, the sequence of Laguerre polynomials (L_n) defined by

$$L_n(x) = \frac{1}{n!}\mathrm{e}^x\frac{d^n}{dx^n}(x^n\mathrm{e}^{-x})$$

is a complete orthonormal system.[21] To minimize the norm of $x \mapsto x^3 - ax^2 - bx - c$, $x \mapsto ax^2 + bx + c$ should be the projection of $x \mapsto x^3$ on the subspace generated by L_0, L_1, and L_2. From

$$\begin{aligned} L_3(x) &= \frac{1}{6}\mathrm{e}^x\frac{d^3}{dx^3}(x^3\mathrm{e}^{-x}) \\ &= 1 - 3x + \frac{3}{2}x^2 - \frac{1}{6}x^3, \end{aligned}$$

---

[21] This definition is slightly different from the definition given in Example 3.4. Defined in this way, the Laguerre polynomials are normed, which is more convenient in this case.

it follows that

$$x^3 + 6L_3(x) = 9x^2 - 19x + 6,$$

i.e., $a = 9$, $b = -18$, $c = 6$.

12. Let $R_n(x) = f(x) \sum_{k=0}^{n} \langle p_n \mid f \rangle p_k(x)$; then, for any k not greater than n, $\langle p_k \mid R_n \rangle = 0$ and, in particular, $\langle p_0 \mid R_n \rangle = \langle 1 \mid R_n \rangle = 0$ which shows that $R_n(x)$ has, at least, one root between a and b. If $x_1, x_2, \ldots, x_j$ are the j roots of $R_n(x)$ ($j > 0$), the sign of $R_n(x)$ is constant in the open intervals $]a, x_1[$,$]x_1, x_2[$, ..., $]x_j, b[$, and it changes from one interval to the following one. Since the polynomial $(x - x_1)(x - x_2) \cdots (x - x_j)$ has the same property, the product $R_n(x)(x-x_1)(x-x_2)\cdots(x-x_j)$ has a constant sign on $]a, b[$, and the scalar product

$$\langle R_n(x)(x-x_1)(x-x_2)\cdots(x-x_j) \mid 1 \rangle = \langle R_n(x) \mid (x-x_1)(x-x_2)\cdots(x-x_j) \rangle$$

cannot be null. But, as a consequence of the orthogonality relations

$$\langle p_k \mid R_n \rangle = 0 \qquad (k = 1, 2, \ldots, n),$$

this cannot be the case if $j \leq n$. Therefore, we necessarily have $j > n$.

13. We have indeed

$$P(x) = \sum_{k=0}^{n} \langle p_k \mid P \rangle p_k(x),$$

hence,

$$\begin{aligned} \langle P(x) \mid K(x,y) \rangle &= \sum_{k=0}^{n} \langle p_k \mid P \rangle \langle p_k(x) \mid p_k(x) \rangle p_k(y) \\ &= \sum_{k=0}^{n} \langle p_k \mid P \rangle p_k(y) \\ &= P(y). \end{aligned}$$

14. To prove unicity, assume that there exist two different polynomials $S_n(x)$ and $R_n(x)$ satisfying the condition; then

$$\begin{aligned} \int_{-1}^{1} \big(S_n(x) - R_n(x)\big)^2\, dx &= \int_{-1}^{1} S_n(x)\big(S_n(x) - R_n(x)\big)\, dx \\ &\quad - \int_{-1}^{1} R_n(x)\big(S_n(x) - R_n(x)\big)\, dx \\ &= \big(S_n(1) - R_n(1)\big) - \big(S_n(1) - R_n(1)\big) \\ &= 0, \end{aligned}$$

i.e., $S_n(x) = R_n(x)$.

To determine $S_n(x)$ by its expansion in Legendre polynomials, put

$$S_n = \sum_{k=0}^{n} s_k P_k \quad \text{and} \quad Q = \sum_{k=0}^{n} q_k P_k.$$

Taking into account the orthogonality relations of the Legendre polynomials, we have

$$\int_{-1}^{1} S_n(x) Q(x)\, dx = \sum_{k=0}^{n} \frac{2}{2k+1} s_k q_k = Q(1) = \sum_{k=0}^{n} q_k,$$

since, for any k, $P_k(1) = 1$. Provided the degree of $Q(x)$ is not greater than n, this relation is true for any value of the coefficient q_k $(k = 1, 2, \ldots, n)$. Hence, $s_k = (2k+1)/2$. Consider now the polynomial $(1-x)S_n(x)$. Its expansion in a series of Legendre polynomials is $\sum_{k=0}^{n+1} c_k P_k(x)$, and if we choose for $Q(x)$ the polynomial $(1-x)x^k$ $(k = 1, 2, \ldots, n-1)$, we obtain

$$(\forall k = 0, 1, 2, \ldots, n-1) \qquad \int_{-1}^{1} (1-x) S_n(x) x^k\, dx = 0,$$

hence, $c_k = 0$ for $k = 1, 2, \ldots, n-1$, and, therefore,

$$(1-x)S_n(x) = c_n P_n(x) + c_{n+1} P_{n+1}(x).$$

From $P_n(1) = 1$, it follows that $c_n + c_{n+1} = 0$, and from $P_n(-1) = (-1)^n$, it follows that $c_n = (-1)^n S_n(-1) = (n+1)/2$. Finally,

$$S_n(x) = \frac{n+1}{2} \frac{P_n(x) - P_{n+1}}{1-x}.$$

The relations

$$\int_{-1}^{1} (1-x) S_n(x) x^k\, dx = 0$$

for $k = 0, 1, 2, \ldots, n-1$ show that S_n is orthogonal to all polynomials of degree less than n (here the weight function is $x \mapsto 1-x$), and we have

$$\begin{aligned}
&\int_{-1}^{1} (1-x)\big(S_n(x)\big)^2\, dx \\
&\quad = \frac{n+1}{2} \int_{-1}^{1} \big(P_n(x) - P_{n+1}(x)\big) \left(\sum_{k=0}^{n} \frac{2k+1}{2} P_k(x) \right) dx \\
&\quad = \frac{n+1}{2} \frac{2n+1}{2} \int_{-1}^{1} \big(P_n(x)\big)^2\, dx \\
&\quad = \frac{n+1}{2}.
\end{aligned}$$

15. r_n is a step function such that, for any nonnegative integer k less than 2^n, we have

$$(\forall x \in]\frac{k}{2^n}, \frac{k+1}{2^n}[) \qquad r_n(x) = (-1)^k.$$

Therefore, on each interval in which r_n keeps a constant sign, any function r_m, where $m > n$, takes an equal number of times the values -1 and 1. This implies that $\langle r_n \mid r_m \rangle = 0$. Furthermore, it is straightforward to see that, for any nonnegative integer n and any real x in $[0,1]$, $(r_n(x))^2 = 1$. This implies $\langle r_n \mid r_n \rangle = 1$.

The sequence of Rademacher functions (r_n) is therefore orthonormal. It is not, however, a Hilbert basis, since there exist nonnull functions in $L^2_m([0,1])$ which are orthogonal to all the functions of the sequence, like, for example, the function $x \mapsto \cos 2\pi x$.

Chapter 4

Distributions

Some physics textbooks make use of "singular functions" whose properties are contradictory. Consider, for example, the so-called "δ function,"or "Dirac function,"defined on **R** by

$$\delta(x) = \begin{cases} 0, & \text{if } x \neq 0; \\ \infty, & \text{if } x = 0; \end{cases}$$

and

$$\int \delta(x)\varphi(x)\,dx = \varphi(0),$$

where φ is any "well-behaved function."In particular, if $\varphi(x) = 1$, this last property implies that

$$\int \delta(x)\,dx = 1.$$

Such a definition is meaningless since, the "δ function" being null almost everywhere, its integral must be equal to 0.

A mathematical object having the properties of the "δ function" can nevertheless be defined. Consider the sequence of functions (f_n) such that

$$f_n = \begin{cases} n, & \text{if } |x| \leq 1/2n; \\ 0, & \text{otherwise.} \end{cases}$$

For any positive integer n, we have

$$\int f_n(x)\,dx = 1,$$

hence,

$$\lim_{n\to\infty} \int f_n(x)\,dx = 1.$$

Furthermore, if φ is a "well-behaved function,"it could be verified that (see Example 3.1)

$$\lim_{n\to\infty} \int f_n(x)\varphi(x)\,dx = \varphi(0).$$

The "δ function" appears to be the limit of the sequence (f_n); but this limit (assuming its existence in a sense to be precise) cannot be a function (note that Lebesgue's dominated convergence theorem does not apply). Therefore, if we want to give a meaning to the symbolic rules satisfied by the "δ function,"we have to define new mathematical objects. ***Distributions***[1] are these new objects, and from the two properties defining the "δ function,"given above, we shall retain the second one, that is, to any "well-behaved function" φ, a distribution associates a number. In other terms, following Laurent Schwartz, distributions will be defined as continuous linear forms on a vector space of sufficiently regular functions. Changing the functional space will change the properties of the corresponding distributions. For most applications, it is sufficient to consider the vector space $C_c^\infty(\mathbf{R})$ of infinitely differentiable functions with bounded support.[2] But to define, for example, Fourier transforms of distributions it is more convenient, as we shall see, to consider a wider functional space.

Distribution theory provides simple and rigorous justifications for a number of formal manipulations often found in the technical literature. In many areas of applied mathematics, especially ones involving differential equations, it supplies powerful tools that greatly simplify the analysis.

1. Test Functions and Distributions

In distribution theory the vector space $C_c^\infty(\mathbf{R})$ of infinitely differentiable functions defined on $\mathbf{R}$ is usually denoted $\mathcal{D}$. The elements of $\mathcal{D}$ are called *test functions.*

It is important to check that $\mathcal{D}$ is indeed not empty. Let φ_a be the function defined by

$$\varphi_a(x) = \begin{cases} \exp -(\frac{a^2}{a^2-x^2}), & \text{if } |x| \le a; \\ 0, & \text{otherwise;} \end{cases}$$

---

[1] Called *generalized functions* by Russian authors.

[2] The *support* of a function φ defined on $\mathbf{R}$, denoted by supp φ, is the closure of the set $\{x \in \mathbf{R} | \varphi(x) \neq 0\}$.

where a is a positive real number. It is clear that φ_a belongs to $\mathcal{D}$ since it has a bounded support, which is the interval $[-a, a]$, and derivatives of all orders. More generally, if f is a continuous function with bounded support, the function f_a defined by

$$f_a(x) = \int f(y)\varphi_a(x-y)\,dy,$$

i.e.,

$$f_a = f * \varphi_a,$$

is in $\mathcal{D}$. Since it is legitimate to take derivatives under the integral sign (Chapter 1, Theorem 14), f_a has derivatives of all orders, and it can be verified that its support is bounded.

In Chapter 2 we used convolution to regularize continuous functions (see Corollary 2 of Theorem 7). The result we obtained shows that $\mathcal{D}$ *is dense in the space of continuous functions with bounded supports.*[3] Regularization will be discussed in more detail in Section 4.

To define continuous linear forms on $\mathcal{D}$, we need the concept of convergence in $\mathcal{D}$.

Definition 1. A sequence of test functions (φ_n) is said to converge in $\mathcal{D}$ to a function φ if

(1) all the supports of the functions of the sequence are contained in a bounded subset of $\mathbf{R}$;

(2) for each positive integer p, the sequence $(\varphi_n^{(p)})$ of the derivatives of order p converges uniformly to the derivative of order p of φ.

It is readily verified that the limit φ is in $\mathcal{D}$ since, from (1), its support is bounded.

For example, the sequence (φ_n) such that

$$\varphi_n(x) = \begin{cases} 0, & \text{if } |x| > n, \\ \frac{1}{n}\exp -\left(\frac{n^2}{n^2-x^2}\right), & \text{if } |x| \le n, \end{cases}$$

does not converge in $\mathcal{D}$, even though it and its derivatives converge uniformly to zero, since there is no bounded subset of $\mathbf{R}$ outside which all the functions of the sequence are null.

---

[3] In the proof, the regularizing functions had integrals over $\mathbf{R}$ equal to 1. In the present case, this condition can easily be satisfied if one considers, instead of φ_a, the function $\frac{1}{C_a}\varphi_a$, where $C_a = \int \varphi_a(x)\,dx$.

Definition 2. A continuous linear form on $\mathcal{D}(\mathbf{R})$ is called a *distribution* on $\mathbf{R}$.

Distributions on $\mathbf{C}$ can be defined in a similar way. A distribution is therefore an element of the *topological dual space* of $\mathcal{D}$. This space is denoted by $\mathcal{D}'$.

If T is a distribution, linearity implies that

$$T(\lambda_1\varphi_1 + \lambda_2\varphi_2) = \lambda_1 T(\varphi_1) + \lambda_2 T(\varphi_2),$$

where φ_1 and φ_2 belong to $\mathcal{D}$, and λ_1 and λ_2 are in $\mathbf{R}$; and from the continuity it follows that

$$\lim_{n\to\infty} T(\varphi_n) = T(\varphi),$$

where (φ_n) is a sequence of test functions that converges to φ in $\mathcal{D}$.

$\mathcal{D}'$ is a vector space, i.e., for any test function φ in $\mathcal{D}$,

$$(\lambda_1 T_1 + \lambda_2 T_2)(\varphi) = \lambda_1 T_1(\varphi) + \lambda_2 T_2(\varphi),$$

where T_1 and T_2 belong to $\mathcal{D}$' and λ_1 and λ_2 are in $\mathbf{R}$.

Example 1. Let f be a function of $\mathbf{R}$ in $\mathbf{R}$ (or $\mathbf{C}$) that is *locally integrable*, that is, integrable on any bounded subset of $\mathbf{R}$;[4] the linear form T_f defined on $\mathcal{D}$ by

$$T_f(\varphi) = \int f(x)\varphi(x)\,dx$$

is a distribution since it is continuous as it can be proved in the following way. Let (φ_n) be a sequence of test functions converging to φ in $\mathcal{D}$, then

$$\begin{aligned}|T_f(\varphi) - T_f(\varphi_n)| &= \Big|\int f(x)\big(\varphi(x) - \varphi_n(x)\big)\,dx\Big| \\ &\le \int_a^b |f(x)|\,|\big(\varphi(x) - \varphi_n(x)\big)|\,dx,\end{aligned}$$

where a and b are such that φ and φ_n are null outside the finite interval $[a, b]$.[5] For any $\epsilon > 0$, there exists $N(\epsilon)$ such that, for all $n \ge N(\epsilon)$,

---

4 The space of locally integrable functions is usually denoted by L^1_{loc}.

5 This is always possible since test functions have bounded supports.

$\left|(\varphi(x)-\varphi_n(x))\right| < \epsilon$. Therefore, taking into account this last inequality and the fact that $\int_a^b |f(x)|\,dx$ is finite, we finally obtain

$$\lim_{n\to\infty} |T_f(\varphi) - T_f(\varphi_n)| = 0,$$

which proves the continuity of T_f.

To any locally integrable function f corresponds a distribution T_f. Such a distribution is said to be *regular*. If two locally integrable functions are equal almost everywhere, the corresponding regular distributions are equal. A distribution that is not regular is *singular*.

Example 2. As a first example of singular distribution we shall consider the *Dirac distribution* δ defined by

$$\delta(\varphi) = \varphi(0),$$

for all φ in $\mathcal{D}$. The form δ is clearly linear and continuous, but it does not represent a regular distribution since there is no locally integrable function δ such that

$$\varphi(0) = \int \delta(x)\varphi(x)\,dx.$$

More generally, the distribution δ_a is defined, for all φ in $\mathcal{D}$, by

$$\delta_a(\varphi) = \varphi(a).$$

These two examples show that the set of all regular distributions is a subspace of $\mathcal{D}'$. Distributions generalize locally integrable functions. But it should be kept in mind that a distribution is not a function, and the notation $\delta(x)$, which is very often found in the literature, is meaningless. The distribution δ does not take the value $\delta(x)$ at point x.

Example 3. The product of two arbitrary distributions cannot be defined in general. For instance, there is no distribution equal to the square of δ. It is, however, possible to define the product of two distributions in some particular cases. For example, if f_1 and f_2 are two locally integrable functions, and if their product $f_1 f_2$ is also a locally integrable function, we can define the product of the regular distributions T_{f_1} and T_{f_2} by

$$T_{f_1}T_{f_2} = T_{f_1 f_2}.$$

It is also possible to define the product of any distribution T by a regular distribution T_ψ, when ψ is an infinitely differentiable function, by

$$(T_\psi T)(\varphi) = T(\psi\varphi).$$

This definition is consistent since, for all φ in $\mathcal{D}$, $\psi\varphi$ is also in $\mathcal{D}$.

Notations. We shall very often use the same symbol to denote a locally integrable function and the corresponding regular distribution. For instance the above definition will be written

$$(\psi T)(\varphi) = T(\psi\varphi).$$

In most cases we shall deal with distributions defined on $\mathbf{R}$. Sometimes, however, we shall also consider distributions defined on $\mathbf{R}^n$. These distributions are continuous linear forms on $\mathcal{D}(\mathbf{R}^n)$, which is the space of infinitely differentiable functions with bounded support defined on $\mathbf{R}^n$.

Distributions are mathematical objects that correctly describe "physical distributions" like distributions of masses, electric charges, etc. These "physical distributions" often have symmetry properties that should be taken into account. This motivates the following definition.

Definition 3. Let u be an invertible linear mapping of $\mathbf{R}^n$ onto itself and let T be a distribution defined on $\mathbf{R}^n$. The *transform* of T by u is the distribution uT defined by

$$(uT)(\varphi) = |J|\, T(u^{-1}\varphi),$$

where $|J|$ is the determinant of the Jacobian of the mapping, and

$$(u^{-1}\varphi)(x) = \varphi(ux).$$

If $uT = T$, T is said to be *invariant* under u.

The consistency of this definition follows from the fact that $x \mapsto \varphi(ux)$ is in $\mathcal{D}(\mathbf{R}^n)$. u can represent a translation, a rotation, a reflection, etc. For example, δ_a is a translate of δ since, if τ_a is a translation, we have

$$\tau_a x = x + a$$
$$\varphi(\tau_a x) = \varphi(x + a),$$

and, therefore,

$$\tau_a \delta = \delta_a.$$

2. Differentiation

A locally integrable function may not have a derivative at certain points or even anywhere at all. On the contrary, a distribution is always differentiable, and its derivative is a distribution. Let f be a locally integrable function which is differentiable everywhere, and let φ be any test function. An integration by parts yields

$$\int f'(x)\varphi(x)\,dx = -\int f(x)\varphi'(x)\,dx.$$

Since φ' is in $\mathcal{D}$, this result suggests the following definition.

Definition 4. Let T be a distribution; its derivative T' is the distribution defined by

$$T'(\varphi) = -T(\varphi').$$

If f is a locally integrable function that is everywhere differentiable, the above discussion shows that the derivative of T_f is $T_{f'}$.

If T is a distribution defined on $\mathbf{R}^n$, the derivative of T with respect to x_i is given by

$$\frac{\partial T}{\partial x_i}(\varphi) = -T(\frac{\partial \varphi}{\partial x_i}).$$

The following results are straightforward to derive.

$$\frac{\partial^2 T}{\partial x_i \partial x_j} = \frac{\partial^2 T}{\partial x_j \partial x_i}$$

$$\frac{\partial^p T}{\partial x_{i_1} \partial x_{i_2} \cdots \partial x_{i_p}}(\varphi) = (-1)^p T(\frac{\partial^p \varphi}{\partial x_{i_1} \partial x_{i_2} \cdots \partial x_{i_p}}),$$

where p is any positive integer.

Example 1. Let ψ be the regular distribution associated to the infinitely differentiable function ψ; then

$$(\psi T)' = \psi' T + \psi T',$$

since

$$\begin{aligned}(\psi T)'(\varphi) &= -(\psi T)(\varphi') \\ &= -T(\psi\varphi') \\ &= -T\big((\psi\varphi)' - \psi'\varphi\big) \\ &= (\psi T')(\varphi) + (\psi' T)(\varphi).\end{aligned}$$

More generally, let D be a linear differential operator of order m such that

$$Dg = \sum_{p=1}^{m} \psi_{i_1 i_2 \cdots i_p} \frac{\partial^p g}{\partial x_{i_1} \partial x_{i_2} \cdots \partial x_{i_p}},$$

where the function g is, at least, m times continuously differentiable, and the functions $\psi_{i_1 i_2 \cdots i_p}$ are infinitely differentiable. If D^+ denotes the differential operator such that

$$D^+ g = \sum_{p=1}^{m} (-1)^p \frac{\partial^p (\psi_{i_1 i_2 \cdots i_p} g)}{\partial x_{i_1} \partial x_{i_2} \cdots \partial x_{i_p}},$$

then, for any distribution T defined on $\mathbf{R}^n$ ($n \geq m$), we have

$$DT(\varphi) = T(D^+ \varphi).$$

The proof is straightforward.

Since $(D^+)^+ = D$, then $D^+ T(\varphi) = T(D\varphi)$.

Example 2. Let θ be the Heaviside function defined on $\mathbf{R}$ by

$$\theta(x) = \begin{cases} 0, & \text{if } x < 0; \\ 1, & \text{if } x \geq 0. \end{cases}$$

The derivative of θ is equal to zero almost everywhere. What is the derivative of the regular distribution T_θ associated to θ? We have

$$T_\theta(\varphi) = \int_0^\infty \varphi(x)\, dx,$$

and

$$T'_\theta(\varphi) = -T_\theta(\varphi') = -\int_0^\infty \varphi'(x)\, dx = \varphi(0).$$

We therefore obtain the important result

$$T'_\theta = \delta,$$

which, when no confusion can arise, will be written

$$\theta' = \delta.$$

We have also

$$T''_\theta(\varphi) = \delta'(\varphi) = -\delta(\varphi') = -\varphi'(0),$$

and more generally

$$T_\theta^{(p)}(\varphi) = (-1)^{p-1}\varphi^{p-1}(0).$$

Similarly, we can find the derivatives of regular distribution associated to a "piecewise smooth" function. Consider, for instance, the case of a regular distribution corresponding to a function f everywhere differentiable except at a finite number of points x_j ($j = 1, 2, \ldots, m$), and let $h_j = f(x_j + 0) - f(x_j - 0)$ be the jump of f at x_j. We have

$$\begin{aligned} T_f'(\varphi) &= -T_f(\varphi') \\ &= -\int_{-\infty}^{x_1} f(x)\varphi'(x)\,dx - \int_{x_1}^{x_2} f(x)\varphi'(x)\,dx - \cdots \\ &\quad \cdots - \int_{x_m}^{\infty} f(x)\varphi'(x)\,dx. \end{aligned}$$

Integrating by parts yields

$$T_f'(\varphi) = \sum_{j=1}^{m} \big(f(x_j + 0) - f(x_j - 0)\big)\varphi(x_j) + \int_{-\infty}^{\infty} f'(x)\varphi(x)\,dx,$$

i.e.,

$$T_f' = T_{f'} + \sum_{j=1}^{m} h_j \delta_{x_j}.$$

T_f' is the derivative of the regular distribution associated to f, whereas $T_{f'}$ is the regular distribution associated to f'.

Example 3. The function $x \mapsto \log|x|$ is locally integrable. The derivative of the corresponding regular distribution is determined as follows[6]

$$\begin{aligned} (\log|x|)'(\varphi) &= -(\log|x|)(\varphi') \\ &= -\int_{-\infty}^{\infty} \log|x|\,\varphi'(x)\,dx. \end{aligned}$$

Since $x \mapsto 1/x$ is not locally integrable, we cannot integrate by parts this last integral. The derivative of $\log|x|$ is not a regular distribution, and

[6] As mentioned before, the regular distribution and the associated locally integrable function are denoted by the same symbol.

to find it we shall use a technique due to Hadamard. The method is very general, and will be systematically used to define singular distributions corresponding to functions that are not locally integrable. Given a divergent integral, the idea is to extract the *finite part* of this integral to define a singular distribution.[7]

From Lebesgue's dominated convergence theorem, it follows that

$$\int \log|x|\,\varphi'(x)\,dx = \lim_{\epsilon\to 0}\int_{|x|\geq\epsilon} \log|x|\,\varphi'(x)\,dx$$

for

$$\chi_\epsilon(x)\big|\log|x|\,\varphi'(x)\big| \leq \big|\log|x|\,\varphi'(x)\big| \in L^1_m(\mathbf{R}),$$

where

$$\chi_\epsilon(x) = \begin{cases} 0, & \text{if } -\epsilon < x < \epsilon; \\ 1, & \text{otherwise.} \end{cases}$$

Furthermore,

$$\int_{|x|\geq\epsilon} \log|x|\,\varphi'(x)\,dx = \int_{-\infty}^{-\epsilon} \log|x|\,\varphi'(x)\,dx + \int_{\epsilon}^{\infty} \log|x|\,\varphi'(x)\,dx,$$

and integrating by parts, each integral yields

$$\begin{aligned}(\log|x|)'(\varphi) = \lim_{\epsilon\to 0}\Big(-(\log\epsilon)\varphi(-\epsilon) + \int_{-\infty}^{-\epsilon} \frac{\varphi(x)}{x}\,dx \\ + (\log\epsilon)\varphi(\epsilon) + \int_{\epsilon}^{\infty} \frac{\varphi(x)}{x}\,dx\Big).\end{aligned}$$

Since the inequality

$$|\varphi(\pm\epsilon) - \varphi(0)| \leq \epsilon \sup|\varphi'|$$

implies

$$\lim_{\epsilon\to 0}(\log\epsilon)\big(\varphi(\pm\epsilon) - \varphi(0)\big) = 0,$$

we may replace $(\log\epsilon)\varphi(\pm\epsilon)$ by $(\log\epsilon)\varphi(0)$ in order to eliminate the divergent terms. We finally obtain

$$(\log|x|)'(\varphi) = \lim_{\epsilon\to 0}\int_{|x|\geq\epsilon} \frac{\varphi(x)}{x}\,dx,$$

[7] For more details, see the remark after Example 5.

i.e.,

$$(\log|x|)'(\varphi) = \mathrm{Pv}\int_{-\infty}^{\infty}\frac{\varphi(x)}{x}\,dx,$$

where the symbol Pv denotes the *Cauchy principal value*. This result is usually written

$$(\log|x|)' = \mathrm{Pv}\,\frac{1}{x}$$

or

$$(\log|x|)' = \mathrm{Pf}\,\frac{1}{x}.$$

The symbol Pf is read *pseudofunction* (see Example 5). Note that $\mathrm{Pv}\,x^{-1}$ is *not* a regular distribution.

Example 4. Consider the principal branch of the logarithm function,[8] and put

$$\log(x+i0) = \lim_{\epsilon\to 0+}\log(x+i\epsilon)$$
$$\log(x-i0) = \lim_{\epsilon\to 0+}\log(x-i\epsilon).$$

Since

$$\log(x+i\epsilon) = \log|x+i\epsilon| + i\arg(x+i\epsilon),$$

as $\epsilon\to 0^+$, we obtain

$$\log(x+i0) = \log|x| + i\pi\theta(-x).$$

The locally integrable function $x\mapsto\log(x+i0)$ defines a regular distribution, and from the preceding results (Examples 2 and 3), we have

$$\big(\log(x+i0)\big)' = \mathrm{Pv}\,\frac{1}{x} - i\pi\delta.$$

Changing i in $-i$, we also have

$$\big(\log(x-i0)\big)' = \mathrm{Pv}\,\frac{1}{x} + i\pi\delta.$$

Example 5. The function $x\mapsto\theta(x)\,x^\alpha$, where $-1<\alpha<0$, is locally integrable. It defines a regular distribution denoted x_+^α. Since the integral $\int x^{\alpha-1}\varphi(x)\,dx$ is not convergent for all test functions φ in $\mathcal{D}$, the derivative

---

[8] The principal branch of the logarithm function is the analytic function defined on $\{z\in\mathbf{C}\mid -\pi<\arg z<\pi\}$ by $\log z = \log|z| + i\arg z$.

of x_+^α is not a regular distribution. To determine this derivative we shall proceed as in Example 3. We have[9]

$$\begin{aligned}(x_+^\alpha)'(\varphi) &= -\int_0^\infty x^\alpha \varphi'(x)\,dx \\ &= -\lim_{\epsilon \to 0} \int_\epsilon^\infty x^\alpha \varphi'(x)\,dx.\end{aligned}$$

Integrating by parts yields

$$\begin{aligned}(x_+^\alpha)'(\varphi) = &-\lim_{\epsilon \to 0} \Big[x^\alpha \big(\varphi(x) + C\big)\Big]_\epsilon^\infty \\ &+ \lim_{\epsilon \to 0} \int_\epsilon^\infty \alpha x^{\alpha-1} \big(\varphi(x) + C\big)\,dx,\end{aligned}$$

where C is *a priori* any constant. The last integral converges if, and only if, $C = -\varphi(0)$. Therefore, we obtain

$$(x_+^\alpha)'(\varphi) = \int_0^\infty \alpha x^{\alpha-1} \big(\varphi(x) - \varphi(0)\big)\,dx.$$

This result is written

$$(x_+^\alpha)' = \alpha \mathrm{Pf}\, x_+^{\alpha-1}.$$

More generally, for any negative noninteger α, the singular distribution $\mathrm{Pf}\, x_+^\alpha$ is defined by

$$(\mathrm{Pf}\, x_+^\alpha)(\varphi) = \int_0^\infty x^\alpha \big(\varphi(x) - \varphi(0) - x\varphi'(0) - \cdots - \frac{x^m}{m!}\varphi^{(m)}(0)\big)\,dx,$$

where m is such that $-m-2 < \alpha < -m-1$; and with this definition we have

$$(\mathrm{Pf}\, x_+^\alpha)' = \alpha \mathrm{Pf}\, x_+^{\alpha-1}.$$

There is an interesting alternative way to define the distribution $\mathrm{Pf}\, x_+^\alpha$. The function

$$\alpha \mapsto \int_0^\infty x^\alpha \varphi(x)\,dx$$

---

[9] As a consequence of Lebesgue's dominated convergence theorem.

is analytic in the complex half-plane $\{\alpha \in \mathbf{C} | \Re\alpha > -1\}$. If we write the integral defining this function under the form

$$\int_0^1 x^\alpha \big(\varphi(x) - \varphi(0)\big)\, dx + \int_1^\infty x^\alpha \varphi(x)\, dx + \frac{1}{\alpha+1}\varphi(0),$$

the first term is defined for $\Re\alpha > -2$, the second one for any α, and the third one for $\alpha \neq -1$. This expression can thus be used to obtain an analytic continuation of the function[10]

$$\alpha \mapsto \int_0^\infty x^\alpha \varphi(x)\, dx.$$

In other terms, the linear form

$$\varphi \mapsto \int_0^\infty x^\alpha \varphi(x)\, dx,$$

which was defined for $\Re\alpha > -1$, is now defined in a larger domain. More generally, the linear form

$$\varphi \mapsto \int_0^1 x^\alpha \Big(\varphi(x) - \varphi(0) - x\varphi'(0) - \cdots - \frac{x^m}{m!}\varphi^{(m)}(0)\Big)\, dx$$
$$+ \int_1^\infty x^\alpha \varphi(x)\, dx + \sum_{k=0}^{m} \frac{\varphi^{(k)}(0)}{k!\,(\alpha+k+1)}$$

is defined for $\Re\alpha > -m-2$ and $\alpha \neq -1, -2, \ldots, -m-1$. Taking into account the relation

$$\int_1^\infty x^{\alpha+k}\, dx = -\frac{1}{\alpha+k+1},$$

which is valid if $\Re\alpha < -k-1$, for $-m-2 < \Re\alpha < -m-1$, the preceding linear form can be written[11]

$$\varphi \mapsto \int_0^\infty x^\alpha \Big(\varphi(x) - \varphi(0) - x\varphi'(0) - \cdots - \frac{x^m}{m!}\varphi^{(m)}(0)\Big)\, dx.$$

---

10 Note that, for $\Re\alpha < -1$, $1/(\alpha+1) = -\int_1^\infty x^\alpha \varphi(0)\, dx$.

11 To check this result, replace $\varphi(x)$ in the integral from 1 to ∞ by $\varphi(0) + x\varphi'(0) + \cdots + \frac{1}{m!}x^m\varphi^{(m)}(0) + \psi(x)$.

Since the function $\alpha \mapsto (\mathrm{Pf}\, x_+^\alpha)(\varphi)$ has poles at $\alpha = -1, -2, \ldots$, we cannot define the distribution $\mathrm{Pf}\, x_+^{-n}$, where n is a positive integer, by taking the limit $\alpha \to -n$ in the expression of $\mathrm{Pf}\, x_+^\alpha$. In order to define $\mathrm{Pf}\, x_+^{-n}$, we shall remove the divergent term of $\mathrm{Pf}\, x_+^\alpha$ when $\alpha \to -n$. We have

$$\begin{aligned}(\mathrm{Pf}\, x_+^\alpha)(\varphi) = \int_0^1 & x^\alpha \big(\varphi(x) - \varphi(0) - x\varphi'(0) - \cdots \\ & - \frac{x^{n-1}}{(n-1)!}\varphi^{(n-1)}(0)\big)\, dx \\ + \int_1^\infty & x^\alpha \big(\varphi(x) - \varphi(0) - x\varphi'(0) - \cdots \\ & - \frac{x^{n-2}}{(n-2)!}\varphi^{(n-2)}(0)\big)\, dx + \frac{\varphi^{(n-1)}(0)}{(n-1)!\,(\alpha+n)}.\end{aligned}$$

The last term of the right-hand side of this equality is the only divergent term when $\alpha \to -n$. Therefore, we shall define $(\mathrm{Pf}\, x_+^{-n})(\varphi)$ as the sum of the first two integrals. This can be written

$$\begin{aligned}(\mathrm{Pf}\, x_+^{-n})(\varphi) = \int_0^\infty & x^{-n} \big(\varphi(x) - \varphi(0) - x\varphi'(0) - \cdots \\ & - \frac{x^{n-2}}{(n-2)!}\varphi^{(n-2)}(0) - \frac{x^{n-1}}{(n-1)!}\varphi^{(n-1)}(0)\theta(1-x)\big)\, dx.\end{aligned}$$

Using

$$(\mathrm{Pf}\, x_+^{-n})'(\varphi) = -(\mathrm{Pf}\, x_+^{-n})(\varphi'),$$

an integration by parts yields

$$(\mathrm{Pf}\, x_+^{-n})' = -n(\mathrm{Pf}\, x_+^{-n-1}) + \frac{(-1)^n}{n!}\delta^{(n)}.$$

This formula differs from the usual differentiation formula.

Similarly, the function $x \mapsto \theta(-x)|x|^\alpha$ is locally integrable if $-1 < \alpha < 0$, and defines a regular distribution denoted by x_-^α such that

$$\begin{aligned}x_-^\alpha(\varphi) &= \int_{-\infty}^0 |x|^\alpha \varphi(x)\, dx \\ &= \int_0^\infty x^\alpha \varphi(-x)\, dx.\end{aligned}$$

More generally, for any negative noninteger α, we can define the singular distribution $\mathrm{Pf}\, x_-^\alpha$ by

$$(\mathrm{Pf}\, x_-^\alpha)(\varphi) = \int_0^\infty x^\alpha \big(\varphi(-x) - \varphi(0) + x\varphi'(0) - \cdots - \frac{(-1)^m x^m}{m!}\varphi^{(m)}(0)\big)\, dx,$$

where m is such that $-m-2 < \alpha < -m-1$; and with this definition we have

$$(\mathrm{Pf}\, x_-^\alpha)' = -\alpha \mathrm{Pf}\, x_-^{\alpha-1}.$$

We could also define the distribution $\mathrm{Pf}\, x_-^{-n}$ by extracting the divergent part of the above integral when $\alpha \to -n$. We shall not go into details.

According to the parity of n, $\mathrm{Pf}\, x^{-n}$ is defined by

$$\mathrm{Pf}\, x^{-n} = \begin{cases} \mathrm{Pf}\, x_+^{-n} + \mathrm{Pf}\, x_-^{-n}, & \text{if } n \text{ is even;} \\ \mathrm{Pf}\, x_+^{-n} - \mathrm{Pf}\, x_-^{-n}, & \text{if } n \text{ is odd.} \end{cases}$$

That is, if $n = 2p$, we have

$$(\mathrm{Pf}\, x^{-2p})(\varphi) = \int_0^\infty x^{-2p}\Big(\varphi(x) + \varphi(-x) - 2\big(\varphi(0) + \frac{x^2}{2}\varphi''(0) + \cdots + \frac{x^{2p-2}}{(2p-2)!}\varphi^{(2p-2)}(0)\big)\Big)\, dx,$$

and, if $n = 2p+1$,

$$(\mathrm{Pf}\, x^{-2p-1})(\varphi) = \int_0^\infty x^{-2p-1}\Big(\varphi(x) - \varphi(-x) - 2\big(x\varphi'(0) + \frac{x^3}{3!}\varphi'''(0) + \cdots + \frac{x^{2p-1}}{(2p-1)!}\varphi^{(2p-1)}(0)\big)\Big)\, dx.$$

In particular,

$$(\mathrm{Pf}\, \frac{1}{x})(\varphi) = \int_0^\infty \frac{\varphi(x) - \varphi(-x)}{x}\, dx,$$

that is (see Example 3),

$$\mathrm{Pf}\, \frac{1}{x} = \mathrm{Pv}\, \frac{1}{x},$$

and

$$(\mathrm{Pf}\, \frac{1}{x^2})(\varphi) = \int_0^\infty \frac{\varphi(x) + \varphi(-x) - 2\varphi(0)}{x^2}\, dx.$$

To conclude, note that

$$(\mathrm{Pf}\, x^{-n})' = -n \mathrm{Pf}\, x^{-n-1}.$$

Remark. In the above example, we made a systematic use of Hadamard's finite part of a divergent integral in order to define correctly the singular distributions corresponding to certain nonlocally integrable functions. In what follows we shall give a more precise definition of the finite part of a divergent integral and study briefly its main properties.

Let f be a function nonlocally integrable on $[a, b]$ but integrable on $[a+\epsilon, b]$, where ϵ is positive. Assume that $f(x)$ can be written

$$f(x) = \sum_{i=1}^{m} \frac{A_i}{(x-\lambda_i)} + g(x),$$

where g is integrable on $[a, b]$ and the exponents λ_i are nonintegers such that $\Re\lambda_i \geq 1$. Then

$$\int_{a+\epsilon}^{b} f(x)\,dx = I(\epsilon) + F(\epsilon),$$

where

$$I(\epsilon) = \sum_{i=1}^{m} \frac{A_i}{\lambda_i - 1} \frac{1}{\epsilon^{\lambda_i - 1}},$$

and

$$F(\epsilon) = -\sum_{i=1}^{m} \frac{A_i}{\lambda_i - 1} \frac{1}{(b-a)^{\lambda_i - 1}} + \int_{a+\epsilon}^{b} g(x)\,dx.$$

When ϵ tends to 0, $I(\epsilon)$ diverges, whereas $F(\epsilon)$ has a finite limit, called the *Hadamard's finite part* of the divergent integral $\int_a^b f(x)\,dx$, denoted by

$$\mathrm{Fp} \int_a^b f(x)\,dx.$$

This generalized integral has the following two important properties

(i) Let $t \mapsto x(t)$ be a infinitely differentiable homeomorphism; then

$$\mathrm{Fp} \int_a^b f(x)\,dx = \mathrm{Fp} \int_{t(a)}^{t(b)} f(x(t))x'(t)\,dt,$$

which means that we can change variables to compute the finite part of a divergent integral.

(ii) Let λ be a complex number whose real part is sufficiently large, then the function $x \mapsto f(x)(x-a)^\lambda$ is integrable on $[a,b]$, and we have

$$\int_a^b f(x)(x-a)^\lambda\,dx = -\sum_{i=1}^m \frac{A_i}{\lambda_i-\lambda-1}\frac{1}{(b-a)^{\lambda_i-\lambda-1}} + \int_a^b g(x)(x-a)^\lambda\,dx,$$

which shows that the finite part of a divergent integral is the analytic continuation of an ordinary integral, and we have

$$\lim_{\lambda\to 0}\int_a^b f(x)(x-a)^\lambda\,dx = \mathrm{Fp}\int_a^b f(x)\,dx.$$

These considerations are valid if no exponent λ_i is equal to 1. If, for instance, $\lambda_1 = 1$, we have

$$f(x) = \frac{A_1}{(x-a)} + \sum_{i=2}^m \frac{A_i}{(x-a)^{\lambda_i}} + g(x),$$

and

$$\mathrm{Fp}\int_a^b f(x)\,dx = A_1\log(b-a) - \sum_{i=2}^m \frac{A_i}{\lambda_i-1}\frac{1}{(b-a)^{\lambda_i-1}}.$$

But, in this case, the previous two results are no more valid.

(i) We cannot change variables. For example,

$$\mathrm{Fp}\int_0^1 \frac{dx}{x} \neq \mathrm{Fp}\int_0^{1/2}\frac{dt}{t},$$

since the first integral is equal to zero, and the second one (obtained by putting $x = 2t$) is equal to $-\log 2$.

(ii) We have not

$$\lim_{\lambda\to 0}\int_a^b f(x)(x-a)^\lambda\,dx = \mathrm{Fp}\int_a^b f(x)\,dx,$$

but

$$\lim_{\lambda\to 0}\Big(\int_a^b f(x)(x-a)^\lambda - \frac{A_1}{\lambda}\Big)\,dx = \mathrm{Fp}\int_a^b f(x)\,dx.$$

If f contains a term in $1/(x-a)^n$, where n is a positive integer, the function $f\varphi$, where φ belongs to $\mathcal{D}$, will, in general, contain a term in $1/(x-a)$,[12] and, in this case, the finite part of $\int_a^b f(x)\varphi(x)\,dx$ will be defined as above.

The linear form

$$\varphi \mapsto \mathrm{Fp} \int_a^b f(x)\varphi(x)\,dx$$

being continuous, it defines the singular distribution $\mathrm{Pf}\, f$.

Example 6. In what follows we shall establish various formulas of *vector analysis* in $\mathbf{R}^3$, and show how they simplify the formulation of classical field theories.

So far, we have considered only *scalar distributions*. In field theories, many physical quantities are vectors or tensors. It is therefore necessary to define *vectorial* and *tensorial distributions*. If φ_i $(i = 1, 2, \ldots, n)$ are the n components of a vectorial test function φ belonging to $\mathcal{D}^n$, the continuous linear form

$$\varphi \mapsto \sum_i T_i(\varphi_i)$$

defines a vectorial distribution whose components are T_i $(i = 1, 2, \ldots, n)$. More generally, if φ_{ij} $(i, j = 1, 2, \ldots, n)$ are the components of a tensorial test function φ, the continuous linear form

$$\varphi \mapsto \sum_{i,j} T_{ij}\varphi_{ij}$$

defines a tensorial distribution of rank two whose components are T_{ij} $(i, j = 1, 2, \ldots, n)$. Tensorial distributions of higher rank are defined in a similar manner.

Let $s(x_1, x_2, x_3) = 0$ be the equation of a surface S. If we assume that the function $s\colon \mathbf{R}^3 \to \mathbf{R}$ is infinitely differentiable, and denote by θ_s the characteristic function of the domain $s(x_1, x_2, x_3) > 0$, that is,

$$\theta_s(x_1, x_2 x_3) = \begin{cases} 1, & \text{if } s(x_1, x_2, x_3) > 0; \\ 0, & \text{if } s(x_1, x_2, x_3) < 0; \end{cases}$$

then we can define a regular distribution θ_s by

$$\begin{aligned} \theta_s(\varphi) &= \int \theta_s(x_1, x_2, x_3)\varphi(x_1, x_2, x_3)\,dx_1\,dx_2\,dx_3 \\ &= \int_{s(x_1,x_2,x_3)>0} \varphi(x_1, x_2, x_3)\,dx_1\,dx_2\,dx_3. \end{aligned}$$

---

12 Because φ is, in general, analytic at a.

In particular, if $s(x_1, x_2, x_3) = x_1$, the surface S is the plane $x_1 = 0$ and the distribution θ_{x_1} is defined by

$$\theta_{x_1}(\varphi) = \int_{x_1>0} \varphi(x_1, x_2, x_3)\, dx_1\, dx_2\, dx_3.$$

The derivative $\partial\theta_{x_1}/\partial x_1$ is then given by

$$\begin{aligned}\frac{\partial\theta_{x_1}}{\partial x_1}(\varphi) &= -\theta_{x_1}\Big(\frac{\partial\varphi}{\partial x_1}\Big)\\ &= -\int_{x_1>0} \frac{\partial\varphi}{\partial x_1}\, dx_1\, dx_2\, dx_3\\ &= \int \varphi(0, x_2, x_3)\, dx_2\, dx_3.\end{aligned}$$

If we put

$$\delta_{x_1}(\varphi) = \int \varphi(0, x_2, x_3)\, dx_2\, dx_3,$$

the previous result may be written

$$\frac{\partial\theta_{x_1}}{\partial x_1} = \delta_{x_1}.$$

More generally, we shall define the distribution δ_s by

$$\frac{\partial\theta_s}{\partial x_i} = \frac{\partial s}{\partial x_i}\delta_s,$$

or, equivalently,

$$\nabla\theta_s = (\nabla s)\delta_s.$$

Note that θ_s and δ_s are often denoted $\theta(s)$ and $\delta(s)$. This notation is particularly convenient when the index does not consist of a single symbol. If, for instance, one considers a propagating surface like a wave front, the distribution corresponding, to the characteristic function of the four-dimensional domain $s(x_1, x_2, x_3) - t > 0$ is usually denoted by $\theta(s(x) - t)$ (see Problem 10).

If $s(x_1, x_2, x_3)$ represents the distance from the point $x = (x_1, x_2, x_3)$ to the surface S,[13] we have

$$\frac{\partial\theta_s}{\partial x_i} = n_i\delta_s,$$

---

[13] Since this distance is equal to zero when x is on S, $s(x_1, x_2, x_3) = 0$ is indeed the equation of S.

where n_i is the i-component of the unit vector $\mathbf{n} = \nabla s$ normal to the surface S at $x = (x_1, x_2, x_3)$.

In that case, let V be an infinitely differentiable scalar field, and $\mathbf{A}$ an infinitely differentiable vector field. The fields $\theta_s V$ and $\theta_s \mathbf{A}$ are null in the domain defined by $s(x) < 0$, and from the formula derived in Example 1, we obtain

$$\nabla(\theta_s V) = \theta_s \nabla V + \delta_s V \mathbf{n}$$
$$\nabla \cdot (\theta_s \mathbf{A}) = \theta_s \nabla \cdot \mathbf{A} + \delta_s \mathbf{n} \cdot \mathbf{A}$$
$$\nabla \times (\theta_s \mathbf{A}) = \theta_s \nabla \times \mathbf{A} + \delta_s \mathbf{n} \times \mathbf{A}.$$

Consider now the fields

$$\theta_s V_1 + (1 - \theta_s) V_2$$

and

$$\theta_s \mathbf{A}_1 + (1 - \theta_s) \mathbf{A}_2,$$

which are discontinuous on S. We have

$$\nabla(\theta_s V_1 + (1-\theta_s)V_2) = \theta_s \nabla V_1 + (1-\theta_s)\nabla V_2 + \delta_s (V_1 - V_2)\mathbf{n}$$
$$\nabla \cdot (\theta_s \mathbf{A}_1 + (1-\theta_s)\mathbf{A}_2) = \theta_s \nabla \cdot \mathbf{A}_1 + (1-\theta_s)\nabla \cdot \mathbf{A}_2 + \delta_s \mathbf{n} \cdot (\mathbf{A}_1 - \mathbf{A}_2)$$
$$\nabla \times (\theta_s \mathbf{A}_1 + (1-\theta_s)\mathbf{A}_2) = \theta_s \nabla \times \mathbf{A}_1 + (1-\theta_s)\nabla \times \mathbf{A}_2 + \delta_s \mathbf{n} \times (\mathbf{A}_1 - \mathbf{A}_2).$$

The derivatives of the singular distribution δ_s are defined by

$$\frac{\partial \delta_s^{(k)}}{\partial x_i} = \frac{\partial s}{\partial x_i} \delta_s^{(k+1)}$$

or

$$\frac{\partial \delta_s^{(k)}}{\partial x_i} = n_i \delta_s^{(k+1)}$$
$$\nabla \delta_s^{(k)} = \mathbf{n} \delta_s^{(k+1)},$$

if $s(x)$ represents the distance from x to S.

In particular, we have

$$\nabla(\delta_s V) = \delta_s \nabla V + \delta_s' V \mathbf{n}$$
$$\nabla \cdot (\delta_s \mathbf{A}) = \delta_s \nabla \cdot \mathbf{A} + \delta_s' \mathbf{n} \cdot \mathbf{A}$$
$$\nabla \times (\delta_s \mathbf{A}) = \delta_s \nabla \times \mathbf{A} + \delta_s' \mathbf{n} \times \mathbf{A}.$$

To show how these results can be used, we shall first establish *Green's formula*. From

$$\nabla^2 \theta_s V = \nabla \times (\theta_s \nabla V + \delta_s V \mathbf{n})$$
$$= \theta_s \nabla^2 V + \delta_s \mathbf{n} \times \nabla V + \nabla \times (\delta_s V \mathbf{n}),$$

it follows that

$$\big(\nabla^2(\theta_s V)\big)(\varphi) = (\theta_s \nabla^2 V)(\varphi) + (\delta_s \mathbf{n} \times \nabla V)(\varphi) + \big(\nabla \times (\delta_s V \mathbf{n})\big)(\varphi),$$

that is,

$$(\theta_s V)(\nabla^2 \varphi) - (\theta_s \nabla^2 V)(\varphi) = (\delta_s \mathbf{n} \times \nabla V)(\varphi) - (\delta_s V \mathbf{n})(\nabla \varphi).$$

Hence,

$$\int_{s(x)>0} (V\nabla^2 - \varphi \nabla V)\, d_3x = \int_{s(x)=0} (\frac{\partial V}{\partial n}\varphi - V\frac{\partial \varphi}{\partial n}\, dS,$$

which is Green's formula. d_3x represents $dx_1\, dx_2\, dx_3$, dS denotes the surface element, and $\frac{\partial}{\partial n} = \mathbf{n} \times \nabla$ is the *normal derivative.*

It is very important to distinguish clearly the distributions δ_a and δ_s. They are respectively defined by

$$\delta_a : \varphi \to \varphi(a)$$
$$\delta_s : \varphi \to \int_{s(x)=0} \varphi(x)\, dS.$$

As an application of Green's formula, we shall calculate the Laplacian of the regular distribution $1/r$, where $r = (x_1^2 + x_2^2 + x_3^2)^{1/2}$. Since the Laplacian of $1/r$ is not locally integrable in $\mathbf{R}^3$, the corresponding distribution must be singular (Examples 3 and 5). The integral on the right-hand side of the formula

$$(\nabla^2 \frac{1}{r})(\varphi) = \int \frac{1}{r} \nabla^2 \varphi\, d_3x$$

cannot therefore be integrated by parts without taking some precautions. From Lebesgue's theorem it follows that

$$\int \frac{1}{r} \nabla^2 \varphi\, d_3x = \lim_{\epsilon \to 0} \int_{r>\epsilon} \frac{1}{r} \nabla^2 \varphi\, d_3x.$$

Applying Green's formula yields

$$\int_{r>\epsilon} (\frac{1}{r}\nabla^2 \varphi - \nabla^2 \frac{1}{r}\varphi)\, d_3x = \int_{r=\epsilon} (-\frac{1}{r^2}\varphi - \frac{1}{r}\frac{\partial \varphi}{\partial r})\, dS.$$

In the complement of the origin in $\mathbf{R}^3$, $\nabla^2 1/r = 0$ and $\partial\varphi/\partial r$ is bounded. Therefore, if we let ϵ go to zero, we finally obtain[14]

$$(\nabla^2 \frac{1}{r})(\varphi) = -4\pi\varphi(0),$$

that is,

$$\nabla^2 \frac{1}{r} = -4\pi\delta.$$

This result is fundamental in classical potential theory (see Example 4.1).

To conclude, we shall show how distributions simplify the formulation of classical field theories. Consider, for example, electrostatics. Let $\mathbf{E}$, $\mathbf{D}$, and ρ be, respectively, the *electric field*, the *electric displacement*, and the *electric charge density*. From the first Maxwell's equation,

$$\nabla \times \mathbf{E} = 0,$$

it follows that there exists a scalar field V, called the *electric potential*, such that

$$\mathbf{E} = -\nabla V.$$

Taking into account the second Maxwell's equation,

$$\nabla \cdot \mathbf{D} = \rho,$$

and the fact that in the vacuum

$$\mathbf{D} = \varepsilon_0 \mathbf{E},$$

where ε_0 is a universal constant (*permittivity of the vacuum*), we obtain the *Poisson's equation*

$$\nabla^2 V + \frac{\rho}{\varepsilon_0} = 0.$$

The electric charge density of a point charge q at the origin is clearly not a function, however, since the potential at a distance r from the origin is

$$V(r) = \frac{1}{4\pi\varepsilon_0}\frac{q}{r}.$$

[14] The area of a sphere of radius ϵ in $\mathbf{R}^3$ is equal to $4\pi\epsilon^2$.

From the Poisson's equation it follows that ρ is the singular distribution $q\delta$. More generally, since the potential, at point x, due to a multipole $q_{ij\cdots k}$ at the origin, is given by[15]

$$V(x) = \frac{1}{4\pi\varepsilon_0} q_{ij\cdots k} \frac{\partial}{\partial x_i} \frac{\partial}{\partial x_j} \cdots \frac{\partial}{\partial x_k} \frac{1}{r},$$

the corresponding density is

$$\rho = q_{ij\cdots k} \frac{\partial}{\partial x_i} \frac{\partial}{\partial x_j} \cdots \frac{\partial}{\partial x_k} \delta.$$

Charges and multipoles distributed on lines or surfaces are also represented by singular distributions. For example, a charge distributed on a surface S with a *surface charge density* σ has a *volume charge density* ρ equal to $\sigma\delta_s$, where, as above, $s(x) = 0$ is the equation of S. Using the relations we have established, the study of the influence on the fields **E** and **D** of such a surface of discontinuity is straightforward. Let the indices 1 and 2 refer, respectively, to the domains defined by $s(x) > 0$ and $s(x) < 0$. We have

$$\begin{aligned} \mathbf{E} &= \theta_s \mathbf{E}_1 + (1-\theta_s)\mathbf{E}_2 \\ \mathbf{D} &= \theta_s \mathbf{D}_1 + (1-\theta_s)\mathbf{D}_2 \\ \rho &= \theta_s \rho_1 + (1-\theta_s)\rho_2 + \sigma\delta_s. \end{aligned}$$

Substituting these fields into Maxwell's equations yields

$$\begin{aligned} &\nabla \times (\theta_s \mathbf{E}_1 + (1-\theta_s)\mathbf{E}_2) = 0 \\ &\nabla \cdot (\theta_s \mathbf{D}_1 + (1-\theta_s)\mathbf{D}_2) = \theta_s \rho_1 + (1-\theta_s)\rho_2 + \sigma\delta_s, \end{aligned}$$

that is to say,

$$\begin{aligned} &\theta_s \nabla \times \mathbf{E}_1 + (1-\theta_s)\nabla \times \mathbf{E}_2) + \delta_s \mathbf{n} \times (\mathbf{E}_1 - \mathbf{E}_2) = 0 \\ &\theta_s \nabla \cdot \mathbf{D}_1 + (1-\theta_s)\nabla \cdot \mathbf{D}_2) + \delta_s \mathbf{n} \cdot (\mathbf{D}_1 - \mathbf{D}_2) = \theta_s \rho_1 + (1-\theta_s)\rho_2 + \sigma\delta_s, \end{aligned}$$

which finally gives

$$\begin{aligned} \nabla \times \mathbf{E}_1 &= 0 & \nabla \cdot \mathbf{D}_1 &= \rho_1 \\ \nabla \times \mathbf{E}_2 &= 0 & \nabla \cdot \mathbf{D}_2 &= \rho_2 \\ \mathbf{n} \times (\mathbf{E}_1 - \mathbf{E}_2) &= 0 & \mathbf{n} \cdot (\mathbf{D}_1 - \mathbf{D}_2) &= \sigma. \end{aligned}$$

---

[15] A 2^m-pole is represented by a tensor of rank m. As is customary with tensors, summations should be performed over all indices repeated twice.

The first four equations are just Maxwell's equations in domains 1 and 2, whereas the last two indicate what are the boundary conditions to be satisfied by the fields, i.e., the tangential component of the electric field is continuous whereas the normal component of the displacement is discontinuous, the jump being equal to the surface charge density.

3. Convergence in $\mathcal{D}'$

Definition 5. A distribution T_λ is said to converge to T_{λ_0} in $\mathcal{D}$' when λ tends to λ_0 if, for any test function φ in $\mathcal{D}$,

$$\lim_{\lambda \to \lambda_0} T_\lambda(\varphi) = T_{\lambda_0}(\varphi).$$

Convergence of sequences and series are defined in a similar manner. It is straightforward to verify that, if the limit exists, it is unique. From the definition it follows immmediately that

(i) If ψ is an infinitely differentiable function then

$$\lim_{\lambda \to \lambda_0} \psi T_\lambda = \psi T_{\lambda_0} \qquad \text{when} \qquad \lim_{\lambda \to \lambda_0} T_\lambda = T_{\lambda_0}.$$

(ii) If the distributions T_λ and S_λ converge, respectively ,to T_{λ_0} and S_{λ_0}, then

$$\lim_{\lambda \to \lambda_0} (T_\lambda + S_\lambda) = T_{\lambda_0} + S_{\lambda_0}.$$

Example 1. *Dirac sequences.* Let f be an integrable function such that

$$\int_{-\infty}^{\infty} f(x)\, dx = 1.$$

The function f_n, defined by

$$f_n(x) = nf(nx),$$

where n is a positive integer, is obviously integrable[16] and, therefore, defines a regular distribution, which will be denoted by the same symbol. We shall prove that, in $\mathcal{D}'$, the sequence (f_n), called the *Dirac sequence associated to* f, converges to δ. That is to say, in the sense of convergence of distributions, we shall show that δ can be regarded as the limit of a sequence of functions.

16 Its integral is actually equal to 1

For any test function φ in $\mathcal{D}$, we have

$$\lim_{n\to\infty}\int_{-\infty}^{\infty} nf(nx)\varphi(x)\,dx = \lim_{n\to\infty}\int_{-\infty}^{\infty} f(x)\big(\varphi(\frac{x}{n}) - \varphi(0)\big)\,dx + \varphi(0).$$

Since φ is bounded, for any $\epsilon > 0$, there exists $A > 0$ such that

$$\Big|\int_{A}^{\infty} f(x)\big(\varphi(\frac{x}{n}) - \varphi(0)\big)\,dx\Big| \le \epsilon$$

and

$$\Big|\int_{-\infty}^{-A} f(x)\big(\varphi(\frac{x}{n}) - \varphi(0)\big)\,dx\Big| \le \epsilon.$$

A being fixed, we can choose n sufficiently large such that

$$(\forall x \in [-A, A]) \qquad |\varphi(\frac{x}{n}) - \varphi(0)| \int_{-A}^{A} |f(t)|\,dt \le \epsilon,$$

which implies

$$\Big|\int_{-A}^{A} f(x)\big(\varphi(\frac{x}{n}) - \varphi(0)\big)\,dx\Big| \le \epsilon.$$

Hence,

$$\lim_{n\to\infty}\int_{-\infty}^{\infty} nf(nx)\varphi(x)\,dx = \varphi(0),$$

which is the result we claimed.

At the beginning of the chapter we said that δ could be regarded as the limit of the sequence f_n such that

$$f_n(x) = \begin{cases} n, & \text{if } |x| < 1/2n; \\ 0, & \text{otherwise.} \end{cases}$$

This follows immediately from the fact that (f_n) is the Dirac sequence associated to the function f such that

$$f(x) = \begin{cases} 1, & \text{if } |x| < 1/2; \\ 0, & \text{otherwise.} \end{cases}$$

Dirac sequences associated to the functions

$$x \mapsto \frac{1}{\pi}\frac{1}{1+x^2} \qquad \text{and} \qquad x \mapsto \frac{1}{\sqrt{2\pi}}\exp(-\frac{x^2}{2})$$

are very frequently used.[17]

Let us indicate another interesting result. Since

$$\frac{1}{\pi}\int_{-\infty}^{\infty}\frac{\sin x}{x}\,dx = 1,$$

we have

$$\lim_{n\to\infty}\frac{1}{\pi}\frac{\sin nx}{x} = \delta.$$

But

$$\frac{1}{\pi}\frac{\sin nx}{x} = \frac{1}{2\pi}\int_{-n}^{n}\exp(itx)\,dt,$$

and, therefore, after having changed t in $2\pi t$, we obtain

$$\lim_{n\to\infty}\int_{-n/2\pi}^{n/2\pi}\exp(2\pi itx)\,dt = \delta.$$

This formula is often written

$$\int_{-\infty}^{\infty}\exp(2\pi itx)\,dt = \delta(x).$$

The left-hand side being meaningless, this is not correct.

Theorem 1. *Let T_λ be a distribution that converges to T_{λ_0} when λ tends to λ_0. Then the derivative T'_λ converges to T'_{λ_0}.*

The proof is very simple. If the limit of T_λ is T_{λ_0}, it means that

$$(\forall\varphi\in\mathcal{D})\qquad \lim_{\lambda\to\lambda_0}T_\lambda(\varphi) = T_{\lambda_0}(\varphi).$$

Then

$$\begin{aligned}\lim_{\lambda\to\lambda_0}T'_\lambda(\varphi) &= -\lim_{\lambda\to\lambda_0}T_\lambda(\varphi')\\ &= -T_{\lambda_0}(\varphi')\\ &= T'_{\lambda_0}(\varphi).\end{aligned}$$

Differentiation is therefore linear and continuous.

Example 2. The trigonometric series $\sum_{n=-\infty}^{\infty} a_n\exp(inx)$, where a_n is bounded by a power of n when n is sufficiently large, converges in $\mathcal{D}'$.

---

[17] The first one has been used in the proof of Theorem 9 of Chapter 2.

More precisely, if $a_n < A|n|^k$, A and k being real constants,[18] the sum of the series is equal to $a_0 + f^{(p)}(x)$, where p is a nonnegative integer not less than $k+2$ and $f(x)$ is the sum of the trigometric series

$$\sum_{\substack{n=-\infty \\ n\neq 0}}^{\infty} \frac{a_n}{(in)^p} \exp(inx).$$

Since $|a_n/(in)^p| < A|n|^2$, the above Fourier series is uniformly convergent. Its sum defines a continuous function f. The associated regular distribution is also equal to the sum of this Fourier series, which, according to Theorem 1, may be differentiated p times, and the result readily follows. For instance, the series

$$\sum_{n=1}^{\infty} \frac{1}{n^2} \cos nx$$

is uniformly convergent, and its sum is equal to

$$\frac{1}{4}(x-\pi)^2 \qquad \text{for} \qquad 0 \leq x \leq 2\pi.$$

Differentiating two times yields

$$\sum_{n=1}^{\infty} \frac{1}{n} \sin nx = \frac{1}{2}(x-\pi) \qquad \text{for} \qquad 0 < x < 2\pi$$

$$\sum_{n=1}^{\infty} \cos nx = -\frac{1}{2} + \pi \sum_{k=-\infty}^{\infty} \delta_{2k\pi}.$$

The second result gives the Fourier expansion of the periodic distribution $\sum_{k=-\infty}^{\infty} \delta_{2k\pi}$, which may also be written (see also Theorem 8)

$$\sum_{k=-\infty}^{\infty} \delta_{2k\pi} = \frac{1}{2\pi} \sum_{n=-\infty}^{\infty} \exp(inx);$$

and, from Theorem 1, it follows that, for any positive integer m,

$$\sum_{k=-\infty}^{\infty} \delta_{2k\pi}^{(m)} = \frac{1}{2\pi} \sum_{n=-\infty}^{\infty} (in)^m \exp(inx).$$

---

18 In this case the coefficients a_n are said to be of *slow growth at infinity*.

Example 3. If ϵ is a real different from zero, we have

$$\begin{aligned}\frac{d}{dx}\log(x+i\epsilon) &= \frac{d}{dx}\log|x+i\epsilon| + i\frac{d}{dx}\arctan\frac{\epsilon}{x}\\ &= \frac{x}{x^2+\epsilon^2} - \frac{i\epsilon}{x^2+\epsilon^2}\\ &= \frac{1}{x+i\epsilon}.\end{aligned}$$

Hence, (see Example 2.4)

$$\lim_{\epsilon\to 0+}\frac{1}{x\pm i\epsilon} = \operatorname{Pv}\frac{1}{x} \mp i\pi\delta,$$

which yields

$$\lim_{\epsilon\to 0}\frac{x}{x^2+\epsilon^2} = \operatorname{Pv}\frac{1}{x},$$

and

$$\lim_{\epsilon\to 0}\frac{1}{\pi}\frac{\epsilon}{x^2+\epsilon^2} = \delta.$$

This last result could have been obtained considering the Dirac sequence associated to the function (see Example 1)

$$x \mapsto \frac{1}{\pi}\frac{1}{1+x^2}.$$

Differentiating n times the distribution $\log(x\pm i\epsilon)$ gives

$$\lim_{\epsilon\to 0+}(x\pm i\epsilon)^{-n} = \operatorname{Pf} x^{-n} \mp \frac{(-1)^{n-1}}{(n-1)!} i\pi\delta^{(n-1)}.$$

The limits of $(x+i\epsilon)^{-n}$ and $(x-i\epsilon)^{-n}$, when ϵ tends to 0^+, are, respectively, denoted $(x+i0)^{-n}$ and $(x-i0)^{-n}$.

Example 4. The derivative T' of a distribution T has been defined by

$$(\forall\varphi\in\mathcal{D})\qquad T'(\varphi) = -T(\varphi').$$

Since (see Definition 3)

$$\begin{aligned}\varphi'(x) &= \lim_{h\to 0}\frac{1}{h}\big(\varphi(x+h)-\varphi(x)\big)\\ &= \lim_{h\to 0}\frac{1}{h}\big((\tau_h^{-1}\varphi)(x)-\varphi(x)\big),\end{aligned}$$

where τ_h is the translation operator defined by

$$(\tau_h \varphi)(x) = \varphi(x-h),$$

we have

$$\begin{aligned} T'(\varphi) &= -\lim_{h\to 0} \frac{1}{h} T\big(\tau_h^{-1}\varphi - \varphi\big) \\ &= -\lim_{h\to 0} \frac{1}{h} (\tau_h T - T)(\varphi) \\ &= \lim_{h\to 0} \frac{1}{h} (\tau_h^{-1} T - T)(\varphi), \end{aligned}$$

that is to say,

$$T' = \lim_{h\to 0} \frac{1}{h} (\tau_h^{-1} T - T).$$

This formula is identical to the relation that defines the derivative of a function.

4. Convolution

In distribution theory, convolution plays an essential role. Various types of differential equations or integral equations are convolution equations. To define the convolution of two distributions, it is necessary first to define the tensor product of two distributions. In some formulas, to avoid ambiguity, we shall attach variables as subscripts to symbols representing spaces or distributions. For example, if we need to specify that the variable is x, we shall denote by $\mathcal{D}_x$ the usual space of test functions.

Definition 6. Let $\mathcal{D}_x$, $\mathcal{D}_y$, and $\mathcal{D}_{x,y}$ be the spaces of test functions defined, respectively, on $\mathbf{R}_x$, $\mathbf{R}_y$, and their Cartesian product $\mathbf{R}_x \times \mathbf{R}_y$. Let S and T be two distributions defined, respectively, on $\mathbf{R}_x$ and $\mathbf{R}_y$. The *tensor product* of S and T is the distribution $S \otimes T$, defined on $\mathbf{R}_x \times \mathbf{R}_y$, such that, for any φ_1 in $\mathcal{D}_x$ and any φ_2 in $\mathcal{D}_y$,

$$(S_x \otimes T_y)\big(\varphi_1(x)\varphi_2(y)\big) = S_x\big(\varphi_1(x)\big) T_y\big(\varphi_2(y)\big).$$

From this definition it follows that the tensor product is commutative.

Since the set of functions $(x,y) \mapsto \varphi_1(x)\varphi_2(y)$ is dense in $\mathcal{D}_{x,y}$, the definition of the tensor product makes sense. That is, the linear form $\varphi \mapsto (S \otimes T)(\varphi)$ is defined for any test function φ in $\mathcal{D}_{x,y}$. The tensor product $S \otimes T$ belongs to $\mathcal{D}'_{x,y}$.

The convolution $f * g$ of the functions f and g is defined by

$$(f * g)(x) = \int_{-\infty}^{\infty} f(y) g(x-y)\, dy.$$

$f * g$ exists if the integral converges. If $f * g$ is locally integrable, then, for any φ in $\mathcal{D}$, we have

$$\int (f * g)(x)\varphi(x)\,dx = \int\int f(y)g(x-y)\varphi(x)\,dx\,dy$$
$$= \int\int f(u)g(v)\varphi(u+v)\,du\,dv.$$

This last formula, in which f and g play a symmetrical role, suggests how to define the convolution of two distributions.

Definition 7. Let S and T be two distributions defined on $\mathbf{R}_x$, the distribution $S * T$ such that, for any φ in $\mathcal{D}_x$,

$$(S * T)_x(\varphi(x)) = (S_\xi \otimes T_\eta)(\varphi(\xi + \eta))$$
$$= S_\xi\Big(T_\eta(\varphi(\xi + \eta))\Big).$$

Since the tensor product is commutative, it follows that the same property holds for convolution: $S * T = T * S$.

The convolution of two arbitrary distributions is, in general, not defined, since the support of the test function $(\xi, \eta) \mapsto \varphi(\xi + \eta)$, which is defined on $\mathbf{R}_\xi \times \mathbf{R}_\eta$, is not bounded. [19] That is, the linear form $\varphi \mapsto (S*T)(\varphi)$ is meaningless if the distributions S and T do not satisfy certain restrictive conditions. To specify these conditions we have to introduce the concept of *support of a distribution.*

Definition 8. If, for any test function φ in $\mathcal{D}$ whose support is contained in the open set Ω of $\mathbf{R}$, the linear form $\varphi \mapsto T(\varphi)$ is null, then the distribution T is said to be *null in* Ω. The complement of the largest open set on which the distribution T is null is called the *support of* T. It is denoted by $\operatorname{supp} T$.

Although it is meaningless to say that "the distribution T is null at x_0," it makes sense to say that "T is null in the neighborhood of x_0" if T is null in any open set containing x_0.

Since δ is null in the complement of the origin, its support is the origin. All the derivatives of δ have the same property. If a distribution T has a support that consists in one point, say zero, this does not mean that $T(\varphi)$ depends only on $\varphi(0)$. For example, $\delta(\varphi) = -\varphi'(0)$, and $\varphi'(0)$ is not

---

[19] If the support of $x \mapsto \varphi(x)$ is, for example, the bounded interval $[a, b]$, then the support of $(\xi, \eta) \mapsto \varphi(\xi + \eta)$ is the strip $\{(\xi, \eta) | a \leq \xi + \eta \leq b\}$, which is not bounded.

determined by $\varphi(0)$. More generally, $T(\varphi)$ does not depend only on the values φ takes on the support of T, but it depends on the values φ takes on a *neighborhood* of the support of T.

It is clear that, if a distribution T has a bounded support, the linear form $\varphi \mapsto T(\varphi)$ is defined even when the support of φ is not bounded. More generally, $T(\varphi)$ is defined when $\operatorname{supp} T \cap \operatorname{supp} \varphi$ is bounded.

The following theorem gives two important cases in which the convolution of two distributions is defined.

Theorem 2. *(1) Let T be a distribution with bounded support; then, for any arbitrary distribution S, the convolution $S * T$ is defined.*

(2) Let S and T be two distributions whose supports are both bounded from below (or both from above); then the convolution $S * T$ is defined.

In the first case the function

$$\psi: \xi \mapsto T_\eta(\varphi(\xi + \eta)$$

is infinitely differentiable (because T is continuous) and has a bounded support (because $\operatorname{supp} T$ is bounded). ψ therefore belongs to $\mathcal{D}$, and $S(\psi)$ is defined.

In the second case the intersection of the support of S and the support of $\psi: \xi \mapsto T_\eta(\varphi(\xi + \eta)$ is bounded.

Theorem 3. *For any distribution T, we have*

$$(1) \qquad \delta * T = T$$

$$(2) \qquad \delta' * T = T'.$$

These two formulas are very important. The first shows that δ is the identity for the binary law $(S, T) \mapsto S * T$, and the second indicates that differentiation can be interpreted as a convolution.

These formulas are valid without any restriction on the support of T since $\operatorname{supp} \delta = \operatorname{supp} \delta' = \{0\}$. They readily follow from Definition 7. We have

$$\begin{aligned}(\delta * T)(\varphi) &= T_\xi\Big(\delta_\eta\big(\varphi(\xi + \eta)\big)\Big)\\ &= T_\xi\big(\varphi(\xi)\big)\\ &= T(\varphi)\end{aligned}$$

and

$$\begin{aligned}(\delta' * T)(\varphi) &= T_\xi\Big(\delta'_\eta\big(\varphi(\xi + \eta)\big)\Big)\\ &= -T_\xi\big(\varphi'(\xi)\big)\\ &= T'(\varphi).\end{aligned}$$

Remark 1. If $\tau_a T$ is the translate of T we have

$$\delta_a * T = \tau_a T$$

for

$$\begin{aligned}(\delta_a * T)(\varphi) &= T_\xi\Big(\delta_{a\,\eta}(\varphi(\xi+\eta))\Big)\\ &= T_\xi(\varphi(\xi+a))\\ &= (\tau_a T)(\varphi).\end{aligned}$$

In particular,

$$\delta_a * \delta_b = \delta_{a+b}.$$

Remark 2. From the second formula it follows that

$$\delta^{(p)} * T = T^{(p)},$$

and more generally

$$D\delta * T = DT,$$

where D is a linear differential operator with constant coefficients.

For distributions with unbounded support, the convolution, when it exists, may not be associative. For example,

$$(\theta * \delta') * 1 \neq \theta * (\delta' * 1)$$

for $(\theta * \delta') * 1 = \delta * 1 = 1$, and $\theta * (\delta' * 1) = \theta * 0 = 0$.

For distributions whose supports are all bounded from below (or all from above) the convolution is always associative.

Theorem 4. *Let S and T be two distributions whose convolution exists; then we have*

$$(S * T)' = S' * T = S * T'.$$

To differentiate a convolution we can differentiate either one of the distributions. This follows from the associativity and commutativity of the convolution.

$$\begin{aligned}(S * T)' &= \delta' * (S * T)\\ &= (\delta' * S) * T = S' * T\\ &= S * (\delta' * T) = S * T'.\end{aligned}$$

As a consequence we have, for instance,

$$(S * T)'' = S'' * T = S' * T' = S * T''.$$

Example 1. The electric potential V due to a charge density ρ is given in $\mathbf{R}^3$ by

$$V = \rho * \frac{1}{4\pi\epsilon_0 r}.$$

Hence (see Example 2.6),

$$\begin{aligned} \nabla^2 V &= \rho * \frac{1}{4\pi\epsilon_0}\nabla^2\frac{1}{r} \\ &= -\frac{\rho}{\epsilon_0}, \end{aligned}$$

which is Poisson's equation.

Example 2. Let $Du = f$ be a linear differential equation with constant coefficients where f is an arbitrary distribution. The distribution g is called a *fundamental solution*[20] if

$$Dg = \delta.$$

A fundamental solution is not unique. If g is a fundamental solution $g + g_0$ is also a fundamental solution provided g_0 satisfies $Dg_0 = 0$. Fundamental solutions are very important in the study of differential equations. If a fundamental solution g is known then the solution of the differential equation is given by

$$u = f * g,$$

since

$$Du = f * Dg = f * \delta = f.$$

For instance, in $\mathbf{R}^3$, the distribution

$$g = -\frac{1}{4\pi r}$$

is a fundamental solution of Poisson's equation

$$\nabla^2 V = -\frac{\rho}{\epsilon_0}.$$

Remark. What has been said concerning fundamental solutions can be extended to any arbitrary convolution equation $A * X = B$, where A and B

---

[20] Also known as an *elementary solution* or, especially in physics, a *Green's function.*

are given distributions and X an unknown distribution. For the convolution to exist we can assume that A has a bounded support, but this is not the only case encountered in practice. Note that a fundamental solution does not always exist. If, for instance, A is in $\mathcal{D}$ then, for any distribution X, $A*X$ is an infinitely differentiable function that, therefore, cannot be equal to δ.

Example 3. In the study of one-dimensional heat flows it is found that the temperature at time t and at position x satisfies a differential equation of the following type

$$\frac{\partial u}{\partial t} = \frac{\partial^2 u}{\partial x^2} \qquad (x \in \mathbf{R}, t > 0).$$

To determine the solution u of this equation we have to know the *initial condition* to be satisfied by u, that is, $u(x,0) = f(x)$. This *initial value problem* is called the *Cauchy problem* for the heat equation. In the sense of distributions, the solution of the equation is a distribution defined on $\mathbf{R}$, which depends upon a parameter t. It is denoted by u_t. From the definition of the convergence in $\mathcal{D}'$, it follows that if, for all φ in $\mathcal{D}_x$, the function $t \mapsto u_t(\varphi)$ is differentiable, then the distribution u_t is differentiable with respect to t, and we have

$$(\forall \varphi \in \mathcal{D}_x) \qquad (\frac{\partial u_t}{\partial t})(\varphi) = \frac{\partial}{\partial t} u_t(\varphi).$$

It is easy to show that the solution of the Cauchy problem is given by

$$u_t = f * g_t,$$

where g_t is the regular distribution associated to the locally integrable function

$$x \mapsto \frac{1}{2\sqrt{\pi t}} \exp -\frac{x^2}{4t} \qquad (x \in \mathbf{R}, t > 0).$$

g_t being a solution of the differential equation, it follows from Theorem 4 that $f * g_t$ is also a solution. Moreover (see Example 3.1)

$$\lim_{t \to 0+} g_t = \delta;$$

therefore,

$$\lim_{t \to 0+} f * g_t = f.$$

The distribution g_t is a *fundamental solution of the Cauchy problem* for the heat equation.

Remark. g_t is an element of $\mathcal{D}'_x$ that depends upon a parameter t. The fundamental solution can also be regarded as an element of $\mathcal{D}'_{x,t}$. Then t plays a role similar to x, and, to emphasize this fact, it is more appropriate simply to denote the fundamental solution by g. In this case, it can be shown that g satisfies the relation

$$(\frac{\partial}{\partial t} - \frac{\partial^2}{\partial x^2})g = \delta,$$

where δ belongs to $\mathcal{D}'_{x,t}$. It represents a unit amount of heat put at $x = 0$ at time $t = 0$. When t is positive, g is a particular solution of the differential equation, and if the distribution $(\partial/\partial t - \partial^2/\partial x^2)g$ is not equal to zero, its support must be the single point $x = 0, t = 0$. This is indeed the case since, for any φ in $\mathcal{D}_{x,t}$, we have

$$\begin{aligned}\Big((\frac{\partial}{\partial t} - \frac{\partial^2}{\partial x^2})g\Big)(\varphi) &= -g(\frac{\partial \varphi}{\partial t}) - g(\frac{\partial^2 \varphi}{\partial x^2})\\ &= \lim_{\epsilon\to 0}\int_{t>\epsilon} -\frac{1}{2\sqrt{\pi t}}\exp -(\frac{x^2}{4t})\left(\frac{\partial \varphi}{\partial t} + \frac{\partial^2 \varphi}{\partial x^2}\right) dx\, dt\\ &= \lim_{\epsilon\to 0}\int \frac{1}{2\sqrt{\pi\epsilon}}\exp -(\frac{x^2}{4\epsilon})\, \varphi(x,\epsilon)\, dx\\ &= \lim_{\epsilon\to 0}\int \frac{1}{2\sqrt{\pi}}\exp -(\frac{x^2}{4})\, \varphi(\sqrt{\epsilon}x,\epsilon)\, dx = \varphi(0,0),\end{aligned}$$

which proves the result we claimed.

In this case the solution u is an element of $\mathcal{D}'_{x,t}$, whereas in the first case the solution u_t is an element of $\mathcal{D}'_x$ that depends upon a parameter t. These two points of view are indeed different. In the first case the function $t \mapsto u_t(\varphi)$ has to be differentiable for any φ in $\mathcal{D}_x$, whereas in the second case u as an element of $\mathcal{D}'_{x,t}$ is always differentiable with respect to t.

Example 4. Let $\mathcal{D}'_+$ (resp. $\mathcal{D}'_-$) be the set of distributions whose support is contained in $[0, \infty[$ (resp. $]-\infty, 0]$). Since the convolution of two distributions belonging to $\mathcal{D}'_+$ (resp. $\mathcal{D}'_-$) is always defined and belongs to $\mathcal{D}'_+$ (resp. $\mathcal{D}'_-$), it follows that $\mathcal{D}'_+$ (resp. $\mathcal{D}'_-$) is a *convolution algebra.*

Let λ be a complex number. The inverse, in $\mathcal{D}'_+$, of $\delta' - \lambda\delta$ is the unique distribution X[21] such that

$$(\delta' - \lambda\delta) * X = \delta,$$

21 The unicity of the inverse is a consequence of associativity. If X_1 and X_2

that is,

$$X' - \lambda X = \delta.$$

$x \mapsto \exp \lambda x$ being infinitely differentiable, we can set $X = Y \exp \lambda x$, where $Y \in \mathcal{D}'_+$ is such that

$$Y' = \delta,$$

that is, $Y = \theta$.

The inverse of $\delta' - \lambda\delta$, in $\mathcal{D}'_+$, is therefore $\theta \exp \lambda x$.[22] More generally, the inverse, in $\mathcal{D}'_+$, of the operator

$$D = (\delta' - \lambda_1\delta) * (\delta' - \lambda_2\delta) * \cdots * (\delta' - \lambda_n\delta)$$

is the distribution

$$\theta \exp \lambda_1 x * \theta \exp \lambda_2 x * \cdots * \theta \exp \lambda_n x.$$

This result is correct only if all the λ_i $(i = 1, 2, \ldots, n)$ are different. If this is not the case, the expression of the inverse has to be modified taking into account the formula

$$\frac{\theta x^{p-1} \exp \lambda x}{(p-1)!} * \frac{\theta x^{q-1} \exp \lambda x}{(q-1)!} = \frac{\theta x^{p+q-1} \exp \lambda x}{(p+q-1)!},$$

which is a consequence of the relation

$$B(p, q) = \frac{\Gamma(p)\Gamma(q)}{\Gamma(p+q)},$$

where B and Γ denote the Eulerian functions.

---

are two inverses of the distribution T, from

$$(X_1 * T) * X_2 = X_1 * (T * X_2),$$

it follows that $X_1 = X_2$. The unicity of the inverse implies that the convolution equation $A * X = B$, where A and B are two given distributions, has the unique solution $X = A^{-1} * B$.

[22] In $\mathcal{D}'_-$ the inverse of $\delta' - \lambda\delta$ would be different. A convolution equation may have one solution in $\mathcal{D}'_+$ and another one in $\mathcal{D}'_-$.

As an illustration of this result consider a simple electrical circuit consisting of a resistor R and a capacitor C connected in series to a generator. If the initial charge of the capacitor is zero, the current $i(t)$, at time $t > 0$, charging the capacitor through the resistor is given by

$$Ri(t) + \frac{1}{C}\int_0^t i(\tau)\,d\tau = e(t),$$

where $e(t)$ is the electromotive force at time t. If we denote, respectively, by i and e the distributions in $\mathcal{D}'_+$ representing the current and the electromotive force, we have

$$(R\delta + \frac{\theta}{C}) * i = e,$$

where we have taken into account that, in $\mathcal{D}'_+$,

$$\int_0^t i(\tau)\,d\tau = (\theta * i)(t).$$

Hence,

$$i = (R\delta + \frac{\theta}{C})^{-1} * e.$$

The distribution $Z = R\delta + \theta/C$ is called the *impedance* of the circuit. Its inverse A is the *admittance*. It is a Green's function. To find the expression of A we can proceed as follows: $A * Z = \delta$ implies that

$$A * Z * \delta' = \delta',$$

hence,

$$A = (Z * \delta')^{-1} * \delta'.$$

But

$$Z * \delta' = R\delta' + \frac{1}{C}\delta,$$

and, from the result found above,

$$(R\delta' + \frac{1}{C}\delta)^{-1} = \frac{1}{R}\theta \exp -\frac{t}{RC}.$$

Finally,

$$A = \frac{1}{R}(\delta - \frac{1}{RC}\theta \exp -\frac{t}{RC}),$$

and

$$i = \frac{1}{R}(\delta - \frac{1}{RC}\theta \exp -\frac{t}{RC}) * e.$$

The convolution can be used to approximate certain functions by more differentiable ones. This technique has been used in Chapter 2 to prove that certain sets of differentiable functions were dense in L^p spaces. It is known as *regularization.* More generally, given a distribution T, it is possible to define a sequence of infinitely differentiable functions whose limit in $\mathcal{D}'$ is T. This result is a consequence of the following theorem.

Theorem 5. *Let T be a distribution and let ψ be a test function; then the convolution $T * \psi$ is an infinitely differentiable function given by*

$$(T * \psi)(x) = T_\xi\big(\psi(x - \xi)\big).$$

The support of ψ being bounded, the convolution $T * \psi$ is defined, and we have

$$\begin{aligned}(T * \psi)(x) &= T_\xi\Big(\psi_\eta\big(\varphi(\xi + \eta)\big)\Big)\\ &= T_\xi\Big(\int \psi(\eta)\varphi(\eta + \xi)\, d\eta\Big)\\ &= T_\xi\Big(\int \psi(x - \xi)\varphi(x)\, dx\Big)\\ &= (\varphi_x \otimes T_\xi)\big(\psi(x - \xi)\big)\\ &= \varphi_x\Big(T_\xi\big(\psi(x - \xi)\big)\Big)\\ &= \int T_\xi\big(\psi(x - \xi)\big)\varphi(x)\, dx.\end{aligned}$$

$T * \psi$ is therefore a regular distribution associated to the function $x \mapsto T_\xi\big(\psi(x - \xi)\big)$. Moreover, T being continuous, this function is infinitely differentiable, and its derivatives are given by

$$\frac{d^m}{dx^m}(T * \psi) = T * \frac{d^m \psi}{dx^m}.$$

This theorem has many interesting applications.

If (ψ_n) is the Dirac sequence associated to the test function ψ, then the sequence of infinitely differentiable functions $(T * \psi_n)$ converges, and its limit is T.

Let f be an infinitely differentiable function and let φ be a function in $\mathcal{D}$ such that, for any x in [-1,1], $\varphi(x) = 1$; then the sequence (f_n) defined by

$$f_n(x) = f(x)\varphi(\frac{x}{n})$$

tends to f in $\mathcal{D}'$. Taking into account the preceding result, it follows that any distribution is the limit in $\mathcal{D}'$ of a sequence of test functions. In other words,

Theorem 6. *The space $\mathcal{D}$, regarded as a subspace of $\mathcal{D}'$, is dense in $\mathcal{D}'$.*

Let ψ and φ be two test functions. The integral $\int \psi(x)\varphi(x)\,dx$ can be approximated by a Riemann sum of the form $\sum_{i\in I} \psi(x_i)\varphi(x_i)\Delta x_i$, where I is a finite set of indices. Consequently, given a function ψ in $\mathcal{D}$, for any $\epsilon > 0$, there exists a finite set of constants $\{a_i\}_{i\in I}$[23] such that

$$(\forall \varphi \in \mathcal{D}) \qquad \left|\psi(\varphi) - \sum_{i\in I} a_i\varphi(x_i)\right| \leq \epsilon.$$

This inequality can also be written

$$\left|(\psi - \sum_{i\in I} a_i\delta_{x_i})(\varphi)\right| \leq \epsilon,$$

which shows that any test function can be approximated, in $\mathcal{D}'$, by a finite linear combination of Dirac distributions.

Taking into account the preceding theorem, this result proves the following:

Theorem 7. *The space of all finite linear combinations of Dirac distributions is dense in $\mathcal{D}'$.*

Example 5. The considerations of Example 4 have been illustrated by a particularly simple electric system for which the *response* i to an *excitation* e was given by $i = A * e$, where A, called the *admittance*, is characteristic of the system.

This relation between response and excitation is very general (see Chapter 2, Examples 4.4, 4.5, 5.8). We shall prove that it is *valid for any linear system.*[24]

---

[23] Take, for example, $\forall i \in I \; a_i = \psi(x_i)\Delta x_i$.

[24] A linear system is such that if i_1 and i_2 are, respectively, the responses to the excitations e_1 and e_2, for any two real numbers λ_1 and λ_2, the response to the excitation $\lambda_1 e_1 + \lambda_2 e_2$ is $\lambda_1 i_1 + \lambda_2 i_2$.

We shall assume that the system satisfies the following conditions

(1) If $e \in \mathcal{D}'_+$, then $i \in \mathcal{D}'_+$.

(2) If i is the response to e, then $\delta_\tau * i$ is the response to $\delta_\tau * e$.

(3) If i_λ is the response to e_λ, $\lim_{\lambda\to\lambda_0} i_\lambda$ is the response to $\lim_{\lambda\to\lambda_0} e_\lambda$.

The first condition expresses *causality* and the second one *time translation invariance.* If i and e are functions, the first condition says that if $e(t) = 0$ when $t < 0$, then $i(t)$ must also be equal to zero when $t < 0$, whereas the second condition indicates that if $i(t)$ is the response to $e(t)$, then $i(t-\tau)$ must be the response to $e(t-\tau)$. The third condition expresses *continuity*: a small variation of the excitation should result in a small variation of the response.

To prove the relation $i = A * e$, we first need to define the admittance. We shall say that *the admittance A of a system is the response to the excitation $e = \delta$.* From condition (1) it follows that A belongs to $\mathcal{D}'_+$. Furthermore, the relation $i = A*e$ being true for $e = \delta$, condition (2) shows that it is also true for $e = \delta_\tau$, and, as a consequence of the linearity of the system, it is true for $e = \sum_{k\in K} a_k \delta_{\tau_k}$, where K is a finite set of indices. Condition (3) and Theorem 7 prove that it is true for any distribution $e \in \mathcal{D}'_+$.

Example 6. We shall establish that any distribution T having a bounded support[25] is the limit of a sequence of polynomials. This result is a generalization of a well-known theorem due to Weierstrass. Note that, since T has a bounded support, the function ψ of Theorem 5 does not need to have a bounded support and may be, in particular, a polynomial. In this case $T * \psi$ is also a polynomial.

Consider the sequence of polynomials (P_n) defined by

$$P_n(x) = \frac{n}{\sqrt{\pi}}\Big(1 - \frac{x^2}{n}\Big)^{n^3}.$$

Its limit is δ. To prove it we have to show that

$$(\forall \varphi \in \mathcal{D}) \qquad \lim_{n\to\infty} \int_{-\infty}^{\infty} \frac{n}{\sqrt{\pi}}\Big(1 - \frac{x^2}{n}\Big)^{n^3} \varphi(x)\,dx = \varphi(0).$$

φ having a bounded support, there exists A such that $\operatorname{supp}\varphi \subset [-A, A]$, and if we denote by I_n the above integral we have

$$I_n = \int_{|x|\le A} \frac{n}{\sqrt{\pi}}\Big(1 - \frac{x^2}{n}\Big)^{n^3} \varphi(x)\,dx.$$

---

[25] The space of distributions having bounded supports is often denoted by $\mathcal{E}'$.

Let $y = nx$; then

$$I_n = \int_{|y|\le nA} \frac{1}{\sqrt{\pi}}\big(1-\frac{y^2}{n^3}\big)^{n^3}\varphi(\frac{y}{n})\,dy$$
$$= \int 1_{[-nA,nA]}(y)\frac{1}{\sqrt{\pi}}\big(1-\frac{y^2}{n^3}\big)^{n^3}\varphi(\frac{y}{n})\,dy.$$

y being fixed, the relation

$$\lim_{n\to\infty}\varphi(\frac{y}{n}) = \varphi(0)$$

implies the existence of $C > 0$ such that, for a sufficiently large n,

$$|\varphi(\frac{y}{n})| \le C.$$

Moreover, for a fixed y and a sufficiently large n, we have

$$\big(1-\frac{y^2}{n^3}\big)^{n^3} \le \exp -y^2,$$

and therefore,

$$1_{[-nA,nA]}(y)\big(1-\frac{y^2}{n^3}\big)^{n^3}\varphi(\frac{y}{n}) \le C\exp -y^2.$$

Thus, from Lebesgue's theorem, it follows that

$$\lim_{n\to\infty} I_n = \frac{\varphi(0)}{\sqrt{\pi}}\int \exp -y^2\,dy = \varphi(0),$$

which proves the result we claimed.

5. Fourier Series

Definition 9. A distribution T is said to be *periodic* if there exists a real positive number a, called a *period* of T, such that $\tau_a T = T$, where $\tau_a T$ is the translate of T.

This definition means that, for any φ in $\mathcal{D}$,

$$(\tau_a T)(\varphi) = T(\tau_{-a}\varphi) = T(\varphi),$$

that is to say,

$$T_x\big(\varphi(x+a)\big) = T_x\big(\varphi(x)\big).$$

Let $\mathcal{D}(S^1)$ be the space of infinitely differentiable functions defined on the circle S^1 whose length is a. To a function ϕ belonging to $\mathcal{D}(S^1)$ we may associate a periodic function $\tilde{\phi}$ defined on $\mathbf{R}$ such that

$$\tilde{\phi}(x) = \phi(s) \qquad (x \in \mathbf{R}) \quad (s \in S^1).$$

Note that ϕ has a bounded support (S^1 is bounded), whereas the support of $\tilde{\phi}$ is not bounded.

Let φ be a function in $\mathcal{D}(\mathbf{R})$, and set

$$\tilde{\phi}(x) = \sum_{l=-\infty}^{\infty} \varphi(x + la).$$

$\tilde{\phi}$ is infinitely differentiable and periodic. Let ϕ be the function defined on S^1 associated to $\tilde{\phi}$; to any distribution T defined on S^1 we can associate a periodic distribution $\tilde{T}$ on $\mathbf{R}$ such that

$$\tilde{T}(\varphi) = T(\phi) \qquad (\varphi \in \mathcal{D}(\mathbf{R})) \quad (\phi \in \mathcal{D}(S^1)).$$

It is indeed easy to verify that $\tilde{T}$ is periodic. For any φ in $\mathcal{D}(\mathbf{R})$, we have

$$\begin{aligned}
\tau_a \tilde{T}(\varphi) &= \tilde{T}(\tau_{-a}\varphi) \\
&= \tilde{T}_x\big(\varphi(x + a)\big) \\
&= T(\phi) \\
&= \tilde{T}(\varphi),
\end{aligned}$$

since changing $\varphi(x)$ in $\varphi(x + a)$ does not change $\phi(x)$.

In the case of a regular distribution associated to a locally integrable periodic function $\tilde{f}$ we have

$$\begin{aligned}
\int_{\mathbf{R}} \tilde{f}(x)\varphi(x)\,dx &= \sum_{l=-\infty}^{\infty} \int_{la}^{(l+1)a} \tilde{f}(x)\varphi(x)\,dx \\
&= \sum_{l=-\infty}^{\infty} \int_0^a \tilde{f}(x)\varphi(x + la)\,dx \\
&= \int_0^a \tilde{f}(x)\tilde{\phi}(x)\,dx \\
&= \int_{S^1} f(s)\phi(s)\,ds,
\end{aligned}$$

which shows that f is the function defined on S^1 associated to the periodic function $\tilde{f}$.

In what follows we shall consider only distributions on S^1, that is elements of $\mathcal{D}'(S^1)$.

Definition 10. Let T be a distribution on the circle S^1 of length a, the numbers

$$c_k(T) = \frac{1}{a} T_s\left(\exp -\frac{2i\pi ks}{a}\right),$$

where $k \in \mathbf{Z}$, are called the *Fourier coefficients* of T, and the series

$$\sum_{k=-\infty}^{\infty} c_k(T) \exp \frac{2i\pi ks}{a}$$

is called the *Fourier series* of T.

Theorem 8. *The Fourier series of δ on S^1 converges, in $\mathcal{D}'(S^1)$, to δ.*

The Fourier coefficients of δ are given by

$$\begin{aligned} c_k(\delta) &= \frac{1}{a} \delta_s\left(\exp -\frac{2i\pi ks}{a}\right) \\ &= \frac{1}{a}, \end{aligned}$$

and we have therefore to establish that

$$\sum_{k=-\infty}^{\infty} \frac{1}{a} \exp \frac{2i\pi ks}{a} = \delta.$$

Since all the coefficients are equal to $1/a$, the series is convergent in $\mathcal{D}'(S^1)$ (see example 3.2). Let T be the sum of this series. We have

$$\left(\exp \frac{2i\pi s}{a}\right)T = \sum_{k=-\infty}^{\infty} \frac{1}{a} \exp \frac{2i\pi(k+1)s}{a} = T,$$

i.e.,

$$\left(\exp \frac{2i\pi s}{a} - 1\right)T = 0.$$

The left-hand side is of the form $sg(s)$ where g is an infinitely differentiable function that, for $s = 0$, is different from zero. We shall show that the solution T of the equation $sgT = 0$ is proportional to δ. For any function ϕ in $\mathcal{D}(S^1)$ we have

$$(sgT)(\phi) = T(sg\phi) = 0,$$

that is, T is null on any function which has a simple zero at the origin.

Let $\psi \in \mathcal{D}$ be such that $\psi(0) = 1$; for any function φ in $\mathcal{D}$ there exists a function $\chi \in \mathcal{D}$, which is defined by

$$\varphi(x) = \varphi(0)\psi(x) + \chi(x)$$

and is such that $\chi(0) = 0$. Hence,

$$\begin{aligned} T(\varphi) &= T\big(\varphi(0)\psi + \chi\big) \\ &= C\varphi(0), \end{aligned}$$

where $C = T(\psi)$. This result being true for any $\varphi \in \mathcal{D}$, it follows that $T = C\delta$.

We have therefore proved that

$$\sum_{k=-\infty}^{\infty} \frac{1}{a} \exp \frac{2i\pi ks}{a} = C\delta.$$

To determine the constant C, consider the particular function $s \mapsto 1$ which belongs to $\mathcal{D}(S^1)$. From

$$\Big(\frac{1}{a} \exp \frac{2i\pi ks}{a}\Big)(1) = \begin{cases} 1, & \text{if } k = 0, \\ 0, & \text{otherwise,} \end{cases}$$

it follows that $C = 1$.

Theorem 9. *The Fourier series of any distribution T on S^1 converges, in $\mathcal{D}'(S^1)$, to T.*

From the preceding theorem and the continuity of the convolution it follows that

$$T = T * \delta = \sum_{k=-\infty}^{\infty} \frac{1}{a} T * \exp \frac{2i\pi ks}{a},$$

but (Theorem 5)

$$\begin{aligned} T * \exp \frac{2i\pi ks}{a} &= T_\sigma\Big(\exp \frac{2i\pi k(s-\sigma)}{a}\Big) \\ &= \exp \frac{2i\pi ks}{a} T_\sigma\Big(\exp -\frac{2i\pi k\sigma}{a}\Big) \\ &= ac_k(T) \exp \frac{2i\pi ks}{a}, \end{aligned}$$

which completes the proof.

Example 1. The Fourier series of the periodic function $\tilde{f}$ defined by

$$\tilde{f}(x) = x \qquad \text{if } x \in]-\pi, \pi[$$

can easily be obtained using the Fourier series of δ. Denote by f the function defined on the unit circle and by T_f the regular distribution associated to f. Since f is discontinuous at $s = \pi$, the jump being equal to 2π, we have

$$T'_f = 1 - 2\pi\delta_\pi,$$

that is,

$$T'_f = 1 - \sum_{k=-\infty}^{\infty} (-1)^k \exp iks.$$

But

$$T'_f = \sum_{k=-\infty}^{\infty} ikc_k(T_f) \exp ikx,$$

hence, for $k \neq 0$,

$$c_k(T_f) = (-1)^k \frac{i}{k}.$$

Since $c_0(T_f) = 0$, the Fourier series of the periodic function $\tilde{f}$ is then given by

$$\tilde{f}(x) = \sum_{k=1}^{\infty} (-1)^{k+1} \frac{2}{k} \sin kx.$$

It converges for $x \neq (2n+1)\pi$, where n is an integer.

The convolution of two distributions S and T on S^1 is always defined. From

$$(S * T)_s(\phi(s)) = S_\xi\Big(T_\eta\big(\phi(\xi + \eta)\big)\Big)$$

it follows that

$$\begin{aligned} c_k(S * T) &= \frac{1}{a}(S * T)_s(\exp -\frac{2i\pi ks}{a}) \\ &= \frac{1}{a} S_\xi(\exp -\frac{2i\pi k\xi}{a}) T_\eta(\exp -\frac{2i\pi k\eta}{a}) \\ &= a c_k(S) c_k(T). \end{aligned}$$

This proves the following.

Theorem 10. *The Fourier coefficients of the convolution of two distributions on the circle S^1 of length a are equal to a times the product of the corresponding Fourier coefficients of the two distributions.*

Example 2. This theorem is a useful tool for solving convolution equations on S^1. Consider, for instance, the equation $A * X = B$ where A and B are given distributions. If a_k and b_k ($k \in \mathbf{Z}$) denote, respectively, the Fourier coefficients of A and B, the Fourier coefficients x_k of X are given by

$$a a_k x_k = b_k.$$

Hence

$$X = \frac{1}{a} \sum_{k=-\infty}^{\infty} \frac{b_k}{a_k} \exp \frac{2i\pi ks}{a}.$$

If, for $k = k_0$, $a_{k_0} = 0$, then, for X to exist, the distribution B must be such that $b_{k_0} = 0$, and in this case x_{k_0} is not defined.

It is also possible to solve convolution equations of a higher degree. For example, the solution of the equation $X * X = A * A$ is determined by

$$(\forall k \in \mathbf{Z}) \qquad x_k^2 = a_k^2,$$

that is

$$(\forall k \in \mathbf{Z}) \qquad x_k = \pm a_k.$$

The set of solutions is uncountable. In the convolution algebra of distributions defined on S^1, d'Alembert's theorem is not valid. This is due to the fact that this algebra contains divisors of zero, i.e., $A * X = 0$, where $A \neq 0$, does not imply $X = 0$.

6. Fourier Transform of Tempered Distributions

In Chapter 2 we defined the Fourier transform $\widehat{f}$ of a function $f \in L^1_m(\mathbf{R})$ by

$$\widehat{f}(t) = \int_{-\infty}^{\infty} f(x) \exp 2i\pi tx \, dx.$$

If φ belongs to $\mathcal{D}$, we have

$$\begin{aligned} \int \widehat{f}(t)\varphi(t)\, dt &= \int\int f(x)\varphi(t) \exp 2i\pi tx \, dx\, dt \\ &= \int f(x)\widehat{\varphi}(x)\, dx, \end{aligned}$$

the application of Fubini's theorem being legitimate since f and φ are integrable, and

$$|f(x)\varphi(t) \exp 2i\pi tx| \leq |f(x)\varphi(t)|.$$

This result suggests the following definition.

Definition 11. The *Fourier transform* $\widehat{T}$ of a distribution T is defined by

$$\widehat{T}(\varphi) = T(\widehat{\varphi}),$$

where $\widehat{\varphi}$ is the Fourier transform of the test function φ.

This definition, however, is meaningless if T is an arbitrary distribution since, except if $\varphi = 0$, $\widehat{\varphi}$ does not belong to $\mathcal{D}$. φ having a bounded support, its Fourier transform $\widehat{\varphi}$ is an entire function[26] and, therefore, can be zero only at isolated points. Arbitrary distributions in $\mathcal{D}'$ have, in general, no Fourier transform in $\mathcal{D}'$. [27] We have, therefore, to define a restricted class of distributions, that is, a wider class of test functions, for which the above definition of the Fourier transform will be acceptable. Such a class of test functions has been defined by L. Schwartz; it is denoted by $\mathcal{S}$. If φ is in $\mathcal{S}$, we want its Fourier transform also to be in $\mathcal{S}$. Since the Fourier transform exchanges differentiation and multiplication, and we want the distributions to be differentiable, $\mathcal{S}$ must be invariant under differentiation and multiplication by polynomials. Therefore, the space $\mathcal{S}$ is the set of all infinitely differentiable functions such that

$$\sup_x |x^m \varphi^{(p)}(x)| < \infty,$$

where m and p are nonnegative integers. In other words, the functions in $\mathcal{S}$ do not necessarily have bounded supports, but are required to go to zero, as $|x| \to \infty$, faster than any inverse power of x, and so are all their derivatives. $\mathcal{S}$ is the space of *rapidly decreasing* test functions.

If $(\varphi)_n$ is a sequence of test functions, it is said to converge to φ in $\mathcal{S}$ if, for all nonnegative integers m and p,

$$\lim_{n\to\infty} \sup_x |x^m (\varphi_n^{(p)}(x) - \varphi^{(p)}(x))| = 0.$$

$\mathcal{D}$ is contained in $\mathcal{S}$, and if $\lim_{n\to\infty} \varphi_n = \varphi$ in $\mathcal{D}$, this is also the case in $\mathcal{S}$.

---

[26] That is, the function $z \mapsto \int \varphi(x) \exp 2i\pi zx \, dx$ is analytic for all finite z.

[27] Note, however, that it is possible to define the Fourier transform of an arbitrary distribution belonging to $\mathcal{D}'$. In general, this Fourier transform is not an ordinary distribution, but an element of a new space (see I. M. Gelfand and G. E. Shilov).

Definition 12. Continuous linear forms on $\mathcal{S}$ are called *tempered distributions.*

The space of tempered distributions is denoted by $\mathcal{S}'$. It is contained in $\mathcal{D}'$.

A real-valued function f defined on $\mathbf{R}$ is said to be of *slow growth at infinity* if there exists a positive integer m and a positive real A such that $|f(x)| \leq A|x|^m$ when $|x|$ goes to infinity. To such a function we can associate a tempered distribution T_f since, for any φ in $\mathcal{S}$, the integral $\int f(x)\varphi(x)\,dx$ is convergent.

The Fourier transform $\widehat{T}$ of a tempered distribution T exists and is a tempered distribution.

Example 1. Let us determine the Fourier transform of $\delta^{(p)}$. We have

$$\begin{aligned}\widehat{\delta^{(p)}}(\varphi) &= \delta^{(p)}(\widehat{\varphi}) \\ &= (-1)^p\delta(\widehat{\varphi}^{(p)}) \\ &= (-1)^p\widehat{\varphi}^{(p)}(0),\end{aligned}$$

that is to say,

$$\widehat{\delta^{(p)}}(\varphi) = (-1)^p \int (2i\pi t)^p \varphi(t)\,dt,$$

which yields

$$\widehat{\delta^{(p)}} = (-2i\pi t)^p.$$

In particular,

$$\widehat{\delta} = 1.$$

If φ is in $\mathcal{S}$, its inverse Fourier transform, denoted by $\check{\varphi}$, is given by[28]

$$\check{\varphi}(x) = \int \varphi(t) \exp -2i\pi tx\,dt.$$

$\check{\varphi}$ is of course in $\mathcal{S}$, and we can define the inverse Fourier transform $\check{T}$ of the tempered distribution T by

$$\check{T}(\varphi) = T(\check{\varphi}).$$

Theorem 11. *The inverse Fourier transform of the Fourier transform $\widehat{T}$ of a tempered distribution T is equal to T.*

---

[28] Note that $\mathcal{S}$ is contained in $L^1_m(\mathbf{R})$, and therefore Theorem 9 of Chapter 2 applies.

Set $\widehat{T} = U$ and $\widehat{\varphi} = \psi$. We have to prove that $\check{U} = T$. Taking into account that $\check{\psi} = \varphi$, we have indeed

$$\check{U}(\psi) = U(\check{\psi}) = \widehat{T}(\varphi) = T(\widehat{\varphi}) = T(\psi).$$

Corollary. *If $\widehat{T} = 0$, then $T = 0$.*

Set $\widehat{T} = U$, then $\check{U} = T = 0$.

Example 2. From $\widehat{\delta} = 1$ it follows that $\check{1} = \delta$. This result is frequently written

$$\int \exp -2i\pi tx \, dt = \delta(x),$$

which is incorrect.

Remark. The relation

$$\check{\varphi}(t) = \widehat{\varphi}(-t)$$

implies

$$\check{T}_t(\varphi(t)) = \widehat{T}_{-t}(\varphi(t)).$$

Thus

$$\widehat{1} = \delta,$$

and more generally, for any nonnegative integer p,

$$\widehat{x^p} = \frac{\delta^{(p)}}{(2i\pi)^p}.$$

Let (T_n) be a sequence of tempered distributions that tends to T in $\mathcal{S}'$; the sequence $(\widehat{T}_n)$ of Fourier transforms tends to the Fourier transform $\widehat{T}$ of the limit, since

$$\begin{aligned} \lim_{n\to\infty} \widehat{T}_n(\varphi) &= \lim_{n\to\infty} T_n(\widehat{\varphi}) \\ &= T(\widehat{\varphi}) \\ &= \widehat{T}(\varphi). \end{aligned}$$

In the same way, if the series of tempered distributions $\sum_{n=1}^{\infty} T_n$ converges to T, then the series of the corresponding Fourier transforms $\sum_{n=1}^{\infty} \widehat{T}_n$ converges to $\widehat{T}$.

Example 3. The Fourier transform of the characteristic function of the interval $[-n, n]$ is easy to derive. We have

$$\begin{aligned} \widehat{1}_{[-n,n]}(t) &= \int_{-n}^{n} \exp 2i\pi tx \, dx \\ &= \frac{\sin 2\pi nt}{\pi t}. \end{aligned}$$

Since $1_{[-n,n]}$ tends to 1 when $n \to \infty$, it follows that

$$\lim_{n\to\infty} \frac{\sin 2\pi nt}{\pi t} = \delta,$$

a result that we have already obtained (see Example 3.1).

Example 4. From the relations

$$\widehat{\delta}_k = \exp 2i\pi kt$$

and

$$\sum_{n=-\infty}^{\infty} \delta_n = \sum_{k=-\infty}^{\infty} \exp 2i\pi kt,$$

it follows that

$$\sum_{n=-\infty}^{\infty} \delta_n = \sum_{k=-\infty}^{\infty} \widehat{\delta}_k.$$

Thus, for any $\varphi \in \mathcal{S}$,

$$\sum_{n=-\infty}^{\infty} \delta_n(\varphi) = \sum_{k=-\infty}^{\infty} \delta_k(\widehat{\varphi}),$$

that is to say,

$$\sum_{n=-\infty}^{\infty} \varphi(n) = \sum_{k=-\infty}^{\infty} \widehat{\varphi}(k).$$

This relation, called *Poisson summation formula*, has various interesting applications.

The function

$$g_\sigma : x \mapsto \frac{1}{\sigma\sqrt{2\pi}} \exp -\frac{x^2}{2\sigma^2}$$

belongs to $\mathcal{S}$, and its Fourier transform $\widehat{g}_\sigma$ is given by (see Chapter 2, Example 4.2)

$$\widehat{g}_\sigma(t) = \exp -2\pi^2\sigma^2 t^2.$$

If σ is small, the series

$$\sum_{n=-\infty}^{\infty} \exp -2\pi^2\sigma^2 n^2$$

converges very slowly. Using thePoisson formula, its sum is equal to

$$\sum_{n=-\infty}^{\infty} \frac{1}{\sigma\sqrt{2\pi}} \exp -\frac{n^2}{2\sigma^2},$$

which converges very rapidly.

The proof of the Poisson formula does not require φ to belong to $\mathcal{S}$. The formula is valid when both series are convergent.

The Fourier transform of

$$l_a : x \mapsto \frac{a}{\pi(x^2 + a^2)}$$

is (see Chapter 2, Example 4.1)

$$\widehat{l}_a : t \mapsto \exp -2\pi a|t|.$$

l_a is not in $\mathcal{S}$, but the two series

$$\sum_{n=-\infty}^{\infty} l_a(n) \qquad \text{and} \qquad \sum_{n=-\infty}^{\infty} \widehat{l}_a(n)$$

being convergent, the Poisson formula can be used to find the sum of the series

$$\sum_{n=-\infty}^{\infty} \frac{a}{\pi(n^2 + a^2)}.$$

Since

$$\begin{aligned} \sum_{n=-\infty}^{\infty} \exp -2\pi a|n| &= 2\sum_{n=0}^{\infty} \exp -2\pi an - 1 \\ &= \coth \pi a, \end{aligned}$$

we find

$$\sum_{n=-\infty}^{\infty} \frac{a}{n^2 + a^2} = \frac{\pi}{2a}\Big(\coth \pi a - \frac{1}{\pi a}\Big).$$

Theorem 12. *The Fourier transform of a distribution T whose support is bounded is an analytic function.*

T being tempered, we have

$$\begin{aligned} \widehat{T}(\varphi) &= T(\widehat{\varphi}) \\ &= T_t\Big(\int_{-\infty}^{\infty} \varphi(x) \exp 2i\pi tx \, dx\Big) \\ &= (T_t \otimes \varphi_x)(\exp 2i\pi tx). \end{aligned}$$

T being linear and continuous, the function $x \mapsto T_t(\exp 2i\pi tx)$ is infinitely differentiable.[29] Denoting by f this function, we have

$$\widehat{T}(\varphi) = \int_{-\infty}^{\infty} f(x)\varphi(x)\,dx,$$

which shows that $\widehat{T}$ is a regular tempered distribution associated to the analytic function f.[30]

Note that a regular distribution does not necessarily have a regular Fourier transform. The Fourier transform of a regular distribution T_f is defined if the associated function f does not grow too fast; this Fourier transform $\widehat{T}_f$ is regular if f is integrable, i.e., if f is sufficiently small at infinity.

The essential property of the Fourier transform is to convert the convolution of two distributions into the product of their Fourier transforms. This is valid, however, only under certain restrictive conditions.

Theorem 13. *Let S be a tempered distribution and let T be a distribution with a bounded support; then*

$$\widehat{S * T} = \widehat{S}\widehat{T}.$$

T having a bounded support, the convolution $S * T$ is defined. Moreover, since $\widehat{S}$ is tempered and $\widehat{T}$ is a regular distribution associated to an analytic function of slow growth (Theorem 12), the product $\widehat{S}\widehat{T}$ makes sense.

We have

$$\begin{aligned}(\widehat{S * T})(\varphi) &= (S * T)(\widehat{\varphi}) \\ &= S_\xi\Big(T_\eta\Big(\int \varphi(x) \exp 2i\pi(\xi + \eta)x)\,dx\Big) \\ &= S_\xi\Big(\int \widehat{T}(x)\varphi(x) \exp 2i\pi\xi x\,dx\Big),\end{aligned}$$

[29] The function $t \mapsto \exp 2i\pi tx$ is infinitely differentiable but does not belong to $\mathcal{S}$; however, since T has a bounded support, the function $x \mapsto T_t(\exp 2i\pi tx)$ is defined.

[30] This means that f is of slow growth.

since $T_\eta(\exp 2i\pi\eta x) = \widehat{T}(x)$. Thus,

$$\begin{aligned}(\widehat{S * T})(\varphi) &= S(\widehat{\widehat{T}\varphi}) \\ &= \widehat{S}(\widehat{T}\varphi) \\ &= (\widehat{S}\widehat{T})(\varphi).\end{aligned}$$

Example 5. The regular distribution θ, associated to the Heaviside function, is tempered. Its Fourier transform may be determined using the preceding theorem. Since $\widehat{\delta'} = -2i\pi t$ and $\widehat{\delta} = 1$, the relation

$$\delta' * \theta = \delta$$

implies that

$$-2i\pi t\widehat{\theta} = 1.$$

Hence,

$$\widehat{\theta} = -\frac{1}{2i\pi}\mathrm{Pv}\,\frac{1}{t} + C\delta,$$

where C is a constant.[31] To find the constant C, note that $\mathrm{Pv}\, t^{-1}$ being odd, its inverse Fourier transform, which is proportional to $\theta - C$, must also be odd; therefore, C must be equal to $1/2$. Finally,

$$\widehat{\theta} = -\frac{1}{2i\pi}\mathrm{Pv}\,\frac{1}{t} + \frac{1}{2}\delta,$$

or

$$\widehat{\theta} = -\frac{1}{2i\pi}\frac{1}{t + i0}.$$

From the relation $\check{T}_t = \widehat{T}_{-t}$ (see the remark following Example 2), after having permuted t and x, it follows that

$$\begin{aligned}\widehat{\mathrm{Pv}\,\frac{1}{x}} &= i\pi(2\theta - 1) \\ \widehat{\frac{1}{x - i0}} &= 2i\pi\theta \\ \widehat{\frac{1}{x + i0}} &= 2i\pi(\theta - 1).\end{aligned}$$

---

[31] Taking into account the definition of $\mathrm{Pv}\, x^{-1}$, it is easy to verify that $x\mathrm{Pv}\, x^{-1} = 1$ (see Problems 2 and 3).

$2\theta(t) - 1$, denoted $\operatorname{sgn} t$, and read *sign of* t, is equal to -1 if $t < 0$, and to 1 if $t > 0$. $\theta(t) - 1 = -\theta(-t)$.

Since θ is the limit, in $\mathcal{S}'$, of the regular distribution associated to the function $x \mapsto \theta(x)\exp(-\epsilon x)$ when $\epsilon > 0$ tends to zero, we may also determine the Fourier transform of θ as follows

$$\begin{aligned}\widehat{\theta}(t) &= \lim_{\epsilon \to 0+} \int_0^\infty e^{-\epsilon x} e^{2i\pi t x}\, dx \\ &= \lim_{\epsilon \to 0+} \frac{1}{\epsilon - 2i\pi t} \\ &= \frac{1}{2i\pi} \frac{1}{t + i0}.\end{aligned}$$

Theorem 14. *Let T be a tempered distribution, and let φ be a test function in $\mathcal{S}$; then $T * \varphi$ is a regular tempered distribution associated to an infinitely differentiable function, and we have*

$$\widehat{T * \varphi} = \widehat{T}\widehat{\varphi}.$$

To prove that $T * \varphi$ is tempered, we have to show that, for any $\psi \in \mathcal{S}$, $(T * \varphi)(\psi)$ is well defined. By definition,

$$\begin{aligned}(T * \varphi)(\psi) &= (T_\eta \otimes \varphi_\xi)(\psi(\xi + \eta)) \\ &= T_\eta\Big(\varphi_\xi(\psi(\xi + \eta))\Big),\end{aligned}$$

where

$$\varphi_\xi(\psi(\xi + \eta)) = \int \varphi(\xi)\psi(\xi + \eta)\, d\xi.$$

Differentiating under the integral sign (which is legitimate), and from the fact that φ and ψ are both in $\mathcal{S}$, one verifies that $\eta \mapsto \int \varphi(\xi)\psi(\xi + \eta)\, d\xi$ belongs to $\mathcal{S}$. T being tempered, $T_\eta\Big((\varphi_\xi)(\psi(\xi + \eta))\Big)$ is well defined, and, therefore, the convolution $T * \varphi$ makes sense. From Theorem 5 it follows that $T * \varphi$ is an infinitely differentiable function. Moreover,

$$\begin{aligned}
(\widehat{T * \varphi})(\psi) &= (T * \varphi)(\widehat{\psi}) \\
&= T_\eta\Big(\varphi_\xi\Big(\int \psi(x) e^{2i\pi(\xi+\eta)x}\,dx\Big)\Big) \\
&= T_\eta\Big(\int\int \varphi(\xi)\psi(x) e^{2i\pi(\xi+\eta)x}\,dx\,d\xi\Big) \\
&= T_\eta\Big(\int \widehat{\varphi}(x)\psi(x) e^{2i\pi\eta x}\,dx\Big) \\
&= T(\widehat{\widehat{\varphi}\psi}) \\
&= (\widehat{T}\widehat{\varphi})(\psi).
\end{aligned}$$

Corollary. *Let T be a tempered distribution, and let φ be a test function in $\mathcal{S}$; then*

$$\widehat{T\varphi} = \widehat{T} * \widehat{\varphi}.$$

This is a straightforward consequence of the inverse Fourier transform theorem and the preceding result.

Remark. It is easy to convince oneself that the formula $\widehat{S * T} = \widehat{S}\widehat{T}$ cannot be valid in general. Even when $S * T$ makes sense, and when its Fourier transform $\widehat{S * T}$ exists, there is no reason whatsoever for the product $\widehat{S}\widehat{T}$ to make sense. For instance, although the convolution $\theta * \theta$ is perfectly defined (it is equal to $x\theta = x_+$), and its Fourier transform exists (Example 7 shows that it is equal to $1/(2i\pi)^2(t + i0)^2$), the product of the Fourier transforms is meaningless.

Example 6. Consider a damped oscillator of unit mass. If ω_0 denotes the angular frequency, γ the damping coefficient, and $F(t)$ $(t > 0)$ the time-depending force, the displacement X is the solution to the differential equation

$$X'' + 2\gamma X' + \omega_0^2 X = F, \quad \text{where for } t < 0 \; X(t) = 0.$$

To solve this equation, we can use the results of Example 4.4 to obtain the inverse in $\mathcal{D}'_+$ of the differential operator

$$\delta'' + 2\gamma + \omega_0^2\delta = \big(\delta' + (\gamma + i\omega_1)\delta\big) * \big(\delta' + (\gamma - i\omega_1)\delta\big),$$

where $\omega_1 = \sqrt{\omega_0^2 - \gamma^2}$ (assuming $\gamma < \omega_0$). We find[32]

$$X(t) = \int_0^t \frac{1}{\omega_1} \sin \omega_1 (t - \tau) e^{-\gamma(t-\tau)} F(\tau)\, d\tau,$$

that is,

$$X = A * F,$$

with

$$A(t) = \frac{1}{\omega_1} \sin \omega_1 t e^{-\gamma t} \theta(t).$$

Note that the displacement X at time t is determined by the values of the force F at times $\tau < t$.[33] This result expresses *causality*. The *effect*, which is the displacement X, *never precedes the cause*, which is the force F.

More generally, for any linear system, we have seen that we can define a distribution A, called the admittance of the system, characterizing its response to an excitation (see Example 4.5). If the system is *causal*, then $A = A\theta$, that is, $A \in \mathcal{D}'_+$.[34]

If A is such that the formula $\widehat{A\theta} = \widehat{A} * \widehat{\theta}$ is valid, then

$$\widehat{A} = \widehat{A} * \Big(-\frac{1}{2i\pi} \operatorname{Pv} \frac{1}{x} + \frac{1}{2}\delta \Big),$$

that is,

$$\widehat{A} = -\frac{1}{i\pi} \widehat{A} * \operatorname{Pv} \frac{1}{x},$$

and, finally,

$$\Re \widehat{A} = -\frac{1}{\pi} \operatorname{Pv} \frac{1}{x} * \Im \widehat{A}$$
$$\Im \widehat{A} = \frac{1}{\pi} \operatorname{Pv} \frac{1}{x} * \Re \widehat{A}.$$

---

32 Note that

$$e^{-\lambda_1 t}\theta * e^{-\lambda_2 t}\theta = \frac{e^{-\lambda_1 t} - e^{-\lambda_2 t}}{\lambda_1 - \lambda_2} \theta.$$

33 This is due to the term θ in the expression of A, which implies that $A(t) = 0$ if $t < 0$.

34 The formula $A = A\theta$ makes sense only if the product of the two distributions A and θ is defined. This is, for instance, the case if A is an infinitely differentiable function.

These relations, which express causality, play a very important role in many areas of physics. The are called *Kramers–Kronig relations.*

Remark 1. The *Hilbert transform* of a distribution T is the distribution $H(T)$ defined, when it exists, by

$$H(T) = -\frac{1}{\pi}\mathrm{Pv}\,\frac{1}{x} * T.$$

Since we have

$$\widehat{\mathrm{Pv}\,\frac{1}{x}} = i\pi\,\mathrm{sgn}\,t,$$

and

$$\mathrm{sgn}\,t\ \mathrm{sgn}\,t = 1,$$

the inverse Hilbert transform is given by

$$H^{-1}(T) = \frac{1}{\pi}\mathrm{Pv}\,\frac{1}{x} * T.$$

Therefore,

$$\begin{aligned}\Re\widehat{A} &= H(\Im\widehat{A})\\ \Im\widehat{A} &= H^{-1}(\Re\widehat{A}).\end{aligned}$$

Remark 2. If the admittance A belongs to $L^1_m(\mathbf{R})$, which is the case in our example, its Fourier transform $\widehat{A}$ is a function, and the Kramers–Kronig relations read

$$\begin{aligned}\Re\widehat{A}(\omega) &= -\frac{1}{\pi}\mathrm{Pv}\int_{-\infty}^{\infty}\frac{\Im\widehat{A}(u)}{\omega - u}\,du\\ \Im\widehat{A}(\omega) &= \frac{1}{\pi}\mathrm{Pv}\int_{\infty}^{\infty}\frac{\Re\widehat{A}(u)}{\omega - u}\,du.\end{aligned}$$

These results can also be derived from Cauchy's formula. If $A(t) = 0$ when $t < 0$, then the function

$$z \mapsto \widehat{A}(z) = \int_0^{\infty} A(t)\mathrm{e}^{izt}\,dt$$

is analytic in the upper half-plane $\{z|\Im z \geq 0\}$, and, from Cauchy's formula, it follows that

$$\widehat{A}(z) = \frac{1}{2i\pi}\int_{-R}^{R}\frac{\widehat{A}(u)}{u - z}\,du + \frac{1}{2i\pi}\int_{\Gamma_1}\frac{\widehat{A}(\zeta)}{\zeta - z}\,d\zeta,$$

where Γ_1 is the path $\theta \mapsto R\exp(i\pi\theta)$ $(\theta \in [0,1])$. When R tends to ∞, according to Jordan's lemma,[35] the second integral tends to zero. Hence,

$$\widehat{A}(z) = \frac{1}{2i\pi} \int_{-\infty}^{\infty} \frac{\widehat{A}(u)}{u - z}\, du.$$

Let $z = \omega + i\epsilon$; if $\epsilon > 0$ tends to 0, we have

$$\widehat{A}(\omega) = \lim_{\epsilon \to 0+} \frac{1}{2i\pi}\Big(\int_{-\infty}^{\omega-\epsilon} \frac{\widehat{A}(u)}{u - \omega}\, du + \int_{\omega+\epsilon} \frac{\widehat{A}(u)}{u - \omega}\, du + \int_{\Gamma_2} \frac{\widehat{A}(\zeta)}{\zeta - \omega}\, d\zeta \Big),$$

where Γ_2 is the path $\theta \mapsto \epsilon \exp i\pi(\theta + 1)$ $(\theta \in [0,1])$. When $\epsilon \to 0$, the last integral tends to $i\pi\widehat{A}(\omega)$, and finally,

$$\widehat{A}(\omega) = \frac{1}{i\pi} \mathrm{Pv} \int_{-\infty}^{\infty} \frac{\widehat{A}(u)}{u - \omega}\, du.$$

Identifying the real and imaginary parts yields the Kramers–Kronig relations.

Theorem 15. *Let T be a tempered distribution; then*

(1) $$\widehat{T^{(n)}} = (-2i\pi t)^n \widehat{T},$$

and

(2) $$\widehat{T}^{(n)} = \widehat{(2i\pi x)^n T}.$$

These formulas are simple consequences of Theorem 13. Indeed.

$$\begin{aligned}\widehat{T^{(n)}} &= \widehat{\delta^{(n)} * T}\\ &= \widehat{\delta^{(n)}}\widehat{T}\\ &= (-2i\pi t)^n \widehat{T}.\end{aligned}$$

[35] If $z = u + iv$, then $\lim_{v\to\infty} |\widehat{A}(u + iv)| = 0$.

The proof of the second formula is similar. We can also establish these formulas using the fact that the test functions do satisfy these relations. For instance,

$$\begin{aligned}\big(\widehat{(2i\pi x)^n T}\big)(\varphi) &= \big((2i\pi x)^n T\big)(\widehat{\varphi})\\ &= T\big((2i\pi x)^n \widehat{\varphi}\big)\\ &= (-1)^n T(\widehat{\varphi^{(n)}})\\ &= (-1)^n \widehat{T}(\varphi^{(n)})\\ &= \widehat{T}^{(n)}(\varphi).\end{aligned}$$

Example 7. We shall use these formulas to find new Fourier transforms. Since

$$\widehat{\frac{1}{x-i0}} = 2i\pi\theta,$$

from (1), it follows

$$\widehat{\Big(\frac{1}{x-i0}\Big)^{(n-1)}} = (-2i\pi t)^{n-1} 2i\pi\theta,$$

that is,

$$\widehat{\frac{1}{(x-i0)^n}} = \frac{(2i\pi)^n}{(n-1)!} t^{n-1}\theta.$$

Similarly,

$$\widehat{\frac{1}{(x+i0)^n}} = \frac{(2i\pi)^n}{(n-1)!} t^{n-1}(\theta - 1).$$

Moreover, the distributions t_+^{n-1} and t_-^{n-1} being, respectively, equal to $t^{n-1}\theta$ and $|t|^{n-1}(1-\theta)$ (Example 2.5), we may write

$$\widehat{\frac{1}{(x \pm i0)^n}} = \frac{(\mp 2i\pi)^n}{(n-1)!} t_{\mp}^{n-1}.$$

The Fourier transforms of x_+^n and x_-^n are easily derived applying formula (2) to the distribution θ. However, it is faster to use the formula $\check{T}_t = \widehat{T}_{-t}$, which, taking into account the preceding result, yields

$$\widehat{x_{\mp}^n} = \frac{n!}{(\mp 2i\pi)^{n+1}} \frac{1}{(t \pm i0)^{n+1}}.$$

7. Laplace Transform

Given a function f whose support is in $\mathbf{R}_+$, the Laplace transform of f is the function (Chapter 2, Section 5)

$$L(f)\colon z \mapsto \int_0^\infty f(x)\mathrm{e}^{-zx}\,dx.$$

This definition can be extended to distributions belonging to $\mathcal{D}'_+$.

Definition 13. Let T be a distribution in $\mathcal{D}'_+$; the *Laplace transform* of T is the function

$$L(T)\colon z \mapsto T_x(\mathrm{e}^{-zx}).$$

As in the case of a function, a distribution has to satisfy certain conditions for its Laplace transform to exist.

Since the function $x \mapsto \mathrm{e}^{-zx}$ does not belong to $\mathcal{D}$, $T_x(\mathrm{e}^{-zx})$ is, in general, meaningless. However, if σ denotes an infinitely differentiable function whose support is bounded from below, and it is such that $\sigma(x) = 1$ for all $x > 0$, then, if $\Re z > 0$, the function $x \mapsto \sigma(x)\mathrm{e}^{-zx}$ belongs to $\mathcal{S}$, and, in this case, $T_x(\sigma(x)\mathrm{e}^{-zx}) = T_x(\mathrm{e}^{-zx})$ makes sense if T is tempered. More generally, let ξ_0 denote a real; then, if $\Re z > \xi_0$, the function $x \mapsto \sigma(x)\mathrm{e}^{-(z-\xi_0)x}$ belongs to $\mathcal{S}$, and the Laplace transform of T exists if $\mathrm{e}^{-\xi_0 x}T$ is tempered, for

$$T_x(\mathrm{e}^{-zx}) = (\mathrm{e}^{-\xi_0 x}T)(\mathrm{e}^{-(z-\xi_0)x}).$$

Note that if $\mathrm{e}^{-\xi_0 x}T$ is tempered, then $\mathrm{e}^{-\xi_0 x}T$ is also tempered for all real $\xi > \xi_0$.

Theorem 16. *Let T be a distribution in $\mathcal{D}'_+$; if there exists a real ξ_0 such that the distribution $\mathrm{e}^{-\xi_0 x}T$ is tempered, then the Laplace transform of T is an analytic function in the half-plane $\{z \mid \Re z > \xi_0\}$.*

To establish this result, it is sufficient to show that the function $z \mapsto T_x(\mathrm{e}^{-zx})$ is differentiable. Note that, from what has been said above, the function does exist for $\Re z > \xi_0$.

If h is such that $\Re(z+h) > \xi_0$, the derivative at z of $L(T)$ is given by

$$\lim_{h\to 0} \frac{1}{h}\Big(T_x\big(\mathrm{e}^{-(z+h)x}\big) - T_x\big(\mathrm{e}^{-zx}\big)\Big) = \lim_{h\to 0} T_x\left(\frac{\mathrm{e}^{-(z+h)x} - \mathrm{e}^{-zx}}{h}\right).$$

Since, in $\mathcal{S}$,

$$\lim_{h\to 0} \frac{\mathrm{e}^{-(z+h)x} - \mathrm{e}^{-zx}}{h} = -x\mathrm{e}^{-zx},$$

and T is continuous, the function $z \mapsto T_x(\mathrm{e}^{-zx})$ is differentiable, and its derivative is $T_x(-x\mathrm{e}^{-zx})$.

Remark 1. If $(\mathrm{e}^{-\xi_0 x}T_n)$ is a sequence of tempered distributions in $\mathcal{D}'_+$ whose limit is the tempered distribution $\mathrm{e}^{-\xi_0 x}T \in \mathcal{D}'_+$, then, for any $\varphi \in \mathcal{S}$, the sequence $\big((\mathrm{e}^{-\xi_0 x}T_n)(\varphi)\big)$ tends to $(\mathrm{e}^{-\xi_0 x}T)(\varphi)$. Therefore, considering the function $x \mapsto \sigma(x)\mathrm{e}^{-(z-\xi_0)x}$, which belongs to $\mathcal{S}$, it is clear that the sequence $\big(L(T_n)\big)$ of Laplace transforms converges to $L(T)$ if, for $\Re z > \xi_0$, all Laplace transforms are defined.

Remark 2. The Laplace transform is closely related to the Fourier transform. As in the case of functions, we shall show that, if $z = \xi + i\eta$, we have

$$T_x\big(\mathrm{e}^{-(\xi+i\eta)x}\big) = \big(F^{-1}(\mathrm{e}^{-\xi x}T)\big)\big(\frac{\eta}{2\pi}\big),$$

where $F^{-1}(\mathrm{e}^{-\xi x}T)$ denotes the inverse Fourier transform of $\mathrm{e}^{-\xi x}T$. The right-hand side is the value at $t = \eta/2\pi$ of $F^{-1}(\mathrm{e}^{-\xi x}T)$.

Let

$$f(x) = \sigma(x)\mathrm{e}^{-(\xi_0-\xi)x} \quad \text{and} \quad S = \mathrm{e}^{-\xi_x}T,$$

where, as above, σ is an infinitely differentiable function whose support is bounded from below, and such that $\sigma(x) = 1$ for all $x > 0$; then

$$\mathrm{e}^{-\xi x}T = fS.$$

Since f is in $\mathcal{S}$, and S in $\mathcal{S}'$, we can apply the corollary of Theorem 14, that is,

$$F^{-1}(\mathrm{e}^{-\xi x}T) = \check{f} * \check{S},$$

and, from theorem 5, it follows that

$$\check{f} * \check{S} = \check{S}_\tau\big(\check{f}(t-\tau)\big).$$

But

$$\begin{aligned}\check{f}(t-\tau) &= \int f(x)\mathrm{e}^{-2i\pi(t-\tau)x}\,dx \\ &= \int f(x)\mathrm{e}^{-2i\pi tx}\mathrm{e}^{2i\pi\tau x}\,dx,\end{aligned}$$

that is, $\check{f}(t-\tau)$ is the Fourier transform of the function $x \mapsto f(x)\mathrm{e}^{-2i\pi tx}$ Hence,

$$\check{S}_\tau\big(\check{f}(t-\tau)\big) = S_x\big(f(x)\mathrm{e}^{-2i\pi tx}\big),$$

and, replacing t by $\eta/2\pi$, and S and f by their expressions, we finally obtain

$$\begin{aligned}\left(F^{-1}(\mathrm{e}^{-\xi x}T)\right)\left(\frac{\eta}{2\pi}\right) &= (\mathrm{e}^{-\xi_0 x}T_x)\left(\sigma(x)\mathrm{e}^{-(\xi_0-\xi)x}\mathrm{e}^{-i\eta x}\right)\\ &= T_x\left(\mathrm{e}^{-(\xi+i\eta)x}\right),\end{aligned}$$

where we have taken into account that, on the support of T, $\sigma(x) = 1$.

Example 1. Let us find the Laplace transform of $\delta^{(p)}$. We have

$$\begin{aligned}\delta_x^{(p)}(\mathrm{e}^{-zx}) &= (-1)^p\delta_x\left((\mathrm{e}^{-zx})^{(p)}\right)\\ &= (-1)^p\delta_x\left((-z)^p\mathrm{e}^{-zx}\right)\\ &= z^p.\end{aligned}$$

In particular,

$$L(\delta) = 1.$$

Theorem 17. *Let S and T be two distributions belonging to $\mathcal{D}'_+$ that have a Laplace transform for, respectively, $\Re z > \xi_1$ and $\Re z > \xi_2$; then*

$$L(S * T) = L(S)\,L(T),$$

for $\Re z > \max(\xi_1, \xi_2)$.

Note that, since S and T both belong to $\mathcal{D}'_+$, the convolution $S * T$ exists and belongs to $\mathcal{D}'_+$.

If $\Re z > \max(\xi_1, \xi_2)$ both Laplace transforms exist, and we have

$$\begin{aligned}(S * T)_x(\mathrm{e}^{-zx}) &= (S_u \otimes T_v)\left(\mathrm{e}^{-z(u+v)}\right)\\ &= S_u(\mathrm{e}^{-zu})\,T_v(\mathrm{e}^{-zv}).\end{aligned}$$

Remark 1. Although the formula $\widehat{\theta * \theta} = \hat{\theta}\hat{\theta}$ is wrong, the right-hand side being meaningless, the formula $L(\theta * \theta) = L(\theta)L(\theta)$ is perfectly correct. If $L(S)$ and $L(T)$ exist, then $L(S * T)$ exists, and we have $L(S * T) = L(S)L(T)$.

Remark 2. If T is a distribution whose bounded support is contained in the interval $[0, a]$, then, for any real ξ, the support of $\mathrm{e}^{-\xi x}T$ is also bounded, and $L(T)$ is an entire function, i.e., analytic for all finite z.

Example 2. Since the series $\sum_{n=0}^{\infty} \delta_{na}$ converges in $\mathcal{D}'_+$,[36] the convolution

$$\tilde{T} = \left(\sum_{n=0}^{\infty} \delta_{na}\right) * T$$

[36] For any $\varphi \in \mathcal{D}$, the series $\sum_{n=0}^{\infty} \delta_{na}(\varphi) = \sum_{n=0}^{\infty} \varphi(na)$ comprises only a finite number of terms, the support of φ being bounded.

is defined. Let us determine $L(\tilde{T})$. The Laplace trasform of δ_{na} being the function $z \mapsto \mathrm{e}^{-naz}$, the Laplace transform of $\sum_{n=0}^{\infty} \delta_{na}$ is the function

$$z \mapsto \sum_{n=0}^{\infty} \mathrm{e}^{-naz} = \frac{1}{1-\mathrm{e}^{-az}} \qquad (\Re z > 0),$$

and, therefore,

$$L(\tilde{T}) = \frac{1}{1-\mathrm{e}^{-az}} L(T).$$

Theorem 18. *Let T be a distribution in $\mathcal{D}'_+$ that has a Laplace transform $L(T)$; then*

$$L(T^{(n)}) = z^n L(T), \tag{1}$$

$$\big(L(T)\big)^{(n)} = L\big((-x)^n T\big). \tag{2}$$

From $T^{(n)} = \delta^{(n)} * T$ it readily follows that

$$\begin{aligned} L(T^{(n)}) &= L(\delta^{(n)}) L(T) \\ &= z^n L(T), \end{aligned}$$

which proves formula (1). To establish formula (2), we may simply repeat the proof of Theorem 16.

Example 3. Let us find the Laplace transform of the regular distribution associated to the function $x \mapsto \log x$. It is a tempered distribution belonging to $\mathcal{D}'_+$, and its Laplace transform is, therefore, an analytic function for $\Re z > 0$.

To calculate the integral $\int_0^\infty \mathrm{e}^{-zx} \log x \, dx$, let us first assume that z is real and positive. Let $zx = u$; then

$$\begin{aligned} \int_0^\infty \mathrm{e}^{-zx} \log x \, dx &= \frac{1}{z} \int_0^\infty \mathrm{e}^{-u} \log \frac{u}{z} \, du \\ &= \frac{1}{z} \int_0^\infty \mathrm{e}^{-u} \log u \, du - \frac{\log z}{z} \int_0^\infty \mathrm{e}^{-u} \, du \\ &= -\frac{\gamma + \log z}{z}, \end{aligned}$$

where γ is the *Euler constant.*[37]

---

[37] To calculate the integral $\int_0^\infty \mathrm{e}^{-u} \log u \, du$, see Chapter 1, Problem 12.

The Laplace transform of $x \mapsto \log x$ being analytic for $\Re z > 0$, by analytic continuation, the result is correct for all z in this half-plane. Here the function $z \mapsto \log z$ is such that $\log z$ is real if z is real and positive.

Knowing the Laplace transform of the distribution $\log x$, we can use Theorem 18(1) and the results of Example 2.5 to derive the Laplace transform of $\operatorname{Pf} x_+^{-n}$, where n is a positive integer. We shall proceed step by step. Let us first consider the derivative of $\log x$. We have

$$\begin{aligned}(\log x)'(\varphi) &= -(\log x)(\varphi') \\ &= -\int_0^\infty \log x \varphi'(x)\,dx \\ &= -\lim_{\epsilon \to)+} \int_\epsilon^\infty \log x \varphi'(x)\,dx,\end{aligned}$$

where the last result follows from the dominated convergence theorem of Lebesgue.[38] An integration by parts yields

$$\begin{aligned}(\log x)'(\varphi) &= \lim_{\epsilon \to)+} \int_\epsilon^\infty \frac{\varphi(x)}{x}\,dx + \log \epsilon \varphi(0) \\ &= \lim_{\epsilon \to)+} \int_\epsilon^1 \frac{\varphi(x) - \varphi(0)}{x}\,dx + \int_1^\infty \frac{\varphi(x)}{x}\,dx \\ &= \int_0^\infty \frac{\varphi(x) - \varphi(0)\theta(1-x)}{x}\,dx \\ &= (\operatorname{Pf} x_+^{-1})(\varphi).\end{aligned}$$

Note that in the first line we have replaced

$$\lim_{\epsilon \to 0+} \log \epsilon \varphi(\epsilon) \quad \text{by} \quad \lim_{\epsilon \to 0+} \log \epsilon \varphi(0)$$

since

$$\lim_{\epsilon \to 0+} \log \epsilon \big(\varphi(\epsilon) - \varphi(0)\big) = 0.$$

Hence,

$$(\log x)' = \operatorname{Pf} x_+^{-1},$$

and, from Theorem 18(1), it follows that

$$L(\operatorname{Pf} x_+^{-1}) = -\gamma - \log z.$$

---

38 $1_{[\epsilon,\infty[}(x)|\log x \varphi'(x)| \leq |\log x \varphi'(x)| \in L^1_m(\mathbf{R}_+)$

To determine $L(\mathrm{Pf}\, x_+^{-n})$, we can use the formula

$$\mathrm{Pf}\, x_+^{-n} = \frac{(-1)^{n-1}}{(n-1)!}(\mathrm{Pf}\, x_+^{-1})^{(n-1)} + \frac{(-1)^{n-1}}{(n-1)!}\delta^{(n-1)}\left(1 + \frac{1}{2} + \cdots + \frac{1}{n-1}\right),$$

where the integer n is greater than 1. This formula is easily established by induction, using (see Example 2.5)

$$(\mathrm{Pf}\, x_+^{-k})' = -k\mathrm{Pf}\, x_+^{-k-1} + \frac{(-1)^k}{k!}\delta^{(k)}.$$

Finally,

$$L(\mathrm{Pf}\, x_+^{-n}) = (-1)^{n-1}\frac{z^{n-1}}{(n-1)!}\left(-\log z - \gamma + 1 + \frac{1}{2} + \cdots + \frac{1}{n-1}\right).$$

In Section 5 of Chapter 2 we have seen that if a function F, analytic in the half-plane $\{z|\Re z > \xi_0\}$, is, as a function of $\eta = \Im z$, integrable on $\mathbf{R}$, it is the Laplace transform of a continuous function f, defined by

$$f(x) = \frac{1}{2i\pi}\int_{\xi_0 - i\infty}^{\xi_0 + i\infty} F(z)\mathrm{e}^{zx}\, dz.$$

Moreover, $f(x) = 0$ for all $x < 0$.

In the case of distributions, we have a similar result valid, however, under weaker conditions.

Theorem 19. *Let F be an analytic function in the half-plane $\{z|\Re z > \xi_0\}$, and such that $|F(z)|$ is, in this half-plane, bounded by a polynomial in $|z|$; then there exists a distribution $T \in \mathcal{D}'_+$ whose Laplace transform is F.*

If n is the degree of the polynomial, the function $z \mapsto F(z)/z^{n+2}$, analytic in the half-plane $\{z|\Re z > \max(\xi_0, 0)\}$, is the Laplace transform of a continuous function f such that $\mathrm{supp}\, f \subset \mathbf{R}_+$ since $\eta \mapsto F(\xi + i\eta)/)(\xi + i\eta)^{n+2}$ is integrable on $\mathbf{R}$. Then, from Theorem 18(2), it follows that F is the Laplace transform of the distribution $T = f^{(n+2)}$.

Example 4. The Laplace transform is a powerful tool to solve, in $\mathcal{D}'_+$, convolution equations of the type $A * X = B$, where A and B are given distributions. If there exists a half-plane $\{z|\Re z > \xi_0\}$ in which $L(A)$ and $L(B)$ are both defined, then X is the unique distribution whose Laplace transform is $L(A)/L(B)$.

This method could be used to obtain very easily the results of Example 4.4.

Problems

1. Solve the differential equation $T' = 0$.

2. Calculate in two different ways the derivative of the distribution $x \log |x|$, and show that

$$x \mathrm{Pv} \frac{1}{x} = 1.$$

3. Show that, for any positive integer n,

$$x^n \mathrm{Pv} \frac{1}{x^n} = 1.$$

4. The function $x \mapsto \cot x$ is not locally integrable, and, therefore, there is no corresponding regular distribution. Show that, it is possible to define, in the neighborhood of the origin, a singular distribution denoted CR $\cot x$, where CR is read "canonical regularization,"by

$$\mathrm{CR}\ \cot x = (x \cot x)\big(\mathrm{Pv} \frac{1}{x}\big).$$

Establish the formula

$$(\tan x)(\mathrm{CR}\ \cot x) = 1,$$

and determine the derivative of CR $\cot x$.

5. Solve the equation $(x - a_1)(x - a_2)T = 0$, where a_1 and a_2 are two different reals.

6. Solve the equation $x^m T = 0$, where m is a positive integer.

7. Solve the equation $(x - a_1)(x - a_2)T = 1$, where a_1 and a_2 are two different reals.

8. Solve the equation $Q(x)T = P(x)$, where $Q(x)$ and $P(x)$ are two polynomials, the roots of $Q(x)$ being all real.

9. Show that the distribution $\delta(r^2 - c^2t^2)$, where $r^2 = x_1^2 + x_2^2 + x_3^2$ and c is a constant, is a solution of the wave equation in $\mathbf{R}^3$

$$\nabla^2 f - \frac{1}{c^2} \frac{\partial^2 f}{\partial t^2} = 0.$$

10. The front of an electromagnetic wave is the surface of discontinuity separating, in space, the domain where the electric field **E** and the magnetic field **H** are null from the domain where they are different from zero. If the equation of this surface at time t is $s(x) - t = 0$, the fields being null in the domain defined by $s(x) - t < 0$, show that, in the vacuum, on the front, we have

$$\nabla s \times \mathbf{E} + \mu_0 \mathbf{H} = 0$$

$$\nabla s \times \mathbf{H} - \epsilon_0 \mathbf{E} = 0$$

What conclusions can be drawn?

11. If, for any $\varphi \in \mathcal{D}$, the function $\lambda \mapsto T_\lambda(\varphi)$ is integrable on $[a, b]$, we shall define the distribution $\int_a^b T_\lambda \, d\lambda$ by

$$\Big(\int_a^b T_\lambda \, d\lambda\Big)(\varphi) = \int_a^b (T_\lambda)(\varphi) \, d\lambda.$$

Show that

$$(2i\pi)^n \int_{-\infty}^{\infty} \lambda^n e^{2i\pi\lambda x} \, d\lambda = \delta^{(n)},$$

where n is a nonnegative integer.

12. Establish the following results:

(1) $$\lim_{\lambda \to 0} |x|^\lambda = 1,$$

(2) $$\lim_{\lambda \to 0} \lambda |x|^{\lambda - 1} = 2\delta.$$

13. Find the solution to the Cauchy problem for the one-dimensional wave equation that satisfies the initial conditions

$$u(x, 0) = f_1(x), \quad \frac{\partial u}{\partial t}(x, 0) = f_2(x),$$

where f_1 and f_2 are locally integrable functions, respectively, of classes C^2 and C^1.

14. Transverse vibrations of a taut string arc governed by the one-dimensional wave equation

$$\frac{\partial^2 u}{\partial t^2} = \frac{\partial^2 u}{\partial x^2}.$$

Show that the solution that satisfies the boundary conditions (fixed ends)

$$u(0, t) = u(1, t) = 0$$

can be written

$$u(x, t) = \sum_{k=1}^{\infty} a_k \sin k\pi x \sin(k\pi t + \phi_k),$$

where the constants a_k and ϕ_k for $k = 1, 2, \ldots$ depend upon the initial conditions. Determine these constants if the initial conditions are (equivalent to plucking the string)

$$u(x,0) = \begin{cases} mx/a, & \text{if } x \leq a; \\ m(x-1)/(a-1), & \text{otherwise;} \end{cases}$$

and

$$\frac{\partial u}{\partial t}(x,0) = 0.$$

Same problem if the initial conditions are (equivalent to striking the string)

$$u(x,0) = 0,$$

and

$$\frac{\partial u}{\partial t}(x,0) = v\delta_a.$$

15. Establish the formula

$$\mathrm{Pf}\,\frac{x_+^{\alpha-1}}{\Gamma(\alpha)} * \mathrm{Pf}\,\frac{x_+^{\beta-1}}{\Gamma(\beta)} = \mathrm{Pf}\,\frac{x_+^{\alpha+\beta-1}}{\Gamma(\alpha+\beta)}.$$

16. Consider the Abel's integral equation

$$\frac{1}{\Gamma(1-\alpha)} \int_0^x \frac{f(t)}{(x-t)^\alpha}\,dt = g(x),$$

where g is a given differentiable function. Show that, if $0 < \alpha < 1$, then

$$f(x) = \frac{1}{\Gamma(\alpha)} \int_0^x \frac{g'(t)}{(x-t)^{1-\alpha}}\,dt.$$

17. The support of the function $\varphi \in \mathcal{D}$ is the interval $[-1/2, 1/2]$. Show that the function $f = \chi * \varphi$, where χ is the characteristic function of the interval $[-1/2, 1/2]$, belongs to $\mathcal{D}$, and that its support is the interval $[-1, 1]$.

Establish that, for all real x,

$$\frac{1}{f(0)} \sum_{n=-\infty}^{\infty} f(x-n) = 1.$$

18. χ being the characteristic function of the interval $[-1, 1]$, determine $F_n = \chi * \chi * \cdots * \chi = \chi^{*n}$.

19. Find the Fourier series expansion of the singular distribution CR $\cot x$ (see Problem 4).

Hint. This distribution is the derivative of the regular distribution $\log|\sin x|$.

20. Show that

$$x^q \delta^{(p+q)} = (-1)^q \frac{(p+q)!}{p!} \delta^{(p)},$$

where p and q are two positive integers.

21. Let f be an analytic function in the neighborhood of the origin such that $f(0) \neq 0$, and let m be a positive integer; under which conditions does the differential equation

$$x^m T' + fT = 0$$

have a solution of the form

$$T = \sum_{p=0}^{n} a_p \delta^{(p)}?$$

22. Find the general solution of the differential equation

$$xT' + T = 0.$$

23. Determine the Fourier transform of the distribution $\mathrm{Pf}\, x^{-n}$, where n is a positive integer.

24. Find the sum of the series

$$\sum_{n=1}^{\infty} \frac{\cos 2\pi n a}{1+n^2},$$

where $0 < a < 1$.

25. Determine the Fourier transform of the distribution

$$\mathrm{Pf}\, \frac{1}{x^2 - a^2},$$

where a is a real.

26. Determine the Fourier transform of the distribution

$$\mathrm{Pf}\, \frac{x_+^{\alpha-1}}{\Gamma(\alpha)}.$$

27. Determine the Laplace transform of the distribution

$$\mathrm{Pf}\, \frac{x_+^{\alpha-1}}{\Gamma(\alpha)}.$$

28. Solve in $\mathcal{D}'_+$ the linear system

$$\delta'' * X_1 + \delta' * X_2 = \delta$$
$$\delta' * X_1 + \delta'' * X_2 = 0.$$

Solutions

1. Since $T'(\varphi) = -T(\varphi') = 0$, for any $\varphi' \in \mathcal{D}$ that is the derivative of a function $\varphi \in \mathcal{D}$, $T' = 0$ implies $T(\varphi') = 0$. A function $\varphi_1 \in \mathcal{D}$ is the derivative of a function $\varphi_0 \in \mathcal{D}$ if, and only if,

$$\int_{-\infty}^{\infty} \varphi_1(x)\, dx = 0.$$

This condition is obviously necessary because, the support of φ_0 being bounded,

$$\int_{-\infty}^{\infty} \varphi_1(x)\, dx = 0 = \Big[\varphi_0\Big]_{-\infty}^{\infty} = 0.$$

It is also sufficient because, if φ_0 is the primitive of φ_1 defined by

$$\varphi_0(x) = \int_{-\infty}^{x} \varphi_1(\xi)\, d\xi,$$

then φ_0, as φ_1, is of class C^∞, and, when x tends to ∞, the condition shows that φ_0 has a bounded support.

Let $\psi_0 \in \mathcal{D}$ be such that $\int_{-\infty}^{\infty} \psi_0(x)\, dx = 1$; for all $\varphi \in \mathcal{D}$, there exists $\varphi_1 \in \mathcal{D}$ such that

$$\varphi(x) = \psi_0(x) \int_{-\infty}^{\infty} \varphi(\xi)\, d\xi + \varphi_1(x),$$

and, since the integral of ψ_0 on $\mathbf{R}$ is equal to 1, we have

$$\int_{-\infty}^{\infty} \varphi_1(x)\, dx = 0.$$

Hence,

$$T(\varphi) = \Big(\int_{-\infty}^{\infty} \varphi(\xi)\, d\xi\Big) T(\psi_0),$$

since $T(\varphi_1) = 0$. If we put $T(\psi_0) = C$, then

$$T' = 0 \Rightarrow T = C,$$

that is, if $T' = 0$, T is a constant distribution.

This result can be obtained much faster using Theorem 15(1):

$$T' = 0 \Rightarrow -2i\pi t\widehat{T} = 0.$$

Hence $\widehat{T} = C\delta$, that is, $T = C$.

2. The function $x \mapsto x$ being infinitely differentiable, the product $x \log|x|$ makes sense (see Example 1.3), and its derivative is given by (see Example 2.1)

$$(x \log|x|)' = \log|x| + x \operatorname{Pv}\frac{1}{x}.$$

But

$$\begin{aligned}(x \log|x|)'(\varphi) &= -(x \log|x|)(\varphi') \\ &= -\int_{-\infty}^{\infty} x \log|x| \varphi'(x)\, dx \\ &= \int_{-\infty}^{\infty} (\log|x| + 1)\varphi(x)\, dx,\end{aligned}$$

which shows that $(x \log|x|)'$ is a regular distribution equal to $\log|x| + 1$. Comparing the two results yields:

$$x \operatorname{Pv}\frac{1}{x} = 1.$$

3. If $n = 2p$, we have

$$\begin{aligned}(\operatorname{Pf} x^{-2p})(\varphi) = \int_0^{\infty} x^{-2p}\Big(\varphi(x) + \varphi(-x) - 2\big(\varphi(0) + \frac{x^2}{2}\varphi''(0) + \cdots \\ + \frac{x^{2p-2}}{(2p-2)!}\varphi^{(2p-2)}(0)\big)\Big)\, dx.\end{aligned}$$

Hence,

$$\begin{aligned}(x^{2p} \operatorname{Pf} x^{-2p})(\varphi) &= (\operatorname{Pf} x^{-2p})(x^{2p}\varphi) \\ &= \int_0^{\infty} \big(\varphi(x) + \varphi(-x)\big)\, dx \\ &= \int_{-\infty}^{\infty} \varphi(x)\, dx.\end{aligned}$$

If $n = 2p + 1$, we have

$$\begin{aligned}(\operatorname{Pf} x^{-2p-1})(\varphi) = \int_0^{\infty} x^{-2p-1}\Big(\varphi(x) - \varphi(-x) - 2\big(x\varphi'(0) \\ + \frac{x^3}{3!}\varphi'''(0) + \cdots + \frac{x^{2p-1}}{(2p-1)!}\varphi^{(2p-1)}(0)\big)\Big)\, dx.\end{aligned}$$

Hence,

$$\begin{aligned}(x^{2p+1}\operatorname{Pf} x^{-2p-1})(\varphi) &= (\operatorname{Pf} x^{-2p-1})(x^{2p+1}\varphi)\\ &= \int_0^{\infty} \big(\varphi(x)+\varphi(-x)\big)\,dx\\ &= \int_{-\infty}^{\infty} \varphi(x)\,dx.\end{aligned}$$

That is, regardless whether n is even or odd, we have

$$x^n \operatorname{Pf} x^{-n} = 1.$$

More generally, it can be shown that, if n and m are two positive integers such that $n > m$, we have

$$\begin{aligned}x^{n+m}\operatorname{Pf} x^{-n} &= x^m\\ x^{n-m}\operatorname{Pf} x^{-n} &= \operatorname{Pf} x^{-m}.\end{aligned}$$

4. $x \mapsto x\cot x$ being an infinitely differentiable function in the neighborhood of the origin, the definition of the distribution CR $\cot x$ makes sense, and we have

$$\begin{aligned}(\text{CR } \cot x)(\varphi) &= \Big(\operatorname{Pv}\frac{1}{x}\Big)(x\cot x\,\varphi)\\ &= \int_0^{\infty} \cot x\big(\varphi(x)-\varphi(-x)\big)\,dx,\end{aligned}$$

where the support of φ is contained in the open interval $]-\pi,\pi[$.

The function $x \mapsto \tan x$ being infinitely differentiable in the neighborhood of the origin, the formula

$$(\tan x)(\text{CR } \cot x) = 1$$

is readily verified.

Moreover,

$$\begin{aligned}(\text{CR } \cot x)' &= \Big(x\cot x \operatorname{Pv}\frac{1}{x}\Big)'\\ &= \frac{\cos x\sin x - x}{\sin^2 x}\operatorname{Pf}\frac{1}{x} - x\cot x \operatorname{Pf}\frac{1}{x^2}.\end{aligned}$$

That is,

$$(\text{CR } \cot x)' = -\frac{x^2}{\sin^2 x}\operatorname{Pf}\frac{1}{x^2}.$$

The function $x \mapsto x^2/\sin^2 x$ being infinitely differentiable in the neighborhood of the origin, the right-hand side defines a singular distribution that is denoted $\mathrm{CR}\, 1/\sin^2 x$, and we have

$$(\mathrm{CR}\,\frac{1}{\sin^2 x})(\varphi) = \int_0^\infty \Big(\frac{\varphi(x)+\varphi(-x)}{\sin^2 x} - \frac{2\varphi(0)}{x^2}\Big)\, dx.$$

Note that $\mathrm{CR}\cot x$ is the derivative of the regular distribution $\log|\sin x|$.

5. Let φ_1 and φ_2 be two test functions such that

$$\begin{aligned} \varphi_1(a_1) &= 1 \quad \varphi_1(a_2) = 0 \\ \varphi_2(a_1) &= 0 \quad \varphi_2(a_2) = 1. \end{aligned}$$

If φ is any arbitrary function in $\mathcal{D}$, then the function $\varphi - \varphi(a_1)\varphi_1 - \varphi(a_2)\varphi_2$, belongs to $\mathcal{D}$, and is null for $x = a_1$ and $x = a_2$. Therefore, there exists a function $\varphi_3 \in \mathcal{D}$ such that

$$\varphi(x) = \varphi(a_1)\varphi_1(x) + \varphi(a_2)\varphi_2(x) + (x-a_1)(x-a_2)\varphi_3(x).$$

Hence,

$$T(\varphi) = \varphi(a_1)T(\varphi_1) + \varphi(a_2)T(\varphi_2),$$

that is

$$T = C_1\delta_{a_1} + C_2\delta_{a_2},$$

where $C_1 = T(\varphi_1)$ and $C_2 = T(\varphi_2)$ are two constants.

6. Let $\varphi_0, \varphi_1, \ldots, \varphi_{m-1}$ be m functions belonging to $\mathcal{D}$ such that

$$\varphi_j^{(k)}(0) = \begin{cases} 1, & \text{if } k = j; \\ 0, & \text{otherwise.} \end{cases}$$

For any $\varphi \in \mathcal{D}$, the function $\varphi - \sum_{j=0}^{m-1} \varphi^{(j)}(0)\varphi_j$, and all its derivatives of order less than m are null at the origin. Therefore, there exists a function $\varphi_m \in \mathcal{D}$ such that

$$\varphi(x) = \sum_{k=0}^{m-1} \varphi^{(k)}(0)\varphi_k(x) + x^m\varphi_m(x),$$

and we have

$$T(\varphi) = \sum_{k=0}^{m-1} \varphi^{(k)}(0)T(\varphi_k),$$

that is

$$T = \sum_{k=0}^{m-1} C_k \delta^{(k)},$$

where C_k $(k = 0, 1, \ldots, m-1)$ are constants.

7. If T_0 is a solution of the equation, then (see Problem 5) $T_0 + C_1\delta_{a_1} + C_2\delta_{a_2}$, where C_1 and C_2 are two constants, is also a solution. To find a solution, we can start from the relation

$$\frac{1}{(x-a_1)(x-a_2)} = \frac{1}{a_1 - a_2}\Big(\frac{1}{x-a_1} - \frac{1}{x-a_2}\Big),$$

which suggests that

$$T_0 = \frac{1}{a_1 - a_2}\Big(\text{Pf}\,\frac{1}{x-a_1} - \text{Pf}\,\frac{1}{x-a_2}\Big)$$

is a particular solution. This is, indeed, readily verified.

8. Proceeding as in the previous problem, the relation

$$\frac{P(x)}{Q(x)} = \sum_{i,j} \frac{p_{ij}(x)}{(x-a_i)^{m_j}}$$

suggests the particular solution

$$T_0 = \sum_{i,j} p_{ij}(x)\text{Pf}\,\frac{1}{(x-a_i)^{m_j}}.$$

Adding a linear combination of Dirac distributions δ_{a_i} and their dervatives yields the general solution. The form of this linear combination is determined as in Problems 5 and 6.

9. Let $s(x_1, x_2, x_3, t) = r^2 - c^2t^2$; then

$$\frac{\partial \delta_s}{\partial x_i} = 2x_i\delta_s' \quad \text{and} \quad \frac{\partial^2 \delta_s}{\partial x_i^2} = 2\delta_s' + 4x_i^2\delta_s''.$$

Hence,

$$\nabla^2\delta_s = 6\delta_s' + 4r^2\delta_s''.$$

Moreover,

$$\frac{\partial \delta_s}{\partial t} = -2c^2t\delta_s' \quad \text{and} \quad \frac{\partial^2 \delta_s}{\partial t^2} = -2c^2\delta_s' + 4c^4t^2\delta_s''.$$

Finally,

$$\left(\nabla^2 - \frac{1}{c^2}\frac{\partial^2}{\partial t^2}\right)\delta_s = (2\delta_s' + s\delta_s'').$$

To show that the right-hand side is null, we can take the derivative of the identity $s\delta_s = 0$, which follows from the fact that $s = 0$ on S. We have

$$\frac{\partial s}{\partial x_i}\delta_s + \frac{\partial s}{\partial x_i}s\delta_s' = 0,$$

that is,

$$\delta_s + s\delta_s' = 0,$$

since, at least, one derivative $\partial s/\partial x_i$ differs from zero. Taking the derivative of the last identity completes the proof.

10. Maxwell's equations in empty space read[39]

$$\nabla \times \theta\big(s(x) - t\big)\mathbf{E} = -\mu_0\frac{\partial}{\partial t}\theta\big(s(x) - t\big)\mathbf{H}$$
$$\nabla \times \theta\big(s(x) - t\big)\mathbf{H} = \epsilon_0\frac{\partial}{\partial t}\theta\big(s(x) - t\big)\mathbf{E}.$$

Using the formulas derived in Example 2.6, the terms in $\delta\big(s(x) - t\big)$ yield the relations

$$\nabla s \times \mathbf{E} + \mu_0\mathbf{H} = 0$$
$$\nabla s \times \mathbf{H} - \epsilon_0\mathbf{E} = 0.$$

Since the vector ∇s is normal to the surface determined by the equation $s(x) - t = 0$, we can set $\nabla s = (1/c)\mathbf{n}$, where c is the velocity of propagation of the wave front,[40] and $\mathbf{n}$ a unit vector normal to the wave front. We have then

$$\frac{1}{c}\mathbf{n} \times \mathbf{E} + \mu_0\mathbf{H} = 0$$
$$\frac{1}{c}\mathbf{n} \times \mathbf{H} - \epsilon_0\mathbf{E} = 0,$$

which show that, on the wave front, $\mathbf{E}$ and $\mathbf{H}$ are orthogonal, and that the plane they determine is tangential to the surface of discontinuity.

[39] We write only the Maxwell's equations we need.

[40] This can be seen differentiating the equation $s(x) - t = 0$.

Furthermore, from these relations it is not difficult to obtain the following results

$$\epsilon_0 \|\mathbf{E}\|^2 = \mu_0 \|\mathbf{H}\|^2$$

$$c = \frac{1}{\sqrt{\epsilon_0 \mu_0}}.$$

11. Let us first consider the distribution $\int_{-\infty}^{\infty} \mathrm{e}^{2i\pi\lambda x}\, d\lambda$. For any $\varphi \in \mathcal{D}$,

$$\begin{aligned}(\mathrm{e}^{2i\pi\lambda x})(\varphi) &= \int_{-\infty}^{\infty} \mathrm{e}^{2i\pi\lambda x} \varphi(x)\, dx \\ &= \widehat{\varphi}(\lambda).\end{aligned}$$

Thus,

$$\begin{aligned}\Big(\int_{-\infty}^{\infty} \mathrm{e}^{2i\pi\lambda x}\, d\lambda\Big)(\varphi) &= \int_{-\infty}^{\infty} \widehat{\varphi}(\lambda)\, d\lambda \\ &= \varphi(0),\end{aligned}$$

that is,

$$\int_{-\infty}^{\infty} \mathrm{e}^{2i\pi\lambda x}\, d\lambda = \delta.$$

Let us now consider the derivative $\Big(\int_{-\infty}^{\infty} \mathrm{e}^{2i\pi\lambda x}\, d\lambda\Big)^{(n)}$. We have

$$\begin{aligned}\Big(\int_{-\infty}^{\infty} \mathrm{e}^{2i\pi\lambda x}\, d\lambda\Big)^{(n)}(\varphi) &= (-1)^n \Big(\int_{-\infty}^{\infty} \mathrm{e}^{2i\pi\lambda x}\, d\lambda\Big)(\varphi^{(n)}) \\ &= (-1)^n \int_{-\infty}^{\infty}\int_{-\infty}^{\infty} \mathrm{e}^{2i\pi\lambda x} \varphi^{(n)}(x)\, dx\, d\lambda\end{aligned}$$

and, after n integrations by parts with respect to x,

$$\Big(\int_{-\infty}^{\infty} \mathrm{e}^{2i\pi\lambda x}\, d\lambda\Big)^{(n)}(\varphi) = \Big(\int_{-\infty}^{\infty} (2i\pi\lambda)^n \mathrm{e}^{2i\pi\lambda x}\, d\lambda\Big)(\varphi).$$

Finally,

$$\int_{-\infty}^{\infty} (2i\pi\lambda)^n \mathrm{e}^{2i\pi\lambda x}\, d\lambda = \delta^{(n)}.$$

12. (1) The proof of this formula is not as simple as it might seem. Although, for any finite and fixed x, the result is valid in the sense of functions, we have to prove it in the sense of distributions. $|x|^\lambda$ being a tempered regular distribution, we have to show that, for any $\varphi \in \mathcal{D}$,

$$\lim_{\lambda \to 0} \int_{-\infty}^{\infty} (|x|^\lambda - 1)\varphi(x)\, dx = 0.$$

For any positive x,

$$\frac{x^\lambda - 1}{x^\lambda + 1} = \frac{\exp(\lambda \log x) - 1}{\exp(\lambda \log x) + 1}$$
$$= \tanh(\frac{1}{2}\lambda \log x).$$

Hence,

$$\left|\frac{x^\lambda - 1}{x^\lambda + 1}\right| \leq \frac{1}{2}|\lambda \log x|.$$

For any real x, we have therefore

$$\left||x|^\lambda - 1\right| \leq \frac{1}{2}|\lambda|(|x|^\lambda + 1)\log|x|.$$

Hence,

$$\lim_{\lambda \to 0} \int_{-\infty}^{\infty} (|x|^\lambda - 1)\varphi(x)\,dx \leq \lim_{\lambda \to 0} \frac{1}{2}|\lambda| \int_{-\infty}^{\infty} (|x|^\lambda + 1)\log|x|\varphi(x)\,dx = 0,$$

which completes the proof.

(2) Proceeding as above, we could show that

$$\lim_{\lambda \to 0} |x|^\lambda \operatorname{sgn} x = \operatorname{sgn} x,$$

where $\operatorname{sgn} x = 2\theta(x) - 1$, and, from Theorem 1, it follows that

$$\lim_{\lambda \to 0} (|x|^\lambda \operatorname{sgn} x)' = (\operatorname{sgn} x)' = 2\delta,$$

that is,

$$\lim_{\lambda \to 0} \lambda |x|^{\lambda - 1} = 2\delta.$$

13. Let u_1 and u_2 be the solutions corresponding to the initial conditions

$$u_1(x,0) = 0 \qquad \frac{\partial u_1}{\partial t}(x,0) = f_1(x)$$
$$u_2(x,0) = 0 \qquad \frac{\partial u_2}{\partial t}(x,0) = f_2(x).$$

The fundamental solution of the one-dimensional wave equation being

$$g_t(x) = \frac{1}{2}\theta(x+t) - \frac{1}{2}\theta(x-t),$$

we have

$$u_1 = f_1 * g_t \quad \text{and} \quad u_2 = f_2 * g_t,$$

that is,

$$u_1(x,t) = \frac{1}{2}\int_{x-t}^{x+t} f_1(\xi)\,d\xi$$

$$u_2(x,t) = \frac{1}{2}\int_{x-t}^{x+t} f_2(\xi)\,d\xi.$$

Then it is not difficult to verify that the solution we are looking for is

$$u = \frac{\partial u_1}{\partial t} + u_2,$$

that is,

$$u(x,t) = \frac{1}{2}\big(f_1(x+t) + f_1(x-t)\big) + \frac{1}{2}\int_{x-t}^{x+t} f_2(\xi)\,d\xi.$$

This result shows that the general solution of the one-dimensional wave equation is

$$u(x,t) = f(x+t) + g(x-t),$$

where f and g are of class C^2.

If we accept solutions that are distributions, it is not necessary to assume that f_1 and f_2 are, respectively, of class C^2 and C^1. In this case, the solution can be written

$$u_t = f_1 * \frac{\partial g_t}{\partial t} + f_2 * g_t.$$

Note, however, that the general solution is always of the form

$$u(x,t) = f(x+t) + g(x-t),$$

where f and g are, for obvious physical reasons, continuous functions. For any $\varphi \in \mathcal{D}$, we have indeed

$$\begin{aligned}
\frac{\partial^2 u_t}{\partial t^2}(\varphi) &= \frac{\partial^2}{\partial t^2}\int \big(f(x+t) + g(x-t)\big)\varphi(x)\,dx \\
&= \int f(u)\varphi''(u-t)\,du + \int g(u)\varphi''(u+t)\,du \\
&= \int \big(f(x+t) + g(x-t)\big)\varphi''(x)\,dx \\
&= \frac{\partial^2 u_t}{\partial x^2}(\varphi).
\end{aligned}$$

14. The general solution of the one-dimensional wave equation is (see preceding problem)

$$u(x,t) = f(x+t) + g(x-t),$$

where f and g are continuous functions. The boundary conditions $u(0,t) = u(1,t) = 0$ imply

$$f(t) + g(t) = 0 \quad \text{and} \quad f(1+t) + g(1-t) = 0,$$

which shows that f and g are periodic functions of period 2, and that the solution may be written

$$\begin{aligned} u(x,t) &= f(x+t) - f(x-t) \\ &= g(x-t) - g(x+t). \end{aligned}$$

Let

$$f(z) = \sum_{k=1}^{\infty} (\alpha_k \sin k\pi z + \beta_k \cos k\pi z);$$

then

$$\begin{aligned} u(x,t) &= \sum_{k=1}^{\infty} \sin k\pi x (2\alpha_k \cos k\pi t - 2\beta_k \sin k\pi t) \\ &= \sum_{k=1}^{\infty} a_k \sin k\pi x \sin(k\pi t + \phi_k), \end{aligned}$$

if we put $(2\alpha_k \cos k\pi t - 2\beta_k \sin k\pi t) = a_k \sin(k\pi t + \phi_k)$.

First case. We have

$$\frac{\partial u}{\partial t}(x,0) = \sum_{k=0}^{\infty} k\pi a_k \sin k\pi x \cos \phi_k = 0;$$

hence, for any $k = 1, 2, \ldots$, $\phi_k = \pi/2$, that is,

$$u(x,0) = \sum_{k=1}^{\infty} a_k \sin k\pi x.$$

The constants a_k are, therefore, given by

$$\begin{aligned} a_k &= 2 \int_0^1 u(x,0) \sin k\pi x \, dx \\ &= 2 \int_0^a \frac{mx}{a} \sin k\pi x \, dx + 2 \int_0^a \frac{m(x-1)}{a-1} \sin k\pi x \, dx \\ &= \frac{2m}{a(1-a)} \frac{\sin k\pi a}{k^2 \pi^2}. \end{aligned}$$

Second case. We have

$$u(x,0) = \sum_{k=1}^{\infty} a_k \sin k\pi x \sin \phi_k = 0;$$

hence, for any $k = 1, 2, \ldots$, $\phi_k = 0$, that is,

$$\frac{\partial u}{\partial t}(x,0) = \sum_{k=0}^{\infty} k\pi a_k \sin k\pi x = v\delta_a.$$

The constants a_k are, therefore, given by

$$a_k = \frac{2v \sin k\pi a}{k\pi}.$$

Note that in the first case the amplitude of the kth harmonic decreases as $1/k^2$, whereas in the second case it decreases as $1/k$. A musical instrument like the piano is richer in harmonics than the harp. If $a = 1/n$, where n is an integer, the amplitude of the harmonics $k = n, 2n, \ldots$ is equal to zero.

15. The convolution is defined since the two distributions have their supports bounded from below. If $\Re\alpha > 0$ and $\Re\beta > 0$, we have

$$\begin{aligned}\frac{x_+^{\alpha-1}}{\Gamma(\alpha)} * \frac{x_+^{\beta-1}}{\Gamma(\beta)} &= \int_0^x \frac{\xi^{\alpha-1}}{\Gamma(\alpha)} \frac{(x-\xi)^{\beta-1}}{\Gamma(\beta)}\, d\xi \quad (x > 0)\\ &= \frac{x^{\alpha+\beta-1}}{\Gamma(\alpha)\Gamma(\beta)} \int_0^1 t^{\alpha-1}(1-t)^{\beta-1}\, dt\\ &= \frac{x_+^{\alpha+\beta-1}}{\Gamma(\alpha+\beta)},\end{aligned}$$

taking into account the well-known result

$$\begin{aligned}B(\alpha,\beta) &= \int_0^1 t^{\alpha-1}(1-t)^{\beta-1}\, dt\\ &= \frac{\Gamma(\alpha)\Gamma(\beta)}{\Gamma(\alpha+\beta)}.\end{aligned}$$

By analytic continuation, the formula remains valid for any α and β provided they differ from nonpositive integers. To extend the formula to nonpositive integers, note that, if n is a positive integer, we have

$$\delta^{(n)} * \frac{x_+^{n-1}}{\Gamma(n)} = \delta,$$

and, therefore,

$$\lim_{\alpha \to 0} \text{Pf}\, \frac{x_+^{\alpha-1}}{\Gamma(\alpha)} = \delta$$

$$\lim_{\alpha \to -n} \text{Pf}\, \frac{x_+^{\alpha-1}}{\Gamma(\alpha)} = \delta^{(n)}.$$

All these results could have been obtained more simply using Fourier or Laplace transforms (see Problems 26 and 27).

16. If $0 < \alpha < 1$, the integral is convergent. If F_α denotes the function $x \mapsto x_+^{\alpha-1}/\Gamma(\alpha)$, the equation can be written

$$F_{1-\alpha} * f = g.$$

But (see Problem 15)

$$F_{1-\alpha} * F_{\alpha-1} = \delta \quad \text{and} \quad \delta' * F_\alpha = F_{\alpha-1}.$$

Hence,

$$f = F_{\alpha-1} * g = F_\alpha * g',$$

that is,

$$f(x) = \frac{1}{\Gamma(\alpha)} \int_0^x (x-t)^{\alpha-1} g'(t)\, dt.$$

The Abel's integral equation could also have been solved using the Laplace transform (see Problem 27).

More generally, it can be shown that the solution of the integral equation

$$\frac{1}{\Gamma(m-\alpha)} \int_0^x \frac{f(t)}{(x-t)^{\alpha+1-m}}\, dt = g(x),$$

where m is a positive integer, is

$$f(x) = \frac{1}{\Gamma(\alpha)} \int_0^x \frac{g^{(m)}(t)}{(x-t)^{1-\alpha}}\, dt.$$

If g is not differentiable, this result has to be interpreted in the sense of distributions.

17. We have

$$f(x) = \int_{-\infty}^{\infty} \chi(x-u)\varphi(u)\, du$$

$$= \int_{x-\frac{1}{2}}^{x+\frac{1}{2}} \varphi(u)\, du,$$

which shows that, taking into account the fact that the support of φ is the interval $[-1/2, 1/2]$, $f(x) = 0$ if $|x| > 1$, thus $\operatorname{supp} f = [-1, 1]$. Since *varphi* $\in \mathcal{D}$, f is infinitely differentiable and, therefore, belongs to $\mathcal{D}$. Furthermore, for all x in the interval $[0, 1]$,

$$\operatorname{supp} \varphi = [-1/2, 1/2] \Rightarrow f(x) + f(x-1) = f(0),$$

and, therefore,

$$\frac{1}{f(0)} \sum_{-\infty}^{\infty} f(x-n) = 1,$$

because, for all x, only two consecutive terms in the sum can be different from zero. Note that this result is valid if $f(0) \neq 0$, that is, if $\int_{-1/2}^{1/2} \varphi(u)\, du \neq 0$.

18. We have

$$\chi' = \delta_{-1} - \delta_1,$$

and, therefore,

$$\begin{aligned} F_n^{(n)} &= (\delta_{-1} - \delta_1)^{*n} \\ &= \delta_{-n} - \binom{n}{1}\delta_{-n+2} + \binom{n}{2}\delta_{-n+4} \cdots (-1)^n \delta_n. \end{aligned}$$

$(x-a)_+^{n-1}/(n-1)!$ being the nth primitive of δ_a, we have

$$F_n(x) = P_{n-1}(x) + \frac{1}{(n-1)!} \sum_{k=0}^{n} (-1)^k \binom{n}{k} \theta(x+n-2k)(x+n-2k)^{n-1},$$

where $P_{n-1}(x)$ is a polynomial of degree $n-1$. But the formula

$$\int F_n(x)\varphi(x)\, dx = \int \chi(x_1)\chi(x_2)\cdots\chi(x_n) \times \varphi(x_1 + x_2 + \cdots + x_n)\, dx_1\, dx_2 \cdots dx_n$$

shows that the support of F_n is the interval $[-n, n]$, and, therefore, $P_{n-1}(x) \equiv 0$.

19. First method. To obtain the Fourier series expansion of the function $x \mapsto \log|\sin x|$, the simplest way is, probably, to consider the Laurent series of the function $z \mapsto \log \frac{1}{2}|z - 1/z|$, which is analytic in an annulus containing the closed path $x \mapsto \exp(ix)$ $(0 < x \leq 2\pi)$ (Chapter 3, Example 3.8, Remark 2). We find

$$\log|\sin x| = -\log 2 - \sum_{k=1}^{\infty} \frac{1}{k} \cos 2kx,$$

and, taking the derivative, we finally obtain

$$\text{CR} \cot x = 2\sum_{k=1}^{\infty} \sin 2kx.$$

Second method. This Fourier series can also be derived from the formula

$$\sin x \, \text{CR} \cot x = \cos x,$$

which follows from the fact that, in the neighborhood of the origin,

$$\text{CR} \cot x = x \cot x \, \text{Pf} \frac{1}{x},$$

and, therefore,

$$\sin x \, \text{CR} \cot x = x \cos x \, \text{Pf} \frac{1}{x} = \cos x.$$

Since CR $\cot x$ is odd, we have

$$\text{CR} \cot x = \sum_{k=1}^{\infty} c_k \sin 2kx.$$

Hence,

$$\begin{aligned}
\sin x \, \text{CR} \cot x &= \sum_{k=1}^{\infty} c_k \sin x \sin 2kx \\
&= \frac{1}{2} \sum_{k=1}^{\infty} c_k \big(\cos(2k-1)x - \cos(2k+1)x\big) \\
&= \cos x,
\end{aligned}$$

which shows that, for all positive integers k, $c_k = 2$.

Note that the formula $\sin x\,\mathrm{CR}\,\cot x = \cos x$ could have been established differently. We have

$$\begin{aligned}(\sin x\,\mathrm{CR}\,\cot x)(\varphi) &= (\mathrm{CR}\,\cot x)(\varphi(x)\sin x)\\ &= \Big(\frac{d}{dx}\log|\sin x|\Big)\big(\varphi(x)\sin x\big)\\ &= -(\log|\sin x|)\big(\varphi(x)\cos x + \varphi'(x)\sin x\big).\end{aligned}$$

An integration by parts shows that

$$-(\log|\sin x|)\big(\varphi'(x)\sin x\big) = (\log|\sin x|)\big(\varphi(x)\cos x\big) + (\cos x)\big(\varphi(x)\big),$$

and the formula follows.

20. We have

$$\begin{aligned}(x^q\delta^{(p+q)})(\varphi) &= \delta^{(p+q)}(x^q\varphi)\\ &= (-1)^{p+q}\delta\big((x^q\varphi)^{(p+q)}\big)\\ &= (-1)^{p+q}\delta\Big(\frac{(p+q)!}{p!q!}q!\varphi^{(p)}\Big),\end{aligned}$$

where, in the derivative of order $p+q$, only the terms different from zero have been kept. Hence,

$$(x^q\delta^{(p+q)})(\varphi) = (-1)^q\frac{(p+q)!}{p!}\delta^{(p)}(\varphi).$$

This formula could also have been derived using Theorem 13. $x^q\delta^{(p+q)}$ is indeed the Fourier transform of the convolution

$$\frac{1}{(-2i\pi)^q}\delta^{(q)} * (2i\pi t)^{p+q} = (-1)^p\frac{(p+q)!}{p!}(2i\pi t)^p,$$

and the right-hand side is precisely the Fourier transform of the distribution $(-1)^q\frac{(p+q)!}{p!}\delta^{(p)}$.

21. Let $f(x) = \sum_{k=0}^{\infty} c_k x^k$, and replace T by $\sum_{p=0}^{n} a_p\delta^{(p)}$ in the differential equation. This yields

$$\sum_{p=0}^{\infty} a_p x^m\delta^{(p+1)} + \sum_{p=0}^{n}\sum_{k=0}^{\infty} a_p c_k x^k\delta^{(p)};$$

but (see Problem 20)

$$x^q\delta^{(\ell)} = \begin{cases}(-1)^q\dfrac{\ell!}{(\ell-q)!}\delta^{(\ell-q)}, & \text{if } \ell\geq q;\\ 0, & \text{otherwise.}\end{cases}$$

Thus,

$$\sum_{p=0}^{n}(-1)^m a_p \frac{(p+q)!}{(p+1-m)!}\delta^{(p+1-m)} = \sum_{p=0}^{n}\sum_{k=0}^{p}(-1)^k a_p c_k \frac{p!}{(p-k)!}\delta^{(p-k)}.$$

The highest order of the derivative of δ being equal to n in the right-hand side whereas it is equal to $n+1-m$ in the left-hand side, m must be equal to 1. This condition being fulfilled, by equating the coefficients of the derivatives of δ of the same order we obtain the following linear equations:

$$a_n(c_0 - n - 1) = 0$$

$$a_{n-1}(c_0 - n) - \frac{n!}{(n-1)!}a_n c_1 = 0$$

$$\cdots\cdots\cdots\cdots\cdots\cdots\cdots\cdots\cdots\cdots$$

$$a_2(c_0 - 3) - \frac{3!}{2!}a_3 c_1 + \frac{4!}{2!}a_4 c_2 - \cdots (-1)^{n-2}\frac{n!}{2!}a_n c_{n-2} = 0$$

$$a_1(c_0 - 2) - 2!\,a_2 c_1 + 3!\,a_3 c_2 - \cdots (-1)^{n-1} n!\,a_n c_{n-1} = 0$$

$$a_0(c_0 - 1) - 1!\,a_1 c_1 + 2!\,a_2 c_2 - \cdots (-1)^n n!\,a_n c_n = 0.$$

If c_0 is not a positive integer, for all p, $a_p = 0$. If c_0 is a positive integer, for $n = c_0 - 1$, a_n is arbitrary, and the coefficients $c_0 - n, c_0 - n + 1, \ldots, c_0 - 2, c_0 - 1$ being all different from zero, the coefficients a_p $(p = 1, 2, \ldots, n-1)$ can be determined step by step. That is, if the conditions $m = 1$ and $c_0 = n + 1$ are fulfilled, the differential equation has a solution of the form $T = \sum_{p=0}^{n} a_p \delta^{(p)}$, where the coefficients a_p $(p = 1, 2, \ldots, n)$ are defined except for a multiplicating factor.

22. Since

$$xT' + T = (xT)',$$

we have

$$xT = \text{const.}.$$

Therefore,

$$T = C_1 \operatorname{Pf}\frac{1}{x} + C_2\delta,$$

where C_1 and C_2 are two constants. Although the differential equation is of the first order, the general solution *depends upon two arbitrary constants.*

23. First method. We have

$$\operatorname{Pf}\frac{1}{x^n} = \frac{(-1)^{n-1}}{(n-1)!}\delta^{(n-1)} * \operatorname{Pf}\frac{1}{x},$$

and from Theorem 13 it follows that

$$\widehat{\operatorname{Pf}\frac{1}{x^n}} = i\pi\frac{(2i\pi t)^{n-1}}{(n-1)!}\operatorname{sgn} t.$$

Second method. This result can also be derived from the following relations

$$\left(\operatorname{Pf}\frac{1}{x}\right)^{(n-1)} = (-1)^{n-1}(n-1)!\operatorname{Pf}\frac{1}{x^n},$$

$$\widehat{\left(\operatorname{Pf}\frac{1}{x}\right)^{(n-1)}} = (-2i\pi t)^{n-1}\widehat{\operatorname{Pf}\frac{1}{x}} \quad \text{(Example 2.5)},$$

and

$$\widehat{\operatorname{Pf}\frac{1}{x}} = i\pi\operatorname{sgn} t \quad \text{(Example 6.5)}.$$

24. The trigonometric series

$$\sum_{n=1}^{\infty}\frac{\cos 2\pi na}{1+n^2}$$

is absolutely convergent, and its sum can be found using Poisson's summation formula. Since

$$\frac{\cos 2\pi na}{1+n^2}$$

is the Fourier transform of

$$\frac{\pi}{2}\left(e^{-2\pi|t+a|} + e^{-2\pi|t-a|}\right),$$

we have

$$\sum_{-\infty}^{\infty}\frac{\cos 2\pi na}{1+n^2} = \frac{\pi}{2}\sum_{-\infty}^{\infty}\left(e^{-2\pi|t+a|} + e^{-2\pi|t-a|}\right),$$

that is,

$$1 + 2\sum_{n=1}^{\infty}\frac{\cos 2\pi na}{1+n^2} = \pi\frac{\cosh\pi(1-2a)}{\sinh\pi},$$

and finally,

$$\sum_{n=1}^{\infty}\frac{\cos 2\pi na}{1+n^2} = \frac{\pi}{2}\frac{\cosh\pi(1-2a)}{\sinh\pi} - \frac{1}{2}.$$

25. The relation

$$\operatorname{Pf}\frac{1}{x^2-a^2} = \frac{1}{2a}\Big(\operatorname{Pf}\frac{1}{x-a} - \operatorname{Pf}\frac{1}{x+a}\Big),$$

which is readily verified, shows that

$$\begin{aligned}\widehat{\operatorname{Pf}\frac{1}{x^2-a^2}} &= \frac{i\pi}{2a}(\mathrm{e}^{2i\pi at} - \mathrm{e}^{-2i\pi at})\operatorname{sgn} t\\ &= -\frac{\pi}{a}\sin at \operatorname{sgn} t.\end{aligned}$$

26. Assume first that $0 < \alpha < 1$, and consider the regular distribution $x_+^{\alpha-1}\mathrm{e}^{-2\pi\tau x}$, where τ is positive. Its Fourier transform is given by the convergent integral

$$I_\tau(t) = \int_0^\infty x^{\alpha-1}\mathrm{e}^{2i\pi(t+i\tau)x}\,dx.$$

Let $2i\pi(t+i\tau)x = -u$. Then, from Cauchy's theorem, it follows that

$$I_\tau(t) = \frac{\mathrm{e}^{i\pi\alpha/2}\Gamma(\alpha)}{(2\pi)^\alpha(t+i\tau)^\alpha},$$

and, if τ tends to zero, we finally obtain

$$\widehat{\frac{x_+^{\alpha-1}}{\Gamma(\alpha)}} = \frac{\mathrm{e}^{i\pi\alpha/2}}{(2\pi)^\alpha(t+i0)^\alpha}.$$

By analytic continuation, this result remains valid for all α.

Note that when α tends to zero,

$$\widehat{\frac{x_+^{\alpha-1}}{\Gamma(\alpha)}} \to 1.$$

Thus,

$$\lim_{\alpha\to 0}\frac{x_+^{\alpha-1}}{\Gamma(\alpha)} = \delta.$$

27. As in the previous problem, we first assume that $0 < \alpha < 1$, and consider the integral

$$I(z) = \int_0^\infty x^{\alpha-1}\mathrm{e}^{-zx}\,dx.$$

Let $zx = u$. Then, from Cauchy's theorem, it follows that

$$I(z) = \frac{\Gamma(\alpha)}{z^\alpha},$$

where the function $f: z \mapsto z^\alpha$ has been defined such that $f(1) = 1$. Thus

$$L\left(\frac{x_+^{\alpha-1}}{\Gamma(\alpha)}\right) = \frac{1}{z^\alpha}.$$

By analytic continuation, this result remains valid for all α.

If n is a positive integer, this result implies that

$$\lim_{\alpha \to -n} \frac{x_+^{\alpha-1}}{\Gamma(\alpha)} = \delta^{(n)}.$$

28. Using the Laplace transformation, we find

$$L(X_1; z) = \frac{1}{z^2 - 1} \quad \text{and} \quad L(X_2; z) = \frac{1}{z(1 - z^2)}.$$

Hence,

$$X_1 = \theta(x) \sinh x \quad \text{and} \quad X_2 = \theta(x)(1 - \cosh x).$$

Chapter 5

Linear Operators

The knowledge of the essential properties of linear operators defined in a Hilbert space is indispensable to the understanding of the mathematical formulation of quantum mechanics. When these operators are continuous their manipulation does not present any major difficulty; but if this is not the case, one has to be very careful, since the domain of definition of the operator then plays a very important role. Since this fact has no analogue in linear algebra, and since linear operators defined on a finite dimensional space are always continuous, one has to be wary of hasty generalizations.

The most important result of this chapter is the spectral theorem, which is proved for compact self-adjoint operators. This is far from being the most general result, which lies outside the rather elementary level of this book; however, this simple spectral theory applies directly to Fredholm integral equations with Hermitian kernel and to Sturm–Liouville operators, which are unbounded differential operators.

The difference between Hermitian and self-adjoint operators is emphasized. Since observables in quantum mechanics are represented by self-adjoint operators, the distinction between these two notions should be clearly stated. This is unfortunately not always the case in some books on quantum mechanics where, sometimes, merely Hermitian operators are referred to as self-adjoint.

1. Preliminaries

Definition 1. Let E be a normed vector space; a linear mapping A of E into itself is called a *linear operator* in E. The subspace of all the elements x of E such that $A(x)$ is defined is called the *domain* of A; it is denoted by D_A. The subspace $A(D_A)$ is called the *range* of A; it is denoted by R_A.

Notation: If x belongs to D_A, the vector $A(x)$ will be denoted by Ax.
Example 1. Consider the Hilbert space $L^2_m([a,b])$. If f belongs to $L^2_m([a,b])$, the operator X defined by $(Xf)(t) = tf(t)$ is called the *position operator*. It is easy to verify that $D_X = L^2_m([a,b])$, i.e., the domain of X coincides with the whole Hilbert space. If we had considered the Hilbert space $L^2_m(\mathbf{R})$ this would not have been the case (see Example 4.2).
Example 2. Consider again the Hilbert space $L^2_m([a,b])$. The operator P defined by $(Pf)(t) = -if'(t)$ is called the *momentum operator*. D_P is the subspace of all differentiable functions that belong to $L^2_m([a,b])$ and whose first derivatives also belong to $L^2_m([a,b])$.

A linear operator cannot be regarded as fully specified until its domain has been specified. If A and B are two linear operators, we write $A = B$ if, and only if, first, $D_A = D_B$ and, second, $Ax = Bx$ for any x in $D_A = D_B$. B is an *extension* of A if $D_A \subset D_B$ and if, for any x in D_A, we have $Ax = Bx$. This is denoted by $A \subset B$. The domain of $A + B$ is $D_A \cap D_B$.[1] If λ is a scalar, $D_{\lambda A} = D_A$. The operator $A \circ B$ is denoted simply by AB; ABx makes sense if, and only if, $Bx \in D_A$, which implies that $D_{AB} \subset D_B$. There is no reason for D_{AB} to be equal to D_{BA} and, even if $D_{AB} \cap D_{BA} \neq \emptyset$, in general, ABx differs from BAx for x in $D_{AB} \cap D_{BA}$. If $ABx = BAx$ for any $x \in D_{AB} \cap D_{BA}$, A and B are said to *commute*, and the operator $AB - BA$, called the *commutator* of A and B, is denoted by $[A, B]$.
Example 3. It is easy to verify that the position and momentum operators defined in $L^2_m([a,b])$ do not commute. Indeed, for any f in $D_{[A,B]}$, we have

$$(XP - PX)f = if.$$

In books on quantum mechanics, this result is often written $[XP - PX] = iI$, where I is the identity operator. This is not correct since $D_{[X,P]} \neq L^2_m([a,b])$.

The *inverse* of a linear operator A is a linear operator B such that $D_B = R_A$ and $BAx = x$ for any $x \in D_A$. In this case $R_B = D_A$, and for any $y \in D_B$ we have $ABy = y$; if B is the inverse of A, A is the inverse of B. We write $B = A^{-1}$. If an operator has an inverse it is said to be *nonsingular*; if this not the case the operator is *singular*. An operator has an inverse if, and only if, the equation $Ax = y$ has only one root $x = A^{-1}y$ or, equivalently, if the equation $Ax = 0$ has the only solution $x = 0$. A is

---

[1] Operator addition defined in this way has certain drawbacks. Addition is not generally associative, and in particular $(A + B) - B$ can be different from A.

said to be *invertible* if, first, it is nonsingular and, second, $D_A = R_A = E$. In this case, for any $x \in E$, $AA^{-1}x = A^{-1}Ax = x$, and it is legitimate to write $AA^{-1} = A^{-1}A = I$. If A and B are invertible, the inverse of AB is $B^{-1}A^{-1}$.

2. Bounded Linear Operators

Definition 2. Let E be a normed vector space; any continuous linear mapping of E into E is called a *bounded* linear operator.

If A is a bounded linear operator (see Chapter 3, Note 1)

$$(\forall x \in D_A) \qquad \|Ax\| \le \|A\|\,\|x\|,$$

the norm of A being defined by

$$\|A\| = \sup_{\|x\|\le 1} \|Ax\| = \sup_{x\in D_A} \frac{\|Ax\|}{\|x\|}.$$

From the Cauchy–Schwarz inequality it follows that $|\langle Ax \mid y\rangle|$ is less than $\|Ax\|$ if $\|y\| \le 1$. If $Ax \neq 0$, $|\langle Ax \mid y\rangle|$ reaches the value $\|Ax\|$ for $y = Ax/\|Ax\|$. We then have

$$\|Ax\| = \sup_{\|y\|\le 1} |\langle Ax \mid y\rangle|,$$

and therefore, taking into account the definition of $\|A\|$, the norm of A is also given by

$$\|A\| = \sup_{\|x\|\le 1, \|y\|\le 1} |\langle Ax \mid y\rangle|.$$

Since A is continuous, if the sequence (x_n) converges to x, the sequence (Ax_n) converges to Ax.[2]

Example 1. Let E be the space of the real-valued functions defined on $[a, b]$. The operator A, defined on E by

$$(Af)(\xi) = \int_a^\xi f(t)\,dt,$$

---

[2] Here the convergence of a sequence (x_n) to x means that $\|x_n - x\|$ tends to zero as n tends to ∞.

is continuous, and it is easy to verify that $\|A\| = b - a$.[3] $R_A \neq E$ since Af is a continuously differentiable function whereas in general f is not.

Example 2. Let E be a Hilbert space and a a vector of unit norm. The operator A, defined on E by

$$Ax = \langle a \mid x \rangle \, a,$$

is bounded, and from the Cauchy–Schwarz inequality it follows that $\|A\| = 1$. A is the projection operator on the subspace generated by a. $D_A = E$, but R_A is the subspace generated by a.

Example 3. The position operator defined on the Hilbert space $L^2_m([a,b])$ is bounded since

$$(\forall f \in L^2_m([a,b])) \qquad \|Xf\|^2 = \int_a^b \xi^2 |f(\xi)|^2 \, d\xi \leq b^2 \|f\|^2.$$

Example 4. The linear operator K, defined on $L^2_m(\mathbf{R})$ by

$$(\forall f \in L^2_m(\mathbf{R})) \qquad (Kf)(x) = \int K(x,y) f(y) \, dy,$$

where the function K, called the *kernel of the integral operator*, belongs to $L^2_m(\mathbf{R}^2)$, is called a *Hilbert–Schmidt operator*. It follows from Fubini's theorem that the functions $y \mapsto K(x,y)$ (for almost all x) and $x \mapsto (\int |K(x,y)|^2 \, dy)^{1/2}$ belong to $L^2_m(\mathbf{R})$; furthermore, as a consequence of the Cauchy–Schwarz inequality, the function $y \mapsto K(x,y)f(y)$ is integrable and the function Kf belongs to $L^2_m(\mathbf{R})$, since

$$|(Kf)(x)| = \left| \int K(x,y) f(y) \, dy \right| \leq \int |K(x,y) f(y)| \, dy,$$

or

$$|(Kf)(x)| \leq \left(\int |K(x,y)|^2 \, dy \right)^{1/2} \left(\int |f(y)|^2 \, dy \right)^{1/2}$$

[3] Here the norm of f is defined by

$$\|f\| = \sup_{\xi \in [a,b]} |f(\xi)|.$$

Hence,

$$\|Kf\|^2 \leq \left(\int |K(x,y)|^2 \, dx \, dy \right) \|f\|^2,$$

which shows that the operator K is bounded.

Theorem 1. *Let A and B be two bounded linear operators and λ a scalar; λA, $A+B$ and AB are bounded linear operators, and we have*

$$\|\lambda A\| = |\lambda| \|A\| \tag{1}$$

$$\|A + B\| \leq \|A\| + \|B\| \tag{2}$$

$$\|AB\| \leq \|A\| \, \|B\| \, . \tag{3}$$

(1) is obvious. (2) follows from $\|Ax + Bx\| \leq \|Ax\| + \|Bx\|$, valid for any $x \in D_{A \cap B}$. To prove (3), it is sufficient to note that, since A is bounded, we have

$$\|ABx\| = \|A(Bx)\| \leq \|A\| \, \|Bx\|,$$

which implies that

$$\|AB\| = \sup_{\|x\| \leq 1} \|ABx\| \leq \|A\| \sup_{\|x\| \leq 1} \|Bx\|.$$

If E is a Banach space, the set of bounded linear operators whose domain is E is also a Banach space.[4] A vector space in which a multiplication has been defined is called an *algebra*. If the vector space is normed, we have a *normed algebra*. If it is complete, and if the elements satisfy relation (3), we have a *Banach algebra*. The set of bounded operators whose domain is E is therefore a Banach algebra. It is denoted by $\mathcal{B}(E)$. The identity operator I, whose norm is equal to 1, is the unit element of this algebra.[5]

Example 5. Let A and B be two bounded linear operators defined on a Hilbert space E; then $[A, B] \neq I$.

---

[4] This assumption allows us to define the sum and the product of two operators without restriction.

[5] The set of complex numbers with the usual norm is the simplest example of a Banach algebra.

Assume that A and B are two bounded linear operators such that

$$[A, B] = I;$$

then, for any positive integer n, we have

$$[A^n, B] = nA^{n-1}.$$

This result is easy to prove by induction. Since

$$[A^2, B] = A[A, B] + [A, B]A = 2,$$

the formula is true for $n = 2$. If we assume that it is true for $n > 2$, we can write

$$A^{n-1}B - BA^{n-1} = (n-1)A^{n-2},$$

and therefore,

$$\begin{aligned} A^nB - BA^n &= A^{n-1}(AB - BA) + (A^{n-1}B - BA^{n-1})A \\ &= A^{n-1} + (n-1)A^{n-1} \\ &= nA^{n-1}. \end{aligned}$$

But this result implies that, for any positive integer n,

$$\begin{aligned} n\|A^{n-1}\| &= \|A^nB - BA^n\| \\ &\leq 2\|A^n\| \, \|B\| \\ &\leq 2\|A^{n-1}\| \, \|A\| \, \|B\|. \end{aligned}$$

Hence,

$$n \leq 2\|A\| \, \|B\|,$$

which contradicts the assumption that A and B are bounded.

Since the position operator X defined on the Hilbert space $L^2_m([a, b])$ is bounded (see Example 3), the result we just proved shows that the momentum operator P cannot be bounded.

Theorem 2. *Let $\mathcal{B}(E)$ be the Banach algebra of the bounded linear operators defined on the Banach space E, and (A_n) and (B_n) be two sequences belonging to $\mathcal{B}(E)$ that, in the norm topology of $\mathcal{B}(E)$, converge, respectively, to A and B; if (λ_n) is a sequence of scalars that converges to λ, then the sequences $(\lambda_n A_n)$, $(A_n + B_n)$, and $(A_n B_n)$ converge, respectively, to λA, $A + B$, and AB.*

The first two results follow directly from the continuity, in a normed vector space, of the multiplication by a scalar and the sum of two elements. To prove the last result, it is sufficient to note that

$$\begin{aligned}\|A_nB_n - AB\| &= \|A_n(B_n - B) + (A_n - A)B\| \\ &\le \|A_n(B_n - B)\| + \|(A_n - A)B\| \\ &\le \|A_n\|\,\|B_n - B\| + \|A_n - A\|\,\|B\|,\end{aligned}$$

and since

$$\lim_{n\to\infty} \|A_n - A\| = \lim_{n\to\infty} \|B_n - B\| = 0,$$

the result follows.

Theorem 3. *Let E be a Banach space and A a bounded linear operator on E such that $\|A\| < 1$; $I - A$ is invertible and bounded, and we have*

$$(I - A)^{-1} = \sum_{n=0}^{\infty} A^n.$$

Let $S_n = \sum_{k=0}^{n} A^k$. Let us first prove that (S_n) is a Cauchy sequence. If $p > q$,

$$\begin{aligned}\|S_p - S_q\| &= \|A^{q+1} + A^{q+2} + \ldots + A^p\| \\ &\le \|A^{q+1}\|\,(1 + \|A\| + \|A\|^2 + \ldots + \|A\|^{p-q-1},\end{aligned}$$

for $\|AB\| \le \|A\|\,\|B\|$ implies that $\|A^k\| \le \|A\|^k$.

Since $\|A\| < 1$, the geometric series $\sum_{k=0}^{\infty} \|A\|^k$ is convergent, and therefore,

$$\|S_p - S_q\| \le \|A\|^{q+1}(1 - \|A\|)^{-1}.$$

Because $\|A\| < 1$, given $\epsilon > 0$, there exists $n_0(\epsilon)$ such that, for $p > q > n_0(\epsilon)$, the right-hand side is less than ϵ, and therefore (S_n) is a Cauchy sequence. Since the set of bounded linear operators on E is a Banach algebra, the Cauchy sequence converges to a bounded linear operator. Finally, we can directly verify that

$$(I - A)\sum_{n=0}^{\infty} A^n = (\sum_{n=0}^{\infty} A^n)(I - A) = I,$$

for, taking into account what precedes, the multiplication term by term is legitimate.

Remark. We can directly verify that the series representing $(I-A)^{-1}$ is bounded. We have indeed

$$\|\sum_{n=0}^{\infty} A^n\| \leq \sum_{n=0}^{\infty} \|A\|^n = (I-A)^{-1}.$$

In physical applications, bounded linear operators on l^2 play an important role. These operators are rather easy to manipulate because they share many properties with linear operators defined on finite-dimensional spaces.

Let (e_n) be a Hilbert basis of l^2; if $x = \sum_{n=1}^{\infty} x_n e_n$ belongs to l^2, and A is a bounded linear operator on l^2, it is legitimate to write

$$Ax = \sum_{n=1}^{\infty} x_n A e_n,$$

and, therefore,

$$\langle e_m \mid Ax\rangle = \sum_{n=1}^{\infty} x_n \langle e_m \mid A e_n\rangle.$$

Denoting $\langle e_m \mid Ae_n\rangle$ by a_{mn}, we have

$$\langle e_m \mid Ax\rangle = \sum_{n=1}^{\infty} a_{mn} x_n$$

and

$$Ax = \sum_{m=1}^{\infty}\sum_{n=1}^{\infty} a_{mn} x_n e_m.$$

The numbers a_{mn} are the *matrix elements of the bounded operator* A. If A and B are two bounded linear operators, the matrix elements of the operators $A+B$ and AB are, respectively, $a_{mn}+b_{mn}$ and $\sum_{p=1}^{\infty} a_{mp} b_{pn}$. If λ is a scalar, the matrix elements of λA are λa_{mn}.

We insisted on the fact that for each operator the domain has to be specified. In the case of bounded linear operators it is, however, always possible to assume that the domain coincides with the whole space.

Theorem 4. *Let A be a linear bounded operator defined on a Hilbert space E, and D_A its domain; then there exists a bounded linear operator $\widehat{A}$ whose domain is E and such that, for any $x \in D_A$,*

$$\widehat{A}x = Ax \quad \text{and} \quad \|\widehat{A}\| = \|A\|.$$

This theorem asserts the existence of an extension $\widehat{A}$ of A to the whole space E, which preserves the norm.

Let us first extend A from D_A to its closure $\overline{D_A}$. For any $x \in \overline{D_A}$, there exists a Cauchy sequence (x_n) of vectors belonging to D_A such that $\lim_{n\to\infty} x_n = x$. Since A is bounded,

$$\|Ax_p - Ax_q\| = \|A(x_p - x_q)\| \le \|A\| \, \|x_p - x_q\|,$$

which shows that (Ax_n) is a Cauchy sequence. Define then the extension A' of A to $\overline{D_A}$ by

$$A'x = \lim_{n\to\infty} Ax_n.$$

It is clear that the extension A' does not depend upon the particular Cauchy sequence (x_n) converging to x. A' is obviously linear, and we have

$$\|A'x\| = \lim_{n\to\infty} \|Ax_n\| \le \|A\| \lim_{n\to\infty} \|x_n\| = \|A\| \, \|x\|;$$

hence, A' is bounded and $\|A'\| \le \|A\|$. But we also have

$$\sup_{x\in\overline{D_A}} \frac{\|A'x\|}{\|x\|} \ge \sup_{x\in D_A} \frac{\|A'x\|}{\|x\|} = \sup_{x\in D_A} \frac{\|Ax\|}{\|x\|},$$

i.e., $\|A'\| \ge \|A\|$. Therefore, $\|A'\| = \|A\|$.

The extension of A' from $\overline{D_A}$ to E is straightforward. Since E is a Hilbert space and $\overline{D_A}$ is a closed subspace of E, $\overline{D_A}$ is also a Hilbert space, and we can write $E = \overline{D_A} \oplus \overline{D_A}^{\perp}$. The extension $\widehat{A}$ of A to E is then defined by

$$\widehat{A}x = \begin{cases} A'x, & \text{if } x \in \overline{D_A}; \\ 0, & \text{if } x \in \overline{D_A}^{\perp}. \end{cases}$$

Indeed, if x belongs to E, we have

$$x = x_1 + x_2,$$

where $x_1 \in \overline{D_A}$ and $x_2 \in \overline{D_A}^{\perp}$, and taking into account the linearity of A, we have

$$\widehat{A}x = A'x_1.$$

Moreover,

$$\sup_{\|x\|=1} \|\widehat{A}x\| = \sup_{\|x_1\|=1} \|A'x_1\| = \sup_{\|x_1\le 1} \|A'x_1\|;$$

hence,

$$\|\widehat{A}\| = \|A'\| = \|A\|.$$

Definition 3. A linear operator A on a Hilbert space E is *Hermitian* or *symmetric* if

$$(\forall x \in E)\,(\forall y \in E) \quad \langle x \mid Ay \rangle = \langle Ax \mid y \rangle.$$

Let x and y be two vectors in l^2 and (e_n) a Hilbert basis in l^2; we have

$$x = \sum_n x_n e_n \quad \text{and} \quad y = \sum_n y_n e_n.$$

If a_{mn} is the matrix element $\langle e_m \mid Ae_n \rangle$ of a Hermitian operator A on l^2, then

$$\langle x \mid Ay \rangle = \sum_{mn} a_{mn} \overline{x_m} y_n = \langle Ax \mid y \rangle = \sum_{mn} \overline{a_{nm} x_m} y_n.$$

Therefore, for any pair n, m of indices, we have

$$a_{mn} = \overline{a_{nm}}.$$

Theorem 5. *Let A be a bounded linear operator on a Hilbert space E; there exists a unique bounded linear operator A^+, called the* adjoint *of A, such that, for any pair x, y in E, we have*

$$\langle y \mid Ax \rangle = \langle A^+ y \mid x \rangle.$$

A and A^+ have the same norm.[6]

The linearity of A^+ is a direct consequence of the sesquilinearity of the form $\langle y \mid Ax \rangle$. From the Cauchy–Schwarz inequality, it follows that

$$|\langle y \mid Ax \rangle| \le \|A\|\,\|x\|\,\|y\|,$$

which implies that

$$|\langle A^+ y \mid x \rangle| \le \|A\|\,\|x\|\,\|y\|.$$

Hence, A^+ is bounded and its norm is equal to $\|A\|$. The unicity of A^+ is obvious.

---

[6] The adjoint of an operator A is also frequently denoted A^*; physicists, however, tend to favor the notation A^+.

If A and B are two bounded linear operators on a Hilbert space E and λ a scalar, we have

$$\begin{aligned} (A^+)^+ &= A; & (\lambda A)^+ &= \overline{\lambda} A^+; \\ (A+B)^+ &= A^+ + B^+; & (AB)^+ &= B^+ A^+; \\ (A^+)^{-1} &= (A^{-1})^+ . \end{aligned}$$

These results readily follow from the definition of the adjoint. The last result supposes that A has an inverse.

In l^2, between the matrix elements of A^+ and A we have the relations

$$(\forall (m,n)) \qquad a^+_{mn} = \overline{a_{nm}},$$

which show that, in l^2, a bounded Hermitian operator is identical to its adjoint.

Example 6. Let A be a linear operator on l^2 defined by $Ae_n = e_{n+1}$, where (e_n) is a Hilbert basis in l^2. It is readily verified that A is bounded. From $\langle A^+ e_{n+1} \mid e_n \rangle = \langle e_{n+1} \mid Ae_n \rangle = 1$ it follows that $A^+ e_{n+1} = e_n$. This relation, however, does not define $A^+ e_1$. Since, for any $x \in l^2$, Ax is orthogonal to e_1, we have $\langle e_1 \mid Ax \rangle = \langle A^+ e_1 \mid x \rangle = 0$ and therefore $A^+ e_1 = 0$. The inverse A^{-1} of A is such that $A^{-1} e_{n+1} = e_n$, but $A^{-1} e_1$ is not defined because its image by A cannot be equal to e_1. A^+ is, therefore, an extension of A^{-1}.

Definition 4. Let A be a bounded linear operator on a Hilbert space E; A is said to be *unitary* if $A^{-1} = A^+$; A is said to be *self-adjoint* if $A = A^+$.

A bounded Hermitian operator is self-adjoint;[7] a self-adjoint operator (bounded or unbounded) is Hermitian.

Theorem 6. *A bounded linear operator on a Hilbert space E is a projection operator if, and only if, it is Hermitian and idempotent.*

Let P be a projection operator on a closed subspace V of E. It is bounded since, for any x in E, $\|Px\| \le \|x\|$, and obviously it is idempotent. It is also Hermitian since, for all x and y belonging to E, we have

$$\langle Px \mid y \rangle = \langle Px \mid Py \rangle = \langle x \mid Py \rangle.$$

Conversely, if P is an idempotent Hermitian operator, for all x and y in E, we have

$$\langle Px \mid y \rangle = \langle P^2 x \mid y \rangle = \langle Px \mid Py \rangle.$$

[7] But there exist unbounded Hermitian operators that are not self-adjoint (see Section 4).

Hence, for any y in E, $y - Py$ is orthogonal to $V = P(E)$. Since V is closed,[8] P is a projection operator.

Corollary 1. *If P is a projection operator on a Hilbert space E, $I - P$ is a projection operator on E.*

It is sufficient to verify that $I - P$ is Hermitian and idempotent. The first property is obvious, and the second one follows from

$$(I - P)^2 = I - 2P + P^2 = I - 2P + P = I - P.$$

Corollary 2. *Let P_1 and P_2 be two projection operators on a Hilbert space E, $P_1 + P_2$ is a projection operator on E if, and only if, $P_1P_2 = 0$.*

The proof is straightforward.

An operator on l^2, represented by a diagonal matrix whose diagonal elements are equal to either zero or 1, is a projection operator since it is Hermitian and idempotent.

Definition 5. A linear operator A in a Hilbert space E is *isometric* (or is an *isometry*) if

$$(\forall x \in D_A) \qquad \|Ax\| = \|x\|.$$

An isometric operator is bounded and its norm is equal to 1.

Theorem 7. *Let A be an isometric operator in a Hilbert space E.*

(1) *For all x and y in D_A, we have*

$$\langle Ax \mid Ay \rangle = \langle x \mid y \rangle.$$

(2) *If $D_A = R_A = E$, A is unitary.*

From

$$\|A(x + y)\|^2 = \|x + y\|^2 \quad \text{and} \quad \|A(x + iy)\|^2 = \|x + iy\|^2,$$

it follows that

$$\begin{aligned} \Re\langle Ax \mid Ay \rangle &= \Re\langle x \mid y \rangle \\ \Im\langle Ax \mid Ay \rangle &= \Im\langle x \mid y \rangle, \end{aligned}$$

which proves (1).

If, for all x and y in E, we have

$$\langle Ax \mid Ay \rangle = \langle x \mid A^+Ay \rangle = \langle x \mid y \rangle,$$

---

[8] Because $V = \ker(I - P)$.

i.e.,

$$\langle x \mid y - A^+ Ay \rangle = 0,$$

then $A^+ A = I$. Moreover, $R_A = E$ implies $D_{A^{-1}} = E$, and, therefore, we have

$$A^+ - A^{-1} = (A^+ - A^{-1})AA^{-1} = (I - A^{-1}A)A^{-1} = 0,$$

which proves (2).

Remark 1. Let A be a unitary operator on a Hilbert space E, if (e_n) is a Hilbert basis, (Ae_n) is also a Hilbert basis. The set of unitary operators on a Hilbert space E is a group under operator multiplication. It is the *group of automorphisms* of the Hilbert space.

Remark 2. The operator A defined in Example 6 is a nonunitary isometric operator.

Example 7. **The Fourier operator**. The Fourier transform of a function f in $L^2_m(\mathbf{R})$ does not exist in general. If, however, f is in $L^1_m(\mathbf{R}) \cap L^2_m(\mathbf{R})$, it turns out that $\widehat{f}$ is in $L^2_m(\mathbf{R})$, and that $\|\widehat{f}\|_2 = \|f\|_2$. This isometry of $L^1_m(\mathbf{R}) \cap L^2_m(\mathbf{R})$ into $L^2_m(\mathbf{R})$ can be extended to an isometry of $L^2_m(\mathbf{R})$ into itself, which will define the Fourier transform of any function in $L^2_m(\mathbf{R})$.

Consider the operator F, called the *Fourier operator*, defined in $L^2_m(\mathbf{R})$ by

$$(Ff)(x) = \frac{1}{\sqrt{2\pi}} \int_{-\infty}^{\infty} e^{ixy} f(y)\, dy.$$

The integral converges if f is integrable, and therefore the domain of F is $L^1_m(\mathbf{R}) \cap L^2_m(\mathbf{R})$. We shall prove that this domain is dense in $L^2_m(\mathbf{R})$,[9] and that F is an isometry.

Let (h_n) be the Hilbert basis that consists of the normed Hermite functions defined by[10]

$$h_n(x) = (\sqrt{\pi} 2^n n!)^{-1/2} (-1)^n e^{x^2/2} \frac{d^n}{dx^n} e^{-x^2},$$

Since h_n is integrable and the sequence (h_n) is complete in $L^2_m(\mathbf{R})$, it follows that $L^1_m(\mathbf{R}) \cap L^2_m(\mathbf{R})$ is dense in $L^2_m(\mathbf{R})$. Furthermore, n integrations by

[9] See also Chapter 2, Problem 5.

[10] See Chapter 3, Example 3.4 and Theorem 9.

parts yield

$$\begin{aligned}\frac{1}{\sqrt{2\pi}}\int_{-\infty}^{\infty} e^{ixy} h_n(y)\,dy &= \frac{(\sqrt{\pi}2^n n!)^{-1/2}}{\sqrt{2\pi}}\int_{-\infty}^{\infty} e^{-y^2}\frac{d^n}{dy^n}e^{(y^2+2ixy)/2}\,dy \\ &= \frac{(\sqrt{\pi}2^n n!)^{-1/2}}{\sqrt{2\pi}}e^{x^2/2}\int_{-\infty}^{\infty} e^{-y^2}\frac{d^n}{dy^n}e^{\left((y+ix)^2/2\right)}\,dy \\ &= \frac{(\sqrt{\pi}2^n n!)^{-1/2}}{i^n\sqrt{2\pi}}e^{x^2/2}\int_{-\infty}^{\infty} e^{-y^2}\frac{d^n}{dx^n}e^{\left((y+ix)^2/2\right)}\,dy \\ &= \frac{(\sqrt{\pi}2^n n!)^{-1/2}}{i^n\sqrt{2\pi}}e^{x^2/2}\frac{d^n}{dx^n}\int_{-\infty}^{\infty} e^{(-x^2-y^2+2ixy)/2}\,dy \\ &= (-i)^n(\sqrt{\pi}2^n n!)^{-1/2}e^{x^2/2}\frac{d^n}{dx^n}e^{-x^2},\end{aligned}$$

i.e.,

$$Fh_n = i^n h_n.$$

Since (h_n) is a Hilbert basis, for any function f in $L^2_m(\mathbf{R})$, we have

$$f = \sum_{n=0}^{\infty} c_n h_n,$$

where $c_n = \langle h_n \mid f\rangle$, and from Parseval's equality it follows that

$$\|f\|_2^2 = \sum_{n=0}^{\infty} |c_n|^2.$$

F being linear, for any f in $L^1_m(\mathbf{R}) \cap L^2_m(\mathbf{R})$, we have

$$Ff = \sum_{n=0}^{\infty} c_n F h_n = \sum_{n=0}^{\infty} i^n c_n h_n,$$

and, here again, from Parseval's equality it follows that

$$\|Ff\|_2 = \|f\|_2.$$

F is therefore isometric, and it is possible to define an extension $\widehat{F}$ of F whose domain will be the whole space $L^2_m(\mathbf{R})$ (Theorem 4).

We can define $\widehat{F}$ by

$$(\forall f \in L^2_m(\mathbf{R})) \qquad \widehat{F}f = \sum_{n=0}^{\infty} c_n F h_n,$$

where $c_n = \langle h_n \mid f \rangle$.

If f and g both belong to $L^2_m(\mathbf{R})$, we have

$$\langle \widehat{F}f \mid \widehat{F}g \rangle = \langle f \mid g \rangle.$$

This important formula, which follows from Theorem 7(1), is called the *Parseval's formula.*[11]

Since $\widehat{F}$ is defined on the whole Hilbert space $L^2_m(\mathbf{R})$, it is an isometry, and we have

$$\widehat{F}^+ = \widehat{F}^{-1},$$

where F^{-1} is given by (Chapter 2, Theorem 9)

$$(F^{-1}f)(x) = \frac{1}{\sqrt{2\pi}} \int_{-\infty}^{\infty} \mathrm{e}^{-ixy} f(y)\, dy.$$

This formula is actually slightly different from the formula given in Chapter 2. It is obtained after a trivial change of variables.

$\widehat{F}$ can also be defined without referring to Hermite functions. If f is in $L^2_m(\mathbf{R})$ but not in $L^1_m(\mathbf{R})$, consider

$$f_n = f 1_{[-n,n]},$$

which , for any positive integer n, is in $L^1_m(\mathbf{R}) \cap L^2_m(\mathbf{R})$. We have

$$\lim_{n\to\infty} \|f - f_n\|_2 = 0,$$

i.e., f is the limit of the sequence (f_n) in the norm topology of $L^2_m(\mathbf{R})$, and we shall define $\widehat{F}$ by

$$\widehat{F}f = \lim_{n\to\infty} F f_n,$$

i.e.,

$$(\widehat{F}f)(x) = \lim_{n\to\infty} \frac{1}{\sqrt{2\pi}} \int_{-n}^{n} \mathrm{e}^{ixy} f(y)\, dy.$$

This result can be written differently. Since, in the norm topology of $L^2_m(\mathbf{R})$,

$$\lim_{n\to\infty} F f_n = \widehat{F}f,$$

---

[11] In general, the extension of F is not denoted by a particular symbol, and even when f or g does not belong to $L^1_m(\mathbf{R})$, Parseval's formula is written $\langle Ff \mid Fg \rangle = \langle f \mid g \rangle$.

then, for any function g in $L^2_m(\mathbf{R})$, we have

$$\lim_{n\to\infty} \langle g \mid Ff_n \rangle = \langle g \mid \widehat{F}f \rangle.$$

Taking $g = 1_{[0,x]}$ yields

$$\lim_{n\to\infty} \int_0^x d\xi \, \frac{1}{\sqrt{2\pi}} \int_{-n}^{n} e^{i\xi y} f(y)\, dy = \int_0^x (\widehat{F}f)(\xi)\, d\xi,$$

and applying Fubini's theorem to the left-hand side (which is legitimate) gives

$$\lim_{n\to\infty} \frac{1}{\sqrt{2\pi}} \int_{-n}^{n} \frac{e^{ixy} - 1}{iy} f(y)\, dy = \int_0^x (\widehat{F}f)(\xi)\, d\xi.$$

Since $y \mapsto (e^{ixy} - 1)/iy$ is in $L^2_m(\mathbf{R})$, as a consequence of the Cauchy–Schwarz inequality, the function $y \mapsto f(y)(e^{ixy} - 1)/iy$ is in $L^1_m(\mathbf{R})$. Therefore, the limit $n \to \infty$ can be taken, and for almost all x in $\mathbf{R}$, we finally have

$$(\widehat{F}f)(x) = \frac{1}{\sqrt{2\pi}} \frac{d}{dx} \int_{-\infty}^{\infty} \frac{e^{ixy} - 1}{iy} f(y)\, dy.$$

The extension $\widehat{F}$ of the Fourier operator F defined in this way is called the *Fourier–Plancherel operator.*

3. Compact Operators

Most statements about linear operators on finite-dimensional spaces have nice generalizations to a particular class of linear operators on infinite-dimensional spaces, namely, to the compact operators.

Definition 6. A linear operator A, defined on a Hilbert space E, is said to be *compact*, or *completely continuous*, if the closure of the image of a bounded subset of E is compact.

Any compact linear operator A is bounded. In fact if $B(0,1) = \{\, x \mid \|x\| \le 1 \,\}$ is the closed ball of unit radius centered at the origin, $\overline{A\big(B(0,1)\big)}$ is, by definition, compact, and therefore

$$\sup_{\|x\|=1} \|Ax\| < \infty,$$

which means that A is bounded.

On the other hand, any bounded operator is not necessarily compact; for an operator to be compact is more restrictive than to be bounded.

Example 1. Although the indentity operator I is bounded, it is not compact. This follows from the fact that the ball $B(0,1)$, which is its own image by I, is not compact. To see this, consider a Hilbert basis (e_n). If $m \neq n$, then $\|e_m - e_n\| = \sqrt{2}$. Therefore, from the covering of $B(0,1)$ by the set of balls $\left(\cup_n B(e_n,\rho)\right) \cup \left(\cup_n B(-e_n,\rho)\right)$, where the radius ρ is such that $1 < \rho < \sqrt{2}$, it is not possible to extract a finite covering.

Remark. Let A and B be two compact operators; the image of any bounded subset Ω by B is bounded (because B is bounded), and since A is compact, the closure of $AB\Omega$ is compact. That is, the product of two compact operators is compact.

This result shows that the inverse A^{-1} of a compact operator A is not compact; otherwise $AA^{-1} = I$ would be compact in contradiction with what has just been proved.

If A is compact and B is bounded, then the product AB is compact. Therefore, if A is compact, its inverse A^{-1} is not even bounded.

Example 2. Any closed bounded subset of a finite-dimensional metric space being compact, any bounded linear operator A, on a Hilbert space E, whose range is finite-dimensional, is compact. Such an operator is said to be of *finite rank*. A projection operator P on a subspace F of E is compact if, and only if, F is finite-dimensional.

The following criterion is very useful to prove that an operator is compact.

Criterion. *Let (x_n) be a bounded sequence in a Hilbert space E; the operator A is compact if, and only if, there is a subsequence (x_{n_k}) such that the sequence (Ax_{n_k}) converges in E.*

This is a direct consequence of the fact that any infinite subset of a compact subset of a metric space has always a limit point.

Theorem 8. *The limit A of convergent sequence (A_n) of compact linear operators, defined on a Hilbert space E, is compact.*

Let (x_n) be a bounded sequence in E. Since A_1 is compact there is a subsequence (x_{n_1}) such that the sequence $(A_1x_{n_1})$ converges in E. A_2 being compact, there is a subsequence (x_{n_2}) of the bounded sequence (x_{n_1}), such that the sequence $(A_2x_{n_2})$ is convergent. Considering successively the operators $A_1, A_2, \ldots, A_m, \ldots$ we can determine the subsequences $(x_n^1), (x_n^2), \ldots, (x_n^m) \ldots$ where (x_n^m) is a subsequence of the sequences (x_n^l) $(l = 1, 2, \ldots, m-1)$. All the sequences $(A_l x_n^m)$ $(l = 1, 2, \ldots, m)$ are convergent. From these considerations it is clear that the particular *diagonal* sequence $(x_n^n) = (x_1^1, x_2^2, \ldots, x_m^m, \ldots)$ is such that, for any $k \in \mathbf{N}$, $(A_k x_n^n)$ converges in E. To prove that A is compact, we shall establish that the sequence (Ax_n^n) is also convergent. Since E is complete, it is sufficient to prove that (Ax_n^n) is a Cauchy sequence. Write

$$\|Ax_n^n - Ax_m^m\| \leq \|Ax_n^n - A_k x_n^n\| + \|A_k x_n^n - A_k x_m^m\| + \|A_k x_m^m - Ax_m^m\|.$$

(x_n) being bounded (i.e., for all n, $\|x_n\| < M$), for any positive ϵ, let k be such that $\|A - A_k\| < \epsilon/3M$, and let $n_0(\epsilon)$ be such that, for $n > n_0(\epsilon)$ and $m > n_0(\epsilon)$, $\|A_k x_n^n - A_k x_m^m\| \leq \epsilon/3$; then, for $n > n_0(\epsilon)$ and $m > n_0(\epsilon)$, $\|Ax_n^n - Ax_m^m\| < \epsilon$, proving what has been claimed.

In particular, the limit of a convergent sequence of finite-rank operators is compact.

Example 3. The Hilbert-Schmidt operator K on the Hilbert space $L_m^2(\mathbf{R}^2)$ (see Example 2.4) is compact. This follows from the preceding theorem. Let (f_n) be a Hilbert basis for $L_m^2(\mathbf{R})$; the function belonging to $L_m^2(\mathbf{R}^2)$, we can write

$$K(x,y) = \sum_{m=1}^{\infty}\sum_{n=1}^{\infty} c_{mn} f_m(x) f_n(y).$$

Let

$$K_r(x,y) = \sum_{m=1}^{r}\sum_{n=1}^{r} c_{mn} f_m(x) f_n(y).$$

The integral operator corresponding to the kernel $(x,y) \mapsto K_r(x,y)$ is finite-rank since, for any f in $L_m^2(\mathbf{R})$,

$$\begin{aligned}(K_r f)(x) &= \int K_r(x,y) f(y)\,dy \\ &= \sum_{m=1}^{r}\sum_{n=1}^{r} c_{mn} f_m(x) \int f_n(y) f(y)\,dy \\ &= \sum_{m=1}^{r}\sum_{n=1}^{r} c_{mn} \langle f_n \mid f\rangle f_m(x).\end{aligned}$$

K_r is therefore compact, and from

$$\lim_{r\to\infty} \int \big(K(x,y) - K_r(x,y)\big)^2\,dx\,dy = 0,$$

it follows that K is compact.

4. Unbounded Linear Operators

If A is a bounded operator on Hilbert space E, we have seen (Theorem 5) that there exists an operator A^+ called the adjoint of A such that

$$(\forall x \in E) \quad (\forall y \in E) \qquad \langle y \mid Ax\rangle = \langle A^+y \mid x\rangle.$$

If A is not bounded, the adjoint A^+ could be defined by

$$(\forall x \in D_A) \quad (\forall y \in D_{A^+}) \qquad \langle y \mid Ax\rangle = \langle A^+y \mid x\rangle.$$

If A^+y is defined, the mapping $x \mapsto \langle A^+y \mid x\rangle$ is continuous. The domain of A^+ is, therefore, the set of all vectors $y \in E$ such that $x \mapsto \langle y \mid Ax\rangle$ is continuous. Moreover, for A^+y to be determined uniquely by y, D_A must be dense in E; otherwise we could find in E a vector $x_0 \neq 0$ orthogonal to D_A such that

$$(\forall x \in D_A) \qquad \langle A^+y + x_0 \mid x\rangle = \langle A^+y \mid x\rangle.$$

Given the importance of adjoint operators in physical applications, we shall consider only densely defined unbounded operators.

The relation $\langle y \mid Ax\rangle = \langle A^+y \mid x\rangle$ shows that, if $x \in \ker A$, then $\langle A^+y \mid x\rangle = 0$; hence,

$$\ker A = (R_{A^+})^\perp.$$

Similarly, if $y \in \ker A^+$, then $\langle y \mid Ax\rangle = 0$; hence,

$$\overline{R_A} = (\ker A^+)^\perp.$$

Note that, since R_A has no reason to be closed, $(\ker A^+)^\perp$ is equal to the closure of R_A.

Example 1. The operator A defined in l^2 by

$$Ae_0 = 0$$
$$Ae_n = \sqrt{n}e_{n-1},$$

where (e_n) $(n \in \{0,1,2,\dots\}$[12] is a Hilbert basis in l^2, is called the *annihilation operator*.

---

[12] Here, to respect the established usage in quantum mechanics, the index n belongs to the set $\{0,1,2,\dots\}$.

The domain of A is the set

$$D_A = \{x | x = \sum_{n=0}^{\infty} c_n(x) e_n, \sum_{n=1}^{\infty} n |c_n(x)|^2 < \infty\}.$$

A is unbounded, but D_A, which contains all finite linear combinations of the basis vectors, is dense in l^2. The adjoint A^+ exists and is defined by

$$\langle y \mid Ax \rangle = \langle A^+ y \mid x \rangle,$$

where

$$x = \sum_{n=0}^{\infty} c_n(x) e_n \in D_A \quad \text{and} \quad y = \sum_{m=0}^{\infty} c_m(y) e_m \in D_{A^+}.$$

Thus,

$$\begin{aligned} \langle A^+ y \mid x \rangle &= \langle \sum_{m=0}^{\infty} c_m(y) e_m \mid \sum_{n=1}^{\infty} c_n(x) \sqrt{n} e_{n-1} \rangle \\ &= \sum_{n=1}^{\infty} c_{n-1}(y) c_n(x) \sqrt{n} \\ &= \langle \sum_{n=0}^{\infty} c_n(y) \sqrt{n+1} e_{n+1} \mid \sum_{m=0}^{\infty} c_m(x) e_m \rangle, \end{aligned}$$

which shows that

$$A^+ y = \sum_{n=0}^{\infty} c_n(y) \sqrt{n+1} e_{n+1},$$

that is,

$$A^+ e_n = \sqrt{n+1} e_{n+1} \qquad (n = 0, 1, 2, \ldots).$$

A^+ is unbounded, and we have $D_{A^+} = D_A$. A^+ is called the *creation operator.*

Definition 7. An operator A in a Hilbert space E is said to be *closed* if, for any sequence (x_n) belonging to D_A, the relations

$$\lim_{n \to \infty} x_n = x \quad \text{and} \quad \lim_{n \to \infty} A x_n = y$$

imply that

$$x \in D_A \quad \text{and} \quad Ax = y.$$

Note that closedness differs from continuity. If A is continuous, then the existence of $\lim_{n \to \infty} x_n$ implies the existence of $\lim_{n \to \infty} A x_n$; if A is only closed, the convergence of the sequence (x_n) does not imply the convergence of the sequence (Ax_n).

The *graph* G_A of an operator A being the set of all the ordered pairs (x, Ax), where $x \in D_A$, A is closed if its graph is closed. If A is not closed, but if the closure $\overline{G_A}$ of its graph is the graph of an operator $\overline{A}$, A is *closable*, and $\overline{A}$, which is the minimal closed extension of A, is called the *closure* of A. To define the closure $\overline{A}$ of A we have, therefore, to add to D_A all the vectors $x \in \overline{D_A}$ that are limits of sequence (x_n) such that (Ax_n) converges and to let

$$\overline{A}x = \lim_{n \to \infty} Ax_n.$$

All bounded operators are closed, but an arbitrary closed operator is, in general, unbounded.

Theorem 9. *If A is a densely defined operator, its adjoint A^+ is closed.*

Let (x_n) be a convergent sequence belonging to D_{A^+} such that the sequence (A^+x_n) is also convergent. Let

$$\lim_{n \to \infty} x_n = x \quad \text{and} \quad \lim_{n \to \infty} A^+x_n = y.$$

For any z in D_A, we have

$$\langle x_n \mid Az \rangle = \langle A^+x_n \mid z \rangle,$$

and, since the scalar product is continuous, we obtain

$$\langle x \mid Az \rangle = \langle y \mid z \rangle,$$

which shows that x belongs to D_{A^+} and that

$$y - \lim_{n \to \infty} A^+x_n = A^+x.$$

Thus A^+ is closed.

Theorem 10. *Let A be a densely defined operator; if $A \subset B$, then $B^+ \subset A^+$.*

If B is an extension of A, for all x in D_A, $Bx = Ax$. Thus $D_A \subset D_B$, which shows that D_B is dense, and that B^+ exists. From the relation

$$(\forall x \in D_B) \quad (\forall y \in D_{B^+}) \qquad \langle y \mid Bx \rangle = \langle B^+y \mid x \rangle,$$

it follows that

$$(\forall x \in D_A) \quad (\forall y \in D_{B^+}) \qquad \langle y \mid Ax \rangle = \langle B^+y \mid x \rangle.$$

If y is in D_{B^+} it is also in D_{A^+}, i.e., $D_{B^+} \subset D_{A^+}$. Moreover, the relation

$$(\forall x \in D_A) \quad (\forall y \in D_{A^+}) \qquad \langle y \mid Ax \rangle = \langle A^+ y \mid x \rangle$$

implies that

$$(\forall y \in D_{B^+} \subset D_{A^+}) \qquad A^+ y = B^+ y,$$

which completes the proof.

Theorem 11. *If the adjoint A^+ of A is densely defined, then A^{++} is an extension of A.*

For all $x \in D_A$ and all $y \in D_{A^+}$, we have

$$\langle y \mid Ax \rangle = \langle A^+ y \mid x \rangle,$$

or, taking the complex conjugates,

$$\langle Ax \mid y \rangle = \langle x \mid A^+ y \rangle.$$

Since D_{A^+} is dense, A^{++} exists, and

$$\langle A^{++} x \mid y \rangle = \langle x \mid A^+ y \rangle,$$

that is,

$$\langle A^{++} x \mid y \rangle = \langle Ax \mid y \rangle.$$

Therefore, if x belongs to D_A it also belongs to $D_{A^{++}}$, and, since the preceding relation is true for all y which belongs to the dense domain of A^+, we have $Ax = A^{++}x$. Thus $A \subset A^{++}$.

Definition 8. (1) An operator A, defined in a Hilbert space E, is *Hermitian* or *symmetric* if, for all x and y in D_A,

$$\langle y \mid Ax \rangle = \langle Ay \mid x \rangle.$$

(2) An operator A in a Hilbert space E is *self-adjoint* if it is densely defined and if $A^+ = A$.

For unbounded operators, it is essential to make a clear distinction between Hermitian and self-adjoint operators. A self-adjoint operator is Hermitian, but, as will be shown in Example 3, a Hermitian operator is not necessarily self-adjoint. In particular, the adjoint of a Hermitian operator, whose domain is not dense, does not even exist. The condition $A^+ = A$, for A to be self-adjoint, carries with it the following requirements: D_A is dense and $D_{A^+} = D_A$.

If A is a densely defined Hermitian operator, A^+ exists and $D_{A^+} \supset D_A$. A^+ is, therefore, densely defined, A^{++} exists, and, as a consequence of Theorems 10 and 11, we have $A \subset A^{++} \subset A^+$.

Since A^+ is closed, the relation $A \subset A^+$ shows that a densely defined Hermitian operator is closable.

Example 2. *The position operator in $L^2_m(\mathbf{R})$.* The position operator X in $L^2_m(\mathbf{R})$ is unbounded, but its domain D_X is dense; that is, for any f in $L^2_m(\mathbf{R})$, there exists a sequence (f_n) in D_X that converges to f. To prove this, it is sufficient to consider the sequence defined by

$$f_n(\xi) = \begin{cases} f(\xi), & \text{if } |\xi| \le n; \\ 0, & \text{otherwise.} \end{cases}$$

For all n, f_n belongs to D_X since $\xi \mapsto \xi f_n(\xi)$ is in $L^2_m(\mathbf{R})$. Moreover, for any positive ϵ, there exists $n_0(\epsilon)$ such that, for any $n \ge n_0(\epsilon)$, we have

$$\|f - f_n\|^2 = \int_{-\infty}^{-n} |f(\xi)|^2 \, d\xi + \int_n^{\infty} |f(\xi)|^2 \, d\xi < \epsilon.$$

X is Hermitian since, for all functions f and g in D_X, we have

$$\begin{aligned} \langle g \mid Xf \rangle &= \int_{-\infty}^{\infty} \overline{g(\xi)} \xi f(\xi) \, d\xi \\ &= \langle Xg \mid f \rangle. \end{aligned}$$

X being Hermitian and densely defined, X^+ exists and is an extension of X. In fact $D_{X^+} = D_X$, that is, X is self-adjoint. To prove this, it is sufficient to show that $D_{X^+} \subset D_X$. Let $h = X^+ g$, where g is in D_{X^+}; then, for any $f \in D_X$,

$$\langle g \mid Xf \rangle = \langle X^+ g \mid f \rangle = \langle h \mid f \rangle,$$

i.e.,

$$\int_{-\infty}^{\infty} \left(\xi \overline{g(\xi)} - \overline{h(\xi)}\right) f(\xi) \, d\xi = 0$$

Since this is true for any $f \in D_X$, and since D_X is dense, then the function $\xi \mapsto \xi g(\xi) - h(\xi)$ is null almost everywhere. Since h belongs to $L^2_m(\mathbf{R})$, so does $\xi \mapsto \xi g(\xi)$, which proves that g belongs to D_X.

Example 3. *The momentum operator in $L^2_m([a,b])$.* The momentum operator P in $L^2_m([a,b])$ is defined by

$$(\forall f \in D_P) \qquad Pf = -if',$$

where the domain D_P is the set of functions $f \in L^2_m([a,b])$, whose derivatives f', in the sense of distributions, are in $L^2_m([a,b])$, and which satisfy certain boundary conditions.

(1) If the boundary conditions have the form $f(a) = f(b) = 0$, as it is often required in quantum mechanics, *P is Hermitian but not self-adjoint.* In fact, since f and f' are in $L^2_m([a,b])$, f is continuous,[13], and the boundary conditions make sense. The integration-by-parts formula being valid for distributions in D_P[14], we have

$$\begin{aligned}\langle g \mid Pf\rangle &= \int_a^b \overline{g(\xi)}\big(-if'(\xi)\big)\,d\xi \\ &= \Big[-i\overline{g(\xi)}f(\xi)\Big]_a^b + \int_a^b \overline{\big(-ig'(\xi)\big)}f(\xi)\,d\xi.\end{aligned}$$

Taking into account the boundary conditions, the first term in the right-hand side is equal to zero. Hence,

$$(\forall f \in D_P)\quad(\forall g \in D_P)\qquad \langle g \mid Pf\rangle = \langle Pg \mid f\rangle,$$

which shows that P is Hermitian. D_P being dense, P^+ exists, and

$$P^+g = -ig',$$

where g' is the derivative of g in the sense of distributions. But, since the above result has been obtained without requiring g to satisfy any boundary conditions, D_P is strictly contained in D_{P^+}. P^+ is an extension of P; P is not self-adjoint. Consider now the operator P^{++}. For any function h in $D_{P^{++}}$, we have

$$\begin{aligned}\langle g \mid P^{++}h\rangle &= \int_a^b \overline{g(\xi)}\big(-ih'(\xi)\big)\,d\xi \\ &= \int_a^b \overline{\big(-ig'(\xi)\big)}h(\xi)\,d\xi. \\ &= \langle P^+g \mid h\rangle\end{aligned}$$

---

13 See appendix Theorem A1.

14 See appendix Theorem A2.

if

$$\left[-i\overline{g(\xi)}h(\xi) \right]_a^b = 0.$$

Since $g(a)$ and $g(b)$ are arbitrary, this condition is satisfied if, and only if, $h(a) = h(b) = 0$, i.e., $P^{++} \subset P$. Since the relation $P \subset P^{++}$ always holds, we have

$$P = P^{++},$$

which implies that P is closed.

(2) If the boundary conditions are $f(a) = f(b) \neq 0$, *P is self-adjoint.* In fact, proceeding as above, we verify that P is Hermitian if

$$\left[-i\overline{g(\xi)}f(\xi) \right]_a^b = 0.$$

This condition is satisfied if $g(a) = g(b)$. In this case,

$$P^+ g = -ig' \quad \text{and} \quad D_{P^+} = D_P,$$

which shows that P is self-adjoint.

(3) If $f(a) = e^{i\theta} f(b) \neq 0$, where θ is real, it is easy to prove that g has to satisfy the same boundary conditions for P to be Hermitian, and that, in this case, *P is self-adjoint.*

The operator P defined in (1) has an infinite number of self-adjoint extensions.

This example illustrates how crucial it is to specify clearly the domain of an unbounded operator.

Example 4. *The momentum operator in $L^2_m(\mathbf{R})$.* The momentum operator P in $L^2_m(\mathbf{R})$ is defined by

$$(\forall f \in D_P) \qquad Pf = -if',$$

where the domain D_P is the set of functions $f \in L^2_m(\mathbf{R})$, whose derivatives f', in the sense of distributions, are in $f \in L^2_m(\mathbf{R})$. *This operator is self-adjoint.* To prove it we have to perform an integration by parts, and we need to know the behavior of $f(\xi)$ when ξ tends to $\pm\infty$. The fact that f belongs to $f \in L^2_m(\mathbf{R})$ does not imply that $f(\xi)$ tends to zero when ξ tends to $\pm\infty$.[15] But if f and f' both belong to $L^2_m(\mathbf{R})$, then $f(\xi) \to 0$ when

---

[15] Consider, for example, the functions $f_p: \mathbf{R} \to \mathbf{R}$, where p is a nonnegative integer, defined by

$$f_p(\xi) = \xi^p \exp -(\xi^{4(p+1)} \sin^2 \xi).$$

$f_p(n\pi)$ does not tend to zero when the integer n tends to $\pm\infty$.

$\xi \to \pm\infty$. From the Cauchy–Schwarz inequality it follows that

$$\left| \int_a^b f(\xi) f'(\xi)\, d\xi \right|^2 = \frac{1}{4} \left| \big(f(b)\big)^2 - \big(f(a)\big)^2 \right|^2$$
$$\leq \int_a^b |f(\xi)|^2\, d\xi \int_a^b |f'(\xi)|^2\, d\xi.$$

If a and b tend, independently, to $+\infty$, both integrals in the right-hand side tend to zero since f and f' belong to $L_m^2(\mathbf{R})$. Thus $\big(f(b)\big)^2 - \big(f(a)\big)^2$ also tends to zero, that is, $f(\xi)$ tends to a constant when $\xi \to \infty$. Since $f \in L_m^2(\mathbf{R})$, this constant must be zero. Similarly we could show that $f(\xi) \to 0$ when $\xi \to -\infty$.

Now it is easy to prove that P is self-adjoint for

$$\langle g \mid Pf \rangle = \int_{-\infty}^{\infty} \overline{g(\xi)} \big(-if'(\xi) \big)\, d\xi$$
$$= \Big[-\overline{g(\xi)} f(\xi) \Big]_{-\infty}^{\infty} + \int_{-\infty}^{\infty} \overline{\big(-ig'(\xi) \big)} f(\xi)\, d\xi.$$

Since the first term in the right-hand side is equal to zero, P^+ is defined by

$$P^+ g = -ig',$$

where g' is the derivative of g in the sense of distributions, and

$$D_{P^+} = \{ g | g \in L_m^2(\mathbf{R}), g' \in L_m^2(\mathbf{R}) \} = D_P.$$

Therefore,

$$P^+ = P.$$

5. Spectral Theory

Given a linear operator A in a Hilbert space E, we shall study the properties of the operator $A - \lambda I$, where λ is a complex number and I the identity operator. For any λ, $D_{A-\lambda I} = D_A$.

The inverse of $A - \lambda I$, when it exists, is called the *resolvent of* A. It is denoted by $R_\lambda(A)$. The spectral theory studies the properties of the resolvent operator $R_\lambda(A)$ as a function of $\lambda \in \mathbf{C}$.

Definition 9. The *resolvent set* of a linear operator A is the set $\rho(A)$ of all complex numbers λ such that $R_\lambda(A)$ exists, is bounded, and is densely defined. The complement of the resolvent set is called the *spectrum* of A and is denoted by $\sigma(A)$.

Note that λ belongs to $\rho(A)$ if, and only if,

(1) $R_\lambda(A)$ exists,

(2) $R_\lambda(A)$ is bounded,

(3) $R_\lambda(A)$ is densely defined.

If one of these conditions is not satisfied, λ belongs to the spectrum.

Definition 10. The set of all complex numbers λ for which $R_\lambda(A)$ does not exist is called the *point spectrum* of A, and is denoted $P\sigma(A)$. The elements of the point spectrum of A are called the *eigenvalues* of A.

The set of all complex numbers λ for which $R_\lambda(A)$ exists and is densely defined but unbounded is called the *continuous spectrum* of A and is denoted by $C\sigma(A)$.

The set of all complex numbers λ for which $R_\lambda(A)$ exists but is not densely defined is called the *residual spectrum* of A and is denoted by $R\sigma(A)$.

So

$$\sigma(A) = P\sigma(A) \cup C\sigma(A) \cup R\sigma(A).$$

Some spectra may be empty. If, for instance, A is defined on a finite-dimensional space, then $C\sigma(A) = R\sigma(A) = \emptyset$, and $\sigma(A) = P\sigma(A)$.

The terminology is not very good. The point spectrum is not necessarily countable (see Problem 11), whereas the continuous spectrum may be countable or even finite (see Problem 12).

If λ is an eigenvalue of A, the kernel of $A - \lambda I$ differs from $\{0\}$, that is, $A - \lambda I$ is not injective. In this case the equation

$$Ax = \lambda x$$

has a solution $x \neq 0$. Such a solution is called an *eigenvector* of A associated to the eigenvalue λ. If the dimension of the vector space $E_\lambda = \ker(A - \lambda I)$ is greater than one, the eigenvalue λ is said to be *degenerate*, and the dimension of E_λ is the *degree of degeneracy* of the eigenvalue.

Let x_1 be an eigenvector associated to the eigenvalue λ_1, and let x_2 be an eigenvector associated to the eigenvalue λ_2; if $\lambda_1 \neq \lambda_2$, then x_1 and x_2 are linearly independent.

λ is an *approximate eigenvalue* if there is no $x \neq 0$ such that $Ax = \lambda x$, but, given a positive ϵ, there is an *approximate eigenvector* $x(\epsilon)$ of unit norm such that

$$\|Ax(\epsilon) - \lambda x(\epsilon)\| < \epsilon.$$

In this case $R_\lambda(A)$ exists but is unbounded. Approximate eigenvalues are, therefore, in $C\sigma(A)$ or in $R\sigma(A)$.

Example 1. The point spectrum of a projection operator P on a Hilbert space E consists only of two eigenvalues: zero and 1. For any x in E we may write:

$$x = x_1 + x_2,$$

where $x_1 \in R_P$ (i.e., $Px = x_1$) and $x_2 \in (R_P)^\perp$. The operator $P - \lambda I$ has an inverse if the equation

$$(P - \lambda I)x = 0$$

has only the trivial solution $x = 0$. But

$$(P - \lambda I)x = x_1 - \lambda x_1 - \lambda x_2,$$

which shows that, if $x \neq 0$ is a solution of the preceding equation, then either $\lambda = 0$, and $x \in (R_P)^\perp$, or $\lambda = 1$, and $x \in R_P$. Zero and 1 are the only eigenvalues of the operator P. Since in most cases R_P and $(R_P)^\perp$ have dimensions greater than 1, the eigenvalues are degenerate.

The continuous spectrum and the residual spectrum of P are empty. To prove this we have to show that, if λ differs from 1 or zero, the resolvent operator $R_\lambda(P) = (P - \lambda I)^{-1}$ is bounded and densely defined. For any vector x in E we have

$$(P - \lambda I)^{-1}x = \frac{x_1}{1 - \lambda} - \frac{x_2}{\lambda},$$

where $x = x_1 + x_2$ with $x_1 \in R_P$ and $x_2 \in (R_P)^\perp$, as can easily be verified by multiplying both sides of this relation by $P - \lambda I$. This shows that the domain of $R_\lambda(P)$ is the whole Hilbert space E. Moreover, from

$$\frac{\|(P - \lambda I)^{-1}x\|}{\|x\|} \leq \frac{\|x_1\|/|1 - \lambda| + \|x_2\|/|\lambda|}{(\|x_1\|^2 + \|x_2\|^2)^{1/2}},$$

it follows that the operator $R_\lambda(P)$ is bounded.

Example 2. Let X be the position operator in $L^2_m(\mathbf{R})$. We shall prove that its point spectrum and its residual spectrum are empty and that its continuous spectrum coincides with the set of all reals $\mathbf{R}$.[16] That is, any complex number λ, such that $\Im\lambda \neq 0$, belongs to the resolvent set $\rho(X)$.

[16] This result is an illustration of Theorem 13, to be proved below. For any self-adjoint operator A, $\sigma(A) = P\sigma(A) \cup C\sigma(A) \subset \mathbf{R}$.

The equation

$$Xf = \lambda f,$$

where $f \in D_X$, can be written

$$(\xi - \lambda) f(\xi) = 0.$$

Its only solution is $f = 0$. For any complex number λ, the operator $R_\lambda(X) = (X - \lambda I)^{-1}$ exists, which proves that the point spectrum of X is empty.

If λ is real, $D_{(X-\lambda I)^{-1}}$ is dense in $L^2_m(\mathbf{R})$. To prove it we have to show that, for any $f \in L^2_m(\mathbf{R})$, there exists a sequence (g_n) of functions in $D_{(X-\lambda I)^{-1}}$ such that

$$\lim_{n\to\infty} \|f - g_n\| = 0.$$

This is indeed the case if we define g_n by

$$g_n(\xi) = \begin{cases} f(\xi), & \text{if } |\xi - \lambda| > 1/n; \\ 0, & \text{otherwise.} \end{cases}$$

For any positive integer n, g_n belongs to $D_{(X-\lambda I)^{-1}}$ since, as a consequence of the Cauchy–Schwarz inequality, the integral

$$\int_{|\xi-\lambda|>1/n} \left|\frac{f(\xi)}{\xi - \lambda}\right|^2 d\xi$$

is convergent. Moreover,

$$\|f - g_n\|^2 = \int_{\lambda-1/n}^{\lambda+1/n} |f(\xi)|^2 \, d\xi$$

goes to zero as n tends to ∞.

Notc that $D_{(X-\lambda I)^{-1}}$ is dense in $L^2_m(\mathbf{R})$ but does not coincide with $L^2_m(\mathbf{R})$ because, for instance, the function f defined by

$$f(\xi) = \begin{cases} 1, & \text{if } |\xi - \lambda| \le \alpha, \\ 0, & \text{otherwise,} \end{cases}$$

where α is a positive real, is in $L^2_m(\mathbf{R})$ but not in $D_{(X-\lambda I)^{-1}}$.

For any real λ, $R_\lambda(X) = (X - \lambda I)^{-1}$ is unbounded. To prove it consider, for instance, the sequence (h_n) of functions belonging to $D_{(X-\lambda I)^{-1}}$ defined by

$$h_n(\xi) = \begin{cases} (3n^3/2)^{1/2}(\xi - \lambda), & \text{if } |xi - \lambda| \le 1/n; \\ 0, & \text{otherwise.} \end{cases}$$

We have

$$(\forall n \in \mathbf{N}) \qquad \|h_n\| = 1$$

and

$$\|(X - \lambda I)^{-1} h_n\| = \sqrt{3}n,$$

which show that $R_\lambda(X)$ is unbounded.

The continuous spectrum is, therefore, the set of all reals $\mathbf{R}$.

If $\Im\lambda \neq 0$, $D_{(X-\lambda I)^{-1}} = L^2_m(\mathbf{R})$ and $R_\lambda(X)$ is bounded. Indeed, λ being real, it is always possible to find a positive number δ such that

$$(\forall \xi \in \mathbf{R}) \qquad |\xi - \lambda| \geq \delta,$$

which implies

$$(\forall f \in L^2_m(\mathbf{R})) \qquad \|(X - \lambda I)^{-1} f\|^2 = \int_{-\infty}^{\infty} \left|\frac{f(\xi)}{\xi - \lambda}\right|^2 d\xi$$
$$\leq \frac{1}{\delta^2}\|f\|^2.$$

This result shows, first, that the function $\xi \mapsto f(\xi)/(\xi - \lambda)$ belongs to $L^2_m(\mathbf{R})$, that is, $D_{(X-\lambda I)^{-1}} = L^2_m(\mathbf{R})$; and second, that $R_\lambda(X)$ is bounded. The resolvent set $\rho(X)$ is, therefore, the complement of $\mathbf{R}$ in $\mathbf{C}$.

Remark 1. In the sense of distributions, $f = \delta_\lambda$ is a solution of the equation $(\xi - \lambda)f = 0$; however, to say, with some authors, that any real λ is, therefore, an eigenvalue of the position operator X, associated to the eigenfunction δ_λ, is not correct because δ_λ is not in $L^2_m(\mathbf{R})$.

Remark 2. Any real λ is an approximate eigenvalue of the position operator X. To prove it, it is sufficient to exhibit a sequence f_n of approximate eigenfunctions, that is, a sequence (f_n) of functions belonging to $L^2_m(\mathbf{R})$ such that

$$(\forall n \in \mathbf{N}) \qquad \|f_n\| = 1 \quad \text{and} \quad \lim_{n\to\infty} \|(X - \lambda I)f_n\| = 0.$$

The sequence defined by

$$f_n(\xi) = \begin{cases} \sqrt{n/2} & \text{if } |\xi - \lambda| \leq 1/n, \\ 0 & \text{otherwise} \end{cases}$$

satisfies these requirements. Note that, in $\mathcal{D}'$, $\lim_{n\to\infty} f_n^2 = \delta_\lambda$.

Theorem 12. *Let A be a linear operator defined in a Hilbert space E.*

(1) *Its resolvent $\rho(A)$ set is open.*

(2) *The function* $\lambda \mapsto R_\lambda(A)$ *is analytic for all* λ *in* $\rho(A)$.

(3) *Its spectrum* $\sigma(A)$ *is closed and nonempty.*

(1) If λ_0 is in $\rho(A)$, the resolvent $R_\lambda(A)$ exists, is bounded, and is densely defined. To establish that $\rho(A)$ is open, it is sufficient to show that, for any λ_0 in $\rho(A)$, there exists a disk of radius r centered at λ_0 and such that, for any λ belonging to this disk, $R_\lambda(A)$ exists, is bounded, and is densely defined.

If the series

$$\sum_{n=0}^{\infty}(\lambda - \lambda_0)^n R_{\lambda_0}^{n+1}(A)$$

converges, it defines a bounded operator whose domain coincides with the domain of $R_{\lambda_0}(A)$ (Theorem 3), and which is, therefore, dense. The series is convergent if

$$|\lambda - \lambda_0| < r = \|R_{\lambda_0}(A)\|^{-1},$$

and, in this case, its sum is equal to

$$\big(I - (\lambda - \lambda_0)R_{\lambda_0}(A)\big)^{-1} R_{\lambda_0}(A) = R_\lambda(A),$$

since

$$\begin{aligned} A - \lambda I &= (A - \lambda_0 I) - (\lambda - \lambda_0)I \\ &= \big(I - (\lambda - \lambda_0)R_{\lambda_0}(A)\big)(A - \lambda I). \end{aligned}$$

(2) The relation

$$R_\lambda(A) = \sum_{n=0}^{\infty}(\lambda - \lambda_0)^n R_{\lambda_0}^{n+1}(A)$$

shows that the operator-valued function $\lambda \mapsto R_\lambda(A)$, defined for all $\lambda \in \rho(A)$, is analytic, and that, in particular, its derivative is equal to $R_\lambda^2(A)$.

This result could have been derived from the *Hilbert relation*

$$R_\lambda(A) - R_\mu(A) = (\lambda - \mu)R_\lambda(A)R_\mu(A),$$

valid for all λ and μ in $\rho(A)$, since

$$\frac{dR_\lambda(A)}{d\lambda} = \lim_{\Delta\lambda} \frac{R_{\lambda_0+\Delta\lambda}(A) - R_{\lambda_0}(A)}{\Delta\lambda} = R_\lambda^2(A).$$

Note that the Hilbert relation, which shows that $R_\lambda(A)$ and $R_\mu(A)$ commute, follows directly from

$$R_\lambda(A) - R_\mu(A) = R_\lambda(A)(A - \mu I)R_\mu(A) - R_\lambda(A)(A - \lambda I)R_\mu(A).$$

(3) $\sigma(A)$ is closed since $\rho(A)$, which is its complement in $\mathbf{C}$, is open. $\sigma(A)$ is not empty because the function $\lambda \mapsto R_\lambda(A)$ would be bounded and analytic in the whole complex plane, and, as a consequence of Liouville's theorem, would be a constant, which is absurd.

Theorem 13. *Let A be a self-adjoint linear operator defined in a Hilbert space E.*

(1) *The eigenvalues are real.*

(2) *The eigenvectors associated to different eigenvalues are orthogonal.*

(3) *Its residual spectrum is empty.*

(4) *Its continuous spectrum is real.*

(1) If λ is an eigenvalue of A,[17] there exists in E a vector $x \neq 0$ such that $Ax = \lambda x$, and, therefore,

$$\lambda = \frac{\langle x \mid Ax\rangle}{\|x\|^2}.$$

But, A being self-adjoint,

$$\begin{aligned}\langle x \mid Ax\rangle &= \langle Ax \mid x\rangle \\ &= \overline{\langle x \mid Ax\rangle};\end{aligned}$$

hence, $\lambda = \overline{\lambda}$.

(2) If x_1 is an eigenvector associated to the eigenvalue λ_1, and x_2 an eigenvector associated to the eigenvalue λ_2, we have

$$(\lambda_1 - \lambda_2)\langle x_1 \mid x_2\rangle = \langle Ax_1 \mid x_2\rangle - \langle x_1 \mid Ax_2\rangle = 0,$$

that is,

$$\langle x_1 \mid x_2\rangle = 0 \quad \text{if} \quad \lambda_1 \neq \lambda_2.$$

(3) If λ is complex, the solution of the equation

$$Ax - \lambda x = y$$

is

$$x = R_\lambda(A)y.$$

---

[17] The point spectrum of a self-adjoint operator may be empty (see Example 2).

$\langle x \mid Ax\rangle$ being real, the relation

$$\langle x \mid Ax\rangle - \lambda\|x\|^2 = \langle x \mid y\rangle$$

implies

$$\begin{aligned} |\Im\lambda|\,\|x\|^2 &= |\Im\langle x \mid y\rangle| \\ &\le |\langle x \mid y\rangle| \\ &\le \|x\|\,\|y\|; \end{aligned}$$

thus,

$$\|x\| = \|R_\lambda(A)y\| \le \frac{\|y\|}{|\Im\lambda|},$$

which shows that $R_\lambda(A)$ is bounded. Therefore, if $\Im\lambda \neq 0$, λ belongs to $\rho(A)$ or to $R\sigma(A)$. Actually it belongs to $\rho(A)$, since

$$\begin{aligned} \overline{D_{R_\lambda(A)}} &= \overline{R_{A-\lambda I}} \\ &= \left(\ker(A - \overline{\lambda}I)\right)^\perp \end{aligned}$$

shows that the domain of $R_\lambda(A)$ is dense in E because, $\overline{\lambda}$ not being an eigenvalue of A, $\ker(A - \overline{\lambda}I) = \{0\}$.

(4) If $\Im\lambda \neq 0$, we found that $\lambda \in \rho(A)$; therefore, $\sigma(A)$, which is the complement of $\rho(A)$ in $\mathbf{C}$, is real.

This theorem is of central importance in quantum mechanics, for any observable (that is, any measurable physical quantity like, for instance, the momentum or the energy of a particle) is represented by a self-adjoint operator.

Theorem 14. *Let A be a bounded linear operator on a Hilbert space E.*

(1) Its spectrum $\sigma(A)$ is contained in a disk of radius $\|A\|$ centered at the origin.

(2) Its resolvent set $\rho(A)$ is not empty.

(1) The Laurent series

$$R_\lambda(A) = -\sum_{n=0}^{\infty} \frac{A^n}{\lambda^{n+1}}$$

being convergent for any $\lambda > \|A\|$, the complement of the disk of radius $\|A\|$ centered at the origin is contained in $\rho(A)$, and therefore $\sigma(A)$, which is the complement of $\rho(A)$, is contained in the disk of radius $\|A\|$ centered at the origin.

(2) $\rho(A)$ cannot be empty since it contains the complement of the disk of radius $\|A\|$ centered at the origin.

Definition 11. Let A be a bounded linear operator; the positive number

$$r(A) = \sup_{\lambda \in \sigma(A)} |\lambda|$$

is called the *spectral radius* of A.

The Laurent series $\sum_{n=0}^{\infty} A^n/\lambda^{n+1}$ being convergent for any $\lambda > \|A\|$, we have

$$r(A) \leq \|A\|.$$

If A is self-adjoint, this inequality becomes an equality, since

$$\begin{aligned} \|A\|^2 &= \sup_{\|x\| \leq 1} |\langle A^2 x \mid x \rangle| \\ &= \sup_{\|x\| \leq 1} |\langle Ax \mid Ax \rangle| \\ &= \|A\|^2, \end{aligned}$$

and therefore, as a consequence of Cauchy's rule,

$$r(A) = \lim_{n \to \infty} \left(\|A\|^{2^n} \right)^{1/2^n} = \|A\|.$$

Theorem 15. *Let A be a bounded self-adjoint linear operator on a Hilbert space E; if, on the set $\{x | x \in E, \|x\| = 1\}$, the mapping $x \mapsto |\langle Ax \mid x \rangle|$ reaches its maximum for $x = x_1$, then x_1 is an eigenvector of A and the associated eigenvalue λ_1 has the greatest absolute value.*

Let y be a vector of E orthogonal to x_1; the vector

$$x = \frac{x_1 + \epsilon y}{\sqrt{1 + |\epsilon|^2 \|y\|^2}},$$

where ϵ is a complex number, is normed, and we have

$$\begin{aligned} \langle Ax \mid x \rangle = \frac{1}{1 + |\epsilon|^2 \|y\|^2} \big(&\langle Ax_1 \mid x_1 \rangle + \epsilon \langle Ax_1 \mid y \rangle \\ &+ \bar{\epsilon} \langle Ay \mid x_1 \rangle + |\epsilon|^2 \langle Ay \mid y \rangle \big). \end{aligned}$$

If ϵ is such that $\epsilon\langle Ax_1 \mid y \rangle$ is real, then

$$\bar{\epsilon} \langle Ay \mid x_1 \rangle = \bar{\epsilon} \langle y \mid Ax_1 \rangle = \overline{\epsilon \langle Ax_1 \mid y \rangle}$$

is also real, and to the first order in ϵ, we have

$$\langle Ax \mid x\rangle = \langle Ax_1 \mid x_1\rangle + 2\epsilon\langle Ax_1 \mid y\rangle,$$

which shows that we could choose ϵ such that

$$|\langle Ax \mid x\rangle| > |\langle Ax_1 \mid x_1\rangle|,$$

in contradiction with the assumption. Hence, any $y \in E$ orthogonal to x_1 is also orthogonal to Ax_1, and consequently,

$$Ax_1 = \lambda_1 x_1,$$

where λ_1 is a real number. In other words, x_1 is an eigenvector of A, and from

$$|\langle Ax_1 \mid x_1\rangle| = |\lambda_1|\,\|x_1\|^2 = |\lambda_1|,$$

it follows that the associated eigenvalue λ_1 has the greatest absolute value.

This proof assumes that λ_1 is not degenerate. If this is not the case, it is not difficult to modify the proof accordingly.

This theorem is of a very practical importance since it provides a method of determining step by step the eigenvalues $\lambda_1, \lambda_2, \ldots, \lambda_k, \ldots$ of a bounded, self-adjoint, linear operator A of decreasing absolute values. With this aim in view, one successively determines the vector spaces $E_1, E_2, \ldots, E_k, \ldots$ maximizing $|\langle Ax \mid x\rangle|$, where x is a vector of unit norm belonging successively to $E, E_1^\perp, E_2^\perp, \ldots, E_k^\perp, \ldots$, where $E_k^\perp$ is the orthogonal supplement of $E_1 \oplus E_2 \oplus \cdots \oplus E_k$. The maximum value of $|\langle Ax \mid x\rangle|$, when x, of unit norm, belongs to $E_{k-1}^\perp$, is equal to $|\lambda_k|$; the eigenvalue λ_k, associated to the eigenspace E_k, is equal to $\langle Ax \mid x\rangle$, where x, of unit norm, belongs to E_k.

Theorem 16. *Let A be a compact linear operator on a Hilbert space E, and let λ be an eigenvalue different from zero; then the vector space $E_\lambda = \ker(A - \lambda I)$ is finite-dimensional.*

To simplify, w shall assume that E is separable.[18] Let (x_n) be a sequence of linearly independent eigenvectors associated to the eigenvalue λ, and denote by E_k the subspace of E_λ generated by $x_1, x_2, \ldots, x_k$. Since x_k does not belong to E_{k-1}, the distance $d(x_k, E_{k-1})$ is different from zero;

[18] This hypothesis is, however, not indispensable.

let α be this distance, and consider $x \in E_{k-1}$ such that $\|x_k - x\|, 2\alpha$. We have

$$d(x_k, E_{k-1}) = d(x_k - x, E_{k-1}),$$

and therefore, the sequence (y_n) defined by

$$y_n = \frac{x_n - x}{\|x_n - x\|}$$

is such that, for any $k \in \mathbf{N}$,

$$y_k \in E_k, \quad \|y_k\| = 1 \quad \text{and} \quad d(y_k, E_{k-1}) > 1/2.$$

Furthermore, from the relation

$$\frac{1}{\lambda} A y_k = y_k$$

it follows that, for $p \neq q$,

$$\|\frac{1}{\lambda} A y_p - \frac{1}{\lambda} A y_q\| = \|y_p - y_q\| > \frac{1}{2},$$

which shows that if the sequence (y_n/λ) were infinite,[19] so would be the sequence $(\frac{1}{\lambda} A y_n)$. But the latter, which is the image of the former by A, cannot contain any convergent subsequence, in contradiction with the fact that A is compact. Thus the sequence (y_n) comprises a finite number of terms; this number is equal to the dimension of E_λ.

The following two corollaries are direct consequences of this theorem.

Corollary 1. *For any positive ϵ, the vector space $\bigoplus_{|\lambda|>\epsilon} E_\lambda$ is finite-dimensional.*

Corollary 2. *For any positive ϵ, the number of eigenvalues whose absolute value is greater than ϵ is finite.*

Theorem 17. Spectral theorem (for compact self-adjoint operators). *Let A be a compact self-adjoint linear operator on a Hilbert space E, and let (λ_n) be the sequence, finite or infinite, of its eigenvalues. We have*

$$A = \sum_n \lambda_n P_{E_{\lambda_n}}, \tag{1}$$

---

[19] Note that this sequence is bounded, since $\lambda \neq 0$.

where $P_{E_{\lambda_n}}$ is the projection operator on the finite-dimensional space $E_{\lambda_n} = \ker(A - \lambda_n I)$, and

$$E = E_0 \oplus_n E_{\lambda_n}, \tag{2}$$

where $E_0 = \ker A$.

(1) *is the* spectral representation *of A, and* (2) *is the* spectral decomposition *of E.*

This theorem is a straightforward consequence of the preceding results. Using Theorem 15, the sequence (λ_n) of eigenvalues is determined step by step. The eigenspace E_{λ_n} is associated to the eigenvalue λ_n. From Theorem 16, it follows that this vector space is finite-dimensional.

If A is different from zero, it has, at least, one nonzero eigenvalue otherwise; for all vectors x in E, $|\langle Ax \mid x\rangle|$ would be equal to zero, in contradiction with the fact that $A \neq 0$.

Once all nonzero eigenvalues of A have been determined, consider the space E_0 orthogonal supplement of $\oplus_n E_{\lambda_n}$.

If E_0 coincides with $\{0\}$, then the sequence (λ_n) must be infinite if the Hilbert space E is infinite-dimensional; and, from Corollary 2 of Theorem 16, it follows that

$$\lim_{n\to\infty} \lambda_n = 0.$$

If E_0 differs from $\{0\}$, zero is an eigenvalue, and, when the sequence (λ_n) is infinite, zero is an accumulation point of this sequence. If the sequence (λ_n) is finite, E_0 is infinite-dimensional when the Hilbert space E is infinite-dimensional.

Therefore, in all cases, we have

$$E = E_0 \oplus E_{\lambda_n},$$

that is,

$$(\forall x \in E) \qquad x = x_0 + \sum_n x_n,$$

where $x_0 \in E_0$ and $x_n \in E_{\lambda_n}$ $(n = 1, 2, \ldots)$. $x_0 + \sum_n x_n$ is called the *canonical representation* of x. This representation is unique. Since

$$P_{E_{\lambda_i}} P_{E_{\lambda_j}} = 0 \quad \text{if} \quad i \neq j,$$

it is seen that

$$P_{E_0} + \sum_n P_{E_{\lambda_n}} = I.$$

The set $\{P_{E_0}, P_{E_{\lambda_1}}, P_{E_{\lambda_2}}, \dots\}$ of projection operators is called a *resolution of the identity*. Moreover, from the relation

$$Ax = \sum_n \lambda_n x_n,$$

it follows that A is given by

$$A = \sum_n P_{E_{\lambda_n}}.$$

The spectral theorem gives a complete characterization of compact self-adjoint operators. Such an operator has a countable spectrum. Each nonzero element of its spectrum is an eigenvalue whose degree of degeneracy is finite. If the sequence (λ_n) of the eigenvalues is finite, zero is necessarily an eigenvalue. If (λ_n) is infinite, zero belongs either to the point spectrum or to the continuous spectrum.

The set of all eigenvectors is total in the range of A. If to this set we add a complete orthogonal system in $\ker A$, we obtain a complete orthogonal system in the Hilbert space E.

Note that if A and B are two compact self-adjoint operators such that $AB = BA$, they have the same spectral decomposition. In fact, if $x_n \in E_{\lambda_n}$, then $Ax_n = \lambda_n x_n$ and $ABx_n = BAx_n = \lambda_n Bx_n$, that is, $Bx_n \in E_{\lambda_n}$.

Theorem 18. *Let A be a compact self-adjoint linear operator on a Hilbert space E, and let y be a vector in E whose canonical representation $y_0 + \sum_n y_n$ is known.*

(1) *If λ is not an eigenvalue of A, the solution of the equation $Ax - \lambda x = y$ is unique, and its canonical representation is*

$$x = -\frac{1}{\lambda} y_0 + \sum_n \frac{1}{\lambda_n - \lambda} y_n.$$

(2) *If $\lambda = \lambda_k$ is an eigenvalue of A, the equation $Ax - \lambda x = y$ has a solution if, and only if, $y_k = 0$. In this case, the solutions are of the form*

$$x = -\frac{1}{\lambda_k} y_0 + \sum_{n \neq k} \frac{1}{\lambda_n - \lambda_k} y_n + x_k,$$

where x_k is an arbitrary vector of E_{λ_k}.

(3) The equation $Ax = y$ has a solution if, and only if, $y_0 = 0$ and the series $\sum_n (1/\lambda_n^2)\|y_n\|^2$ is convergent; in this case the solutions are given by

$$x = \sum_n \frac{1}{\lambda_n} y_n + x_0,$$

where x_0 is an arbitrary vector of E.

The proof of all these results is straightforward. It is a direct consequence of the unicity of the canonical representation.

Results (1) and (2) are known as the *Fredholm alternative.*

Example 3. Consider the integral equation

$$\int_0^\pi \cos(x+y) f(y)\, dy - \lambda f(x) = \cos 3x.$$

It is an equation of the form $Kf - \lambda f = g$, where K is a Hilbert–Schmidt operator. The kernel of the operator being symmetric, the operator is self-adjoint. The eigenvalues equation reads

$$\int_0^\pi \cos(x+y)\phi(y)\, dy = \lambda\phi(x).$$

Expanding $\cos(x+y)$ we find that K has only 2 nonzero eigenvalues $\pi/2$ and $-\pi/2$. They are nondegenerate, and the corresponding eigenvectors are, respectively, $x \mapsto \cos x$ and $x \mapsto \sin x$. In $L^2_m([0,\pi])$, the function $x \mapsto \cos 3x$ being orthogonal to the two eigenvectors, three cases can be distinguished:

(1) If λ differs from 0 $\pi/2$ and $-\pi/2$, then the solution is $f\colon x \mapsto -\lambda^{-1}\cos 3x$.

(2) If $\lambda = \pi/2$, then the solution is $f\colon x \mapsto C\cos x - (2/\pi)\cos 3x$, where C is a constant.

(3) If $\lambda = -\pi/2$, then the solution is $f\colon x \mapsto C\sin x + (2/\pi)\cos 3x$, where C is a constant.

Most unbounded self-adjoint operators encountered in physics are differential operators; therefore, we shall not try to establish a more general spectral theorem valid for any self-adjoint operator but rather study various examples to show how the spectrum of a differential operator can be determined.

Example 4. *Spectrum of the momentum operator.* Let P be the momentum operator defined in $L^2_m(\mathbf{R})$ by

$$Pf = -if',$$

where f' is the derivative of f in the sense of distributions, and

$$D_P = \{f | f \in L^2_m(\mathbf{R}), f' \in L^2_m(\mathbf{R})\}.$$

We know already that P is self-adjoint (see Example 4.4), and that its spectrum is real (Theorem 13).

The solution of the eigenvalues equation

$$-if' = \lambda f \qquad (\lambda \in \mathbf{R})$$

is $f: x \mapsto \exp(i\lambda x)$; it does not belong to $L^2_m(\mathbf{R})$. The point spectrum is, therefore, empty. On the other hand, any real λ belongs to the continuous spectrum, since $R_\lambda(P) = (P - \lambda I)^{-1}$ exists (because $\lambda \notin P\sigma(P)$), its domain is dense, and it is unbounded.[20] All these results could be established following the method used in Example 2. There is, however, a more interesting method. If f is a tempered distribution that is a solution of the equation $-if' - \lambda f = g$, its Fourier transform $\hat{f}$ is a tempered distribution such that $\xi \hat{f}(\xi) - \lambda \hat{f}(\xi) = \hat{g}(\xi)$. The spectrum of the momentum operator P coincides, therefore, with the spectrum of the position operator X. If F denotes the unitary Fourier–Plancherel operator, it can be shown that

$$D_P = F(D_X) \quad \text{and} \quad P = FXF^{-1}.$$

More generally, if two operators A and B are such that

$$D_B = U(D_A) \quad \text{and} \quad B = UAU^{-1},$$

where U is a unitary operator, they are said to be *unitarily equivalent.* In this case, it can be shown that if A is self-adjoint, so is B, and that the two operators have the same spectrum.

Example 5. *Operators having a compact resolvent.* Let A be a self-adjoint linear operator in a Hilbert space E. A is said to have a *compact resolvent* if there is a λ_0 in $\rho(A)$ for which the resolvent $R_{\lambda_0}(A)$ is compact. From the

[20] If $\Im\lambda \neq 0$, the solution of the equation $Pf - \lambda f = g$ is $f = R_\lambda(P)g$, where

$$\big(R_\lambda(P)g\big)(x) = \begin{cases} i\int_{-\infty}^{x} \mathrm{e}^{i\lambda(x-\xi)} g(\xi)\, d\xi & \text{if } \Im\lambda > 0; \\ -i\int_{x}^{\infty} \mathrm{e}^{i\lambda(x-\xi)} g(\xi)\, d\xi & \text{if } \Im\lambda < 0. \end{cases}$$

relation $R_\lambda(A) = R_{\lambda_0}(A)\big(I+(\lambda-\lambda_0)R_\lambda(A)\big)$ it follows that, if the resolvent is compact for $\lambda_0 \in \rho(A)$, it is compact for all $\lambda \in \rho(A)$ because $I + (\lambda - \lambda_0)R_\lambda(A)$ is bounded. A large class of differential operators have resolvents that may be expressed as integral operators. In this form, it is easy to check whether the resolvent is compact or not. From Theorem 17 it follows that the spectrum of $R_\lambda(A)$ is a bounded countable set with no accumulation point except at the origin, and that any element of the spectrum of $R_\lambda(A)$ different from zero is an eigenvalue whose degree of degeneracy is finite. That leads immediately to the knowledge of the spectrum of A.

Let us first assume that $\mu \neq 0$ belongs to the resolvent set of $R_\lambda(A)$. Then, for any $y \in E$, the equation

$$R_\lambda(A)x - \mu x = y$$

has a solution. If we replace y by $-\mu R_\lambda(A)z$, where z is any element of E, the equation

$$R_\lambda(A)x - \mu x = -\mu R_\lambda(A)z$$

also has a solution. Hence, if x is in the domain of $A - \lambda I$, that is, in the range of $R_\lambda(A)$, the equation

$$x - \mu(A - \lambda I)x = -\mu z,$$

which may be written

$$Ax - (\lambda + \frac{1}{\mu})x = z,$$

has a solution, for any $z \in E$. $\lambda + 1/\mu$ is, therefore, in the resolvent set of A.

Now, let us assume that $\mu \neq 0$ is an eigenvalue of $R_\lambda(A)$, and that x is an associated eigenvector. The equation

$$R_\lambda(A)x - \mu x = 0,$$

which may be written

$$x - \mu(A - \lambda I)x = 0$$

or

$$Ax - (\lambda + \frac{1}{\mu})x = 0,$$

has a solution $x \neq 0$. $\lambda + 1/\mu$ is, therefore, an eigenvalue of A, and x is an associated eigenvector.

In conclusion, an operator with a compact resolvent has a pure point spectrum, its eigenvalues have finite degrees of degeneracy, and they accumulate only at ∞.

Consider, for example, the operator A defined in $L^2_m([0,1])$ by

$$Af = -f'',$$

where the derivative f'' is in the sense of distributions, and

$$D_A = \{f | f \in L^2_m([0,1]), f'' \in L^2_m([0,1]), f(0) = f(1) = 0\}.$$

Since zero is not an eigenvalue of A, we can consider the resolvent corresponding to $\lambda = 0$, which is the inverse A^{-1} of A. It is not difficult to verify that A^{-1} is a Hilbert–Schmidt integral operator whose kernel K is given by

$$K(x,y) = \begin{cases} x(1-y), & \text{if } 0 \le x \le y \le 1; \\ (1-x)y, & \text{if } 0 \le y \le x \le 1. \end{cases}$$

Indeed, the solution of the equation $-f'' = g$ that satisfies the boundary conditions is $g = Kf$, since

$$\begin{aligned} \int_0^1 K(x,y)g(y)\,dy &= \int_0^x (1-x)yg(y)\,dy + \int_x^1 (1-y)xg(y)\,dy \\ &= \int_0^x yg(y)\,dy + \int_x^1 xg(y)\,dy - \int_0^1 xyg(y)\,dy, \end{aligned}$$

which shows that $f(0) = f(1) = 0$, and that

$$\begin{aligned} f'(x) &= \int_x^1 g(y)\,dy - \int_0^1 yg(y)\,dy \\ f''(x) &= -g(x). \end{aligned}$$

The spectrum of A is, consequently, the set of reals λ such that $1/\lambda$ is an eigenvalue of K. To determine these eigenvalues, it is easier to solve the differential equation

$$-f'' = \lambda f$$

than the equation

$$Kf = \frac{1}{\lambda} f.$$

It is found that (see Problem 18)

$$\lambda_n = n^2\pi^2 \quad \text{and} \quad f_n(x) = \sin n\pi x,$$

where, $n = 1, 2, \ldots$. 0 not being an eigenvalue of K, the sequence (λ_n) is infinite, and the set of all the eigenvectors of A is a complete orthogonal system.

The differential operator $-d^2/dx^2$ belongs to a large family of second-order differential operators that arise in the study of a wide class of physical systems like vibrating strings, wave guides, or one-dimensional quantum-mechanical models.

Example 6. *Sturm–Liouville operators.* Let A be a differential operator defined in $L^2_m([a,b])$ by

$$Af = f'' - qf,$$

where q is a continuous function on $[a,b]$. The domain of A is the set of functions in $L^2_m([a,b])$ such that $Af \in L^2_m([a,b])$,[21] and which satisfy the boundary conditions

$$h_1 f(a) + k_1 f'(a) = 0 \tag{1}$$

$$h_2 f(b) + k_2 f'(b) = 0, \tag{2}$$

where h_1, h_2, k_1, k_2 are real constants, of which h_1 and k_1 are not both zero, and h_2 and k_2 are not both zero. A is called a *Sturm–Liouville operator.* It will be shown that A has a pure point spectrum, the set of eigenvalues being countable with no accumulation point, except at ∞. In order to prove this result, we shall first show that the resolvent of A is a compact operator of the Hilbert–Schmidt type.

Since we can always change q in $q + \mu$, where μ is not an eigenvalue of A, we may assume that zero is not an eigenvalue of A. Denote by u_1 (resp. u_2) a solution of the equation $Af = 0$ that satisfies the boundary condition (1) (resp. (2)). u_1 and u_2 are not proportional; otherwise they would satisfy both conditions (1) and (2), and zero would then be an eigenvalue of A in contradiction with what has been assumed. u_1 and u_2 being not proportional,

$$W = u_1' u_2 - u_1 u_2',$$

and, since

$$W' = u_1'' u_2 - u_1 u_2'' = 0,$$

we have

$$W(x) = const. = W(a).$$

---

[21] As usual, derivatives are taken in the sense of distributions.

Let K be the Hilbert–Schmidt operator on $L^2_m([a,b])$ whose kernel is

$$K(x,y) = \begin{cases} -u_1(x)u_2(y)/W(a), & \text{if } a \le x \le y \le b; \\ -u_1(y)u_2(x)/W(a), & \text{if } a \le y \le x \le b. \end{cases}$$

K is a self-adjoint compact operator, and it is straightforward to verify that the solution of the equation $Af = g$, which satisfies the boundary conditions (1) and (2), is $f = Kg$. Indeed, from

$$f(x) = -\frac{u_2(x)}{W(a)} \int_a^x u_1(y)g(y)\,dy - \frac{u_1(x)}{W(a)} \int_x^b u_2(y)g(y)\,dy,$$

it follows that

$$f'(x) = -\frac{u_2'(x)}{W(a)} \int_a^x u_1(y)g(y)\,dy - \frac{u_1'(x)}{W(a)} \int_x^b u_2(y)g(y)\,dy$$

and

$$f''(x) = -\frac{u_2''(x)}{W(a)} \int_a^x u_1(y)g(y)\,dy - \frac{u_1''(x)}{W(a)} \int_x^b u_2(y)g(y)\,dy + g(x).$$

Replacing $u_1''(x)$ (resp. $u_2''(x)$) by $qu_1(x)$ (resp. $qu_2(x)$), we find $f'' - qf = g$. Taking into account the expressions of $f(x)$ and $f'(x)$, it can be seen that the boundary conditions are satisfied.

If zero were an eigenvalue of K, there would exist a function $g \neq 0$ such that $Kg = 0$, that is,

$$u_2(x) \int_a^x u_1(y)g(y)\,dy + u_1(x) \int_x^b u_2(y)g(y)\,dy = 0.$$

Differentiating with respect to x, we obtain

$$u_2'(x) \int_a^x u_1(y)g(y)\,dy + u_1'(x) \int_x^b u_2(y)g(y)\,dy = 0.$$

u_1 and u_2 being linearily independent, for all $x \in [a,b]$, we have

$$\int_a^x u_1(y)g(y)\,dy = 0 \quad \text{and} \quad \int_x^b u_2(y)g(y)\,dy = 0,$$

that is, $g = 0$, and, therefore, zero cannot be an eigenvalue of K.

Since 0 is not an eigenvalue of K, A has a countable point spectrum with no accumulation point, except at ∞, and the set of its eigenvectors constitutes a complete orthogonal system.

Note that all the eigenvalues of a Sturm–Liouville operator are nondegenerate, because, if the equation $f'' - (q + \lambda f) = 0$ had two independent solutions f_1 and f_2 satisfying the boundary conditions, then any function $f = c_1 f_1 + c_2 f_2$, where c_1 and c_2 are arbitrary constants, would be a solution satisfying the boundary conditions. This is absurd for, in this case, $f(a)$ and $f'(a)$ could take arbitrary values.

In quantum mechanics, the Hamiltonian of a particle of unit mass moving in the interval $[a, b]$ is $H = \frac{1}{2}P^2 + V$, where $\frac{1}{2}P^2$ and V are, respectively the operators representing the kinetic energy and the potential energy defined by

$$(\frac{1}{2}P^2 f)(x) = -\frac{1}{2}f''(x) \quad \text{and} \quad (Vf)(x) = V(x)f(x).$$

The results we have obtained show that if the domain of H is such that the boundary conditions (1) and (2) are satisfied, which means that the particle cannot go outside the interval $[a, b]$, then the spectrum of H is a pure point spectrum, its eigenvalues are real and nondegenerate, and the set of the normed eigenvectors is a Hilbert basis.

A more general Sturm–Liouville operator can be defined as follows. Let

$$Af = (pf')' - qf,$$

where p is a continuously differentiable function that has no zero on $[a, b]$, and q is a continuous function on $[a, b]$. The domain of A is such that the boundary conditions (1) and (2) have to be satisfied. Here again, proceeding as above, it is shown that A has a compact resolvent. If we assume that zero is not an eigenvalue of A, and if we denote by u_1 (resp. u_2) the solution of the equation $Af = 0$ that satisfies the boundary condition (1) (resp. (2)), as above, u_1 and u_2 are not proportional, and, therefore, $W = u_1' u_2 - u_1 u_2' \neq 0$; but $(pW)' = 0$, that is,

$$p(x)W(x) = const. = p(a)W(a).$$

Let K be the Hilbert–Schmidt operator on $L^2_m([a, b])$ whose kernel is given by

$$K(x, y) = \begin{cases} -u_1(x)u_2(y)/p(a)W(a), & \text{if } a \leq x \leq y \leq b; \\ -u_1(y)u_2(x)/p(a)W(a), & \text{if } a \leq y \leq x \leq b. \end{cases}$$

K is a self-adjoint compact operator, and, as above, it is straightforward to verify that the solution of the equation $Af = g$, which satisfies the boundary conditions (1) and (2), is $f = Kg$.

All the conclusions reached previously remain valid:

(1) A has a pure point spectrum, the set of eigenvalues being countable with no accumulation point, except at ∞.

(2) All the eigenvalues are real and nondegenerate.

(3) The set of the normed eigenvectors of A is a Hilbert basis of $L^2_m([a,b])$.

Note, however, that these results are valid only if the Sturm–Liouville operator satisfies the following conditions:

(1) The interval $[a,b]$ is bounded.

(2) p is a continuously differentiable function which has no zero on $[a,b]$.

(3) q is a continuous function on $[a,b]$.

(4) The boundary conditions are conditions (1) and (2).

A Sturm–Liouville operator satisfying these conditions is said to be *regular*; if one of these conditions is not satisfied, the operator is *singular*. The spectral theory of a singular operator is more complicated, and certain results obtained for a regular operator may be not true.

Consider, for instance, the operator A of the preceding example, that is, the operator A in $L^2_m([0,1])$ such that $Af = -f''$, but modify the domain of A requiring the functions f to satisfy the boundary condition

$$f(0) = f(1).$$

Note that, for a regular operator, each boundary condition refers to one endpoint; here, this is not the case.

The eigenvalues of A are

$$\lambda_n = 4n^2\pi^2 \qquad (n = 0, 1, 2, \ldots)$$

zero is nondegenerate, and the corresponding eigenfunction is $x \mapsto 1$; but, for $n \neq 0$, the eigenvalues λ_n are twice degenerate, the corresponding eigenfunctions being $x \mapsto \sin 2n\pi x$ and $x \mapsto \cos 2n\pi x$.[22]

[22] For more details on singular Sturm–Liouville operators see R. D. Richtmyer.

Appendix

Theorem A1. *Let f belong to $L^2_m(\mathbf{R})$; then $F: x \mapsto \int_{x_0}^{x} f(\xi)\, d\xi$ is a continuous function, and, in the sense of distributions, $F' = f$.*

Let (φ_k) be a sequence of functions in $\mathcal{D}$ that converges to $f \in L^2_m(\mathbf{R})$.[23] $L^2_m(\mathbf{R})$ being complete, (φ_k) is a Cauchy sequence, and, as a consequence of the Cauchy–Schwarz inequality, we have

$$\begin{aligned}\left|\int_{x_0}^{x} (\varphi_k(\xi) - \varphi_l(\xi))\, d\xi\right|^2 &\leq \int_{x_0}^{x} \left|\varphi_k(\xi) - \varphi_l(\xi)\right|^2 d\xi \int_{x_0}^{x} d\xi \\ &\leq \|\varphi_k - \varphi_l\|^2\, |x - x_0|.\end{aligned}$$

Thus the sequence (Φ_k), such that

$$\Phi_k(x) = \int_{x_0}^{x} \varphi_k(\xi)\, d\xi,$$

converges to a function F; the convergence being uniform on any finite interval, F is continuous.

Furthermore, for any $\psi \in \mathcal{D}$, we have

$$\begin{aligned}F'(\psi) &= -F(\psi') \\ &= -\lim_{k\to\infty} \Phi_k(\psi') \\ &= \lim_{k\to\infty} \varphi_k(\psi) \\ &= f(\psi),\end{aligned}$$

which completes the proof.

Theorem A2. Integration by parts formula. *Let f and g be two functions belonging to $L^2_m(\mathbf{R})$ such that, as distributions, their derivatives f' and g' also belong to $L^2_m(\mathbf{R})$; then*

$$\int_a^b (f(\xi)g'(\xi) + f'(\xi)g(\xi))\, d\xi = \Big[f(\xi)g(\xi)\Big]_a^b.$$

[23] $\mathcal{D}$ is dense in $L^2_m(\mathbf{R})$.

Note that the right-hand side of the formula is well defined since, f' and g' belonging to $L^2_m(\mathbf{R})$, f and g are continuous (Theorem A1). As a consequence of the Cauchy–Schwarz inequality, the integral on the left-hand side is convergent since

$$\int_a^b \big(f(\xi)g'(\xi) + f'(\xi)g(\xi)\big)\,d\xi \leq \|f\|\,\|g'\| + \|f'\|\,\|g\|.$$

To prove the validity of the integration by parts formula, we have only to show that, as distributions,

$$(fg)' = fg' + f'g.$$

Since neither f nor g is infinitely differentiable, the proof given in Example 2.1 of Chapter 4 does not apply. We have here to use the formula derived in Example 3.4 of the same chapter, that is, if T is a distribution, its derivative T' is given by

$$T' = \lim_{h\to 0} \frac{1}{h}(\tau_h^{-1}T - T),$$

where, for a function f,

$$(\tau_h^{-1}f)(x) = f(x+h).$$

Hence, we have to show that, in $\mathcal{D}'$,

$$(fg' + f'g)(x) = \lim_{h\to 0} \frac{1}{h}\big(f(x+h)g(x+h) - f(x)g(x)\big).$$

This result follows immediately from the fact that the right-hand side can be written

$$\lim_{h\to 0} \frac{1}{h} f(x+h)\big(g(x+h) - g(x)\big) + \lim_{h\to 0} \frac{1}{h}\big(f(x+h) - f(x)\big)g(x).$$

Problems

1. Let (e_n) be a Hilbert basis in l^2, and let A be a linear operator in l^2 defined by

$$(\forall n \in \mathbf{N}) \qquad Ae_n = \frac{1}{n} e_n.$$

Show that A is bounded but that its inverse A^{-1} is not.

2. Let A and B be two bounded linear operators on a Banach space; if A is invertible and $\|A^{-1}B\| < 1$, show that $A - B$ is invertible, and find the expression of its inverse.

3. Let (f_n) be a Hilbert basis in the Hilbert space $L^2_m(\mathbf{R})$; if k_{mn} denotes the matrix element $\langle f_m \mid Kf_n \rangle$ of a Hilbert–Schmidt operator K, show that

$$\int\int |K(x,y)|^2 \, dx \, dy = \sum_{n=1}^{\infty} \sum_{m=1}^{\infty} |k_{mn}|^2.$$

4. Find the adjoint K^+ of a Hilbert–Schmidt operator K on the Hilbert space $L^2_m(\mathbf{R})$.

5. Show that any bounded linear operator A on a complex Hilbert space E can be written $A = A_1 + iA_2$, where A_1 and A_2 are self-adjoint bounded linear operators on E.

6. Let A and B be two self-adjoint bounded linear operators on a complex Hilbert space E. If A and B do not commute, show that $AB + BA$ and $i(AB - BA)$ are self-adjoint bounded linear operators on E.

7. If P and Q are two projection operators on a Hilbert space E, which conditions have to be satisfied for

(1) $P + Q$ to be a projection operator?

(2) PQ to be a projection operator?

8. Let A be a linear operator on $L^2_m(\mathbf{R})$ defined by

$$(\forall f \in L^2_m(\mathbf{R})) \quad (Af)(\xi) = \begin{cases} f(\xi), & \text{if } \xi \geq a; \\ 0, & \text{otherwise.} \end{cases}$$

What can be said of A?

9. Let A be a linear operator on $L^2_m(\mathbf{R}_+)$ defined by

$$(\forall f \in L^2_m(\mathbf{R}_+)) \quad (Af)(\xi) = \begin{cases} f(\xi - a), & \text{if } \xi \geq a, \\ 0, & \text{otherwise,} \end{cases}$$

where $a > 0$. What can be said of A?

10. Show that the annihilation and creation operators A and A^+ defined in Example 4.1 satisfy the commutation relation

$$[A, A^+] = I.$$

11. Let P be the momentum operator defined in $L^2_m(\mathbf{R}_+)$ by:

$$Pf = -if',$$

where f' is the derivative of f in the sense of distributions, and

$$D_P = \{f | f \in L^2_m(\mathbf{R}_+), f' \in L^2_m(\mathbf{R}_+), f(0) = 0\}.$$

(1) Show that P is Hermitian but not self-adjoint.
(2) Determine the eigenvalues of P and P^+.

12. Let (e_n) be a Hilbert basis in l^2, and let A be a linear operator in l^2 defined by

$$(\forall n \in \mathbf{N}) \qquad Ae_n = \lambda_n e_n,$$

where the sequence (λ_n) is such that

$$(\forall n \in \mathbf{N}) \qquad \lambda_n \neq a$$

and

$$\lim_{n\to\infty} = a.$$

Determine the spectrum of A.

13. Let A be the linear operator defined in l^2 by

$$Ax_n = \lambda_n x_n,$$

where the sequence (x_n) belongs to l^2, and $\lambda_n \in \mathbf{C}$. Show that if

$$\lim_{n\to\infty} \lambda_n = 0,$$

A is compact.

14. Let A be the linear operator defined in l^2 by

$$Ax_1 = 0 \quad \text{and} \quad Ax_n = \frac{x_{n-1}}{n-1} \quad \text{if } n > 1,$$

where the sequence (x_n) belongs to l^2. Show that A is compact and that its point spectrum is empty.

15. Show that the spectrum of a unitary operator lies on the unit circle, and that its residual spectrum is empty.

16. Solve in $L_m^2([0,1])$ the Fredholm integral equation

$$\int_0^1 K(x,y)f(y)\,dy - \lambda f(x) = g(x)$$

in the following two cases
(1) $K(x,y) = xy$,
(2) $K(x,y) = \exp(x+y)$.

17. Let K be a self-adjoint Hilbert–Schmidt operator on the Hilbert space $L_m^2([a,b])$, and denote by (λ_n) the sequence of its eigenvalues and by (ϕ_n) the sequence of the corresponding normed eigenfunctions. Assuming that no eigenvalue is degenerate, show that

$$K(x,y) = \sum_n \lambda_n \phi_n(x)\overline{\phi_n(y)}.$$

This result is known as *Mercer's theorem.*

18. Determine the eigenvalues and the eigenfunctions of the Hilbert–Schmidt operator defined on $L_m^2([0,1])$ by

$$(Kf)(x) = \int_0^x yf(y)\,dy + x\int_x^1 f(y)\,dy.$$

19. Let A be the linear operator defined in $L_m^2(\mathbf{R})$ by

$$Af = -f'',$$

where f'' is understood as a distribution derivative, and

$$D_A = \{f \mid f \in L_m^2(\mathbf{R}), f'' \in L_m^2(\mathbf{R})\}.$$

Show that A is self-adjoint, determine its spectrum, and find the expression of the resolvent operator $R_\lambda(A)$.

20. Determine the eigenvalues and the eigenfunctions of the integral operator K defined in $L_m^2([0,\pi])$ by

$$(Kf)(x) = \frac{1}{2\pi}\int_{-\pi}^{\pi} \frac{f(y)\,dy}{1 - 2r\cos(x-y) + r^2},$$

where $0 < r < 1$.

21. What can be said of the spectrum of the Sturm–Liouville operator A defined in $L_m^2([0,1])$ by

$$(Af)(x) = \big(xf'(x)\big)',$$

with

$$D_A = \{f \mid f \in L_m^2([0,1]), Af \in L_m^2([0,1]), f(0) < \infty, f(1) = 0\}?$$

Solutions

1. For any $x = \sum_n \alpha_n e_n$ in l^2, we have

$$Ax = \sum_n \alpha_n Ae_n = \sum_n \frac{1}{n}\alpha_n e_n,$$

and, since the series $\sum_n |\alpha_n|^2$ is convergent, the series $\sum_n |\alpha_n|^2/n^2$ is also convergent. Therefore Ax belongs to l^2, and $D_A = l^2$. Moreover, from

$$\begin{aligned}\|Ax\|^2 &= \sum_n \frac{|\alpha_n|^2}{n^2}\\ &\leq \sum_n |\alpha_n|^2 \sum_n \frac{1}{n^2}\end{aligned}$$

it follows that A is bounded, and that $\|A\|^2 \leq \sum_n 1/n^2$. The range of A differs from l^2, it is the subspace

$$\{y|y = \sum_n \beta_n e_n, \sum_n n^2|\beta_n|^2 < \infty\}.$$

The inverse A^{-1} of A is defined by

$$(\forall n \in \mathbf{N}) \qquad A^{-1}e_n = ne_n.$$

Its domain is, of course, the range of A, and it is easy to check that it is unbounded since

$$\|A^{-1}e_n\| = n\|e_n\| = n.$$

2. It is sufficient to write $A - B = A(I - A^{-1}B)$ to prove that $A - B$ is invertible, and that we have

$$(A-B)^{-1} = A^{-1} + A^{-1}BA^{-1} + A^{-1}BA^{-1}BA^{-1} + \cdots.$$

3. We have

$$\begin{aligned}(Kf_n)(x) &= \int K(x,y)f_n(y)\,dy\\ &= \sum_m \langle f_m \mid Kf_n\rangle f_m(x)\\ &= \sum_m k_{mn}f_m(x),\end{aligned}$$

and the Parseval's identity yields

$$\int \left| \int K(x,y) f_n(y)\, dy \right|^2 dx = \sum_m |k_{mn}|^2.$$

Furthermore, since the sequence $(\overline{f_n})$ is also a Hilbert basis, we may write

$$K(x,y) = \sum_n \Big(\int K(x,y) f_n(y)\, dy \Big) \overline{f_n(y)},$$

and

$$\int |K(x,y)|^2\, dy = \sum_n \left| \int K(x,y) f_n(y)\, dy \right|^2,$$

from which the result, we are looking for, readily follows.

4. If f and g both belong to $L^2_m(\mathbf{R})$, we have

$$\begin{aligned} \langle g \mid Kf \rangle &= \int \overline{g(x)} \Big(\int K(x,y) f(y)\, dy \Big)\, dx \\ &= \int \int K(x,y) \overline{g(x)} f(y)\, dy\, dx \end{aligned}$$

and

$$\begin{aligned} \langle K^+ g \mid f \rangle &= \int \overline{\Big(\int K^+(x,y) g(y)\, dy \Big)} f(x)\, dx \\ &= \int \int \overline{K^+(y,x) g(x)} f(y)\, dx\, dy, \end{aligned}$$

where, in the last integral, we have permuted x and y. Hence, from

$$\langle g \mid Kf \rangle = \langle K^+ g \mid f \rangle,$$

it follows that

$$K^+(x,y) = \overline{K(y,x)}.$$

5. Denote by A^+ the adjoint of A, and let

$$\begin{aligned} A_1 &= \frac{1}{2}(A + A^+) \\ A_2 &= \frac{1}{2i}(A - A^+). \end{aligned}$$

A_1 and A_2 are self-adjoint, and $A = A_1 + iA_2$. To prove the unicity of this decomposition, assume that there exists another pair of self-adjoint operators A_1' and A_2' such that $A = A_1' + iA_2'$; then $A_1 - A_1' = i(A_2' - A_2)$, where $A_1 - A_1'$ and $A_2' - A_2$ are self-adjoint. If B is a self-adjoint operator, for all $x \in E$, $\langle x \mid Bx \rangle$ is real since

$$\langle x \mid Bx \rangle = \langle Bx \mid x \rangle = \overline{\langle x \mid Bx \rangle}.$$

Therefore, for all $x \in E$,

$$\alpha_1 = \langle x \mid (A_1 - A_1')x \rangle \quad \text{and} \quad \alpha_2 = \langle x \mid (A_2' - A_2)x \rangle$$

are two reals such that $\alpha_1 = i\alpha_2$, that is $\alpha_1 = \alpha_2 = 0$, which proves the unicity of the decomposition.

6. These results readily follow from the relation

$$(AB)^+ = B^+A^+ = BA.$$

7. A linear operator is a projection operator if, and only if, it is Hermitian and idempotent.

(1) P and Q being Hermitian, $P + Q$ is Hermitian, and from

$$(P+Q)^2 = P^2 + PQ + QP + Q^2 = P + PQ + QP + Q,$$

it follows that $P + Q$ is idempotent if $PQ + QP = 0$, that is $PQ = -QP$. But, in this case,

$$PQ = P^2Q = -PQP = QP^2 = QP,$$

and, therefore, $PQ = QP = 0$. This condition, which is necessary for $P+Q$ to be a projection operator, is also sufficient, since

$$(P+Q)^2 = P^2 + Q^2 = P + Q.$$

(2) P and Q being Hermitian, for all x and y in E, we have

$$\langle PQx \mid y \rangle = \langle Qx \mid Py \rangle = \langle x \mid QPy \rangle.$$

Hence PQ is Hermitian if, and only if, $PQ = QP$. If this condition is satisfied, then PQ is idempotent for

$$(PQ)^2 = PQPQ = P^2Q^2 = PQ.$$

8. A is clearly idempotent. It is also Hermitian since, for all functions f and g belonging to $L_m^2(\mathbf{R})$, we have

$$\langle g \mid Af \rangle = \int_a^\infty \overline{g(\xi)} f(\xi)\, d\xi = \langle Ag \mid f \rangle.$$

A is, therefore, a projection operator.

9. A is an isometry since

$$\|Af\|^2 = \int_a^\infty |f(\xi - a)|^2\, d\xi = \int_0^\infty |f(\xi)|^2\, d\xi = \|f\|^2.$$

The range of A being the subspace $L_m^2([a, \infty[)$ of $L_m^2(\mathbf{R}_+)$, A is not unitary.

10. For any $x = \sum_{n=0}^\infty c_n(x) e_n \in D_A = D_{A^+}$, we have

$$\sum_{n=0}^\infty A^+ A c_n(x) e_n = \sum_{n=0}^\infty n c_n(x) e_n$$

and

$$\sum_{n=0}^\infty A A^+ c_n(x) e_n = \sum_{n=0}^\infty (n+1) c_n(x) e_n.$$

Hence, $AA^+x - A^+Ax = x$, that is,

$$[A, A^+] = I.$$

Recall (see Example 1.3) that this last relation is not entirely correct since $D_{[A,A^+]} = D_A \neq l^2$.

11 (1) If f and g belong to D_P, we have

$$\begin{aligned}
\langle g \mid Pf \rangle &= \int_0^\infty \overline{g(\xi)}\big(- \imath f'(\xi)\big)\, d\xi \\
&= \Big[- i\overline{g(\xi)} f(\xi) \Big]_o^\infty + \int_0^\infty \big(\overline{-ig'(\xi)}\big) f(\xi)\, d\xi \\
&= \langle Pg \mid f \rangle;
\end{aligned}$$

since f and g are equal to zero for $\xi = 0$ and $\xi = \infty$, $\big[- i\overline{g(\xi)} f(\xi) \big]_0^\infty = 0$. P is, therefore, Hermitian.

D_P being dense in $L_m^2(\mathbf{R}_+)$, P^+ is well defined, and it is found that

$$P^+ g = -ig' \quad \text{with} \quad D_{P^+} = \{g | g \in L_m^2(\mathbf{R}_+), g' \in L_m^2(\mathbf{R}_+)\},$$

where g' is, as usual, understood as a distribution derivative. Since g does not have to satisfy any boundary condition, P^+ is an extension of P; that is, P is not self-adjoint.

(2) The solution of the differential equation

$$-if' = \lambda f$$

is

$$f(\xi) = \exp(i\lambda\xi).$$

If f has to satisfy the boundary condition $f(0) = 0$, the only solution is $f = 0$: the point spectrum of P is empty.

If f does not satisfy any boundary condition, for $\Im\lambda > 0$, $\xi \mapsto a\exp(i\lambda\xi)$ is in $L^2_m(\mathbf{R}_+)$: the point spectrum of P^+ is the upper complex half-plane.

In quantum mechanics, P represents the radial momentum operator. It is Hermitian, but cannot be made self-adjoint by any choice of its domain.

12. It is clear that A is bounded and that $D_A = l^2$. For any n, $\lambda = \lambda_n$ is a nondegenerate eigenvalue of A, and e_n is the corresponding eigenvector. If $\lambda \neq \lambda_n$, we have

$$(A - \lambda I)x = \sum_n (\lambda_n - \lambda)x_n e_n,$$

where $x = \sum_n x_n e_n \in l^2$. Since the right-hand side is equal to zero only if, for all n, $x_n = 0$, it follows that $R_\lambda(A) = (A - \lambda I)^{-1}$ exists if $\lambda \neq \lambda_n$. The point spectrum of A is the set of all λ_n.

Since any finite linear combination of basis vectors is in the domain of $R_\lambda(A)$, its domain is dense, and the residual spectrum is, therefore, empty.

To determine the continuous spectrum, we have to find the set of all λ such that $R_\lambda(A)$ is unbounded. From

$$R_\lambda(A)x = \sum_n \frac{x_n e_n}{\lambda_n - \lambda},$$

where $x = \sum_n x_n e_n$ is any vector in the domain of $R_\lambda(A)$, it follows that

$$\|R_\lambda(A)x\|^2 = \sum_n \frac{|x_n|^2}{|\lambda_n - \lambda|^2}.$$

If $\lambda \neq a$, $\lambda_n - \lambda$ does not tend to zero, and there exists $\delta > 0$ such that, for all n, $|\lambda_n - \lambda| \geq \delta$. In this case,

$$\|R_\lambda(A)x\|^2 \leq \frac{\|x\|^2}{\delta^2},$$

which shows that $R_\lambda(A)$ is bounded, and that the resolvent set is given by

$$\rho(A) = \{\lambda \in \mathbf{C} | (\forall n)\, \lambda \neq \lambda_n, \lambda \neq a\}.$$

$R_\lambda(A)$ is unbounded only if $\lambda_n - \lambda$ may go to zero, that is, if $\lambda = a$. The continuous spectrum contains, therefore, only one element, namely, $\lambda = a$.

13. Let A_p be the finite-rank, hence compact, operator defined by

$$A_p x_n = \begin{cases} \lambda_n x_n, & \text{if } p \geq n; \\ 0, & \text{otherwise.} \end{cases}$$

Then

$$(A - A_p)x_n = \begin{cases} 0, & \text{if } p \geq n; \\ \lambda_n x_n, & \text{otherwise.} \end{cases}$$

Hence,

$$\lim_{p\to\infty} \|A - A_p\| = \lim_{p\to\infty} \sup_{n>p} |\lambda_n| = 0.$$

A as the limit of a sequence of compact operators is, therefore, compact.

14. Let S and D be the operators on l^2 defined by

$$Sx_n = x_{n-1} \quad \text{and} \quad Dx_n = \frac{x_n}{n}.$$

S is an isometry, hence bounded, and D is compact (Problem 13). Thus, $A = DS$ is compact.

If λ is an eigenvalue of A, there should exist a nonzero $x = (x_n) \in l^2$ such that $Ax = \lambda x$; that is,

$$Ax_1 = \lambda x_1 = 0,$$

and

$$Ax_n = \lambda x_n = \frac{x_{n-1}}{n-1} \quad \text{if } n > 1.$$

However, it is readily seen that these relations imply, for all n, $x_n = 0$.

This result is interesting. It shows that, contrary to what happens in the finite-dimensional case, a compact operator in an infinite-dimensional Hilbert space may have an empty point spectrum.

15. If λ is an eigenvalue of the unitary operator U, we have

$$\|Ux\|^2 = \|x\|^2 = |\lambda|^2 \|x\|^2 \quad (x \neq 0).$$

Any eigenvalue is, therefore, of the form $\exp(i\theta)$, where θ is real.

If λ is not an eigenvalue ($|\lambda| \neq 1$), the equation

$$Ux - \lambda x = y,$$

whose solution is $x = R_\lambda(U)y$, implies

$$\big||\lambda| \|x\| - \|y\|\big| \leq \|Ux\| \leq |\lambda| \|x\| + \|y\|;$$

that is, taking into account the relation $\|Ux\| = \|x\|$,

$$\|R_\lambda(U)y\| \leq \frac{\|y\|}{\big|1 - |\lambda|\big|}.$$

If $|\lambda| \neq 1$, $R_\lambda(U)$ is bounded, and the relation

$$\overline{R_{U-\lambda I}} = \left(\ker(U^+ - \overline{\lambda} I)\right)^\perp$$

shows that $R_{U-\lambda I}$ is dense, because $\overline{\lambda}$ not being an eigenvalue of U^+, $\ker(U^+ - \overline{\lambda} I) = \{0\}$. Therefore, if $|\lambda| \neq 1$, $R_\lambda(U)$ exists, is bounded, and is densely defined. In conclusion, the residual spectrum is empty, and the resolvent set is the complement in **C** of the unit circle.

Note that we have not proved that any complex number of the form $\exp(i\theta)$ (θ real) is an eigenvalue, but that any eigenvalue is of this form. Indeed, the elements of the continuous spectrum are also of this form.

16. In both cases, the operator K is a Hilbert-Schmidt operator, which is self-adjoint since the kernels, of the form $f(x)f(y)$, are obviously symmetric. Moreover, this particular form of the kernels shows immediately that the eigenvalue equation $K\phi = \lambda\phi$ has the solution

$\phi(x) = x$ with $\lambda = 1/3$ in the first case, and

$\phi(x) = \exp x$ with $\lambda = (e^2 - 1)/2$ in the second one.[24]

---

[24] These results can also be obtained by differentiating twice the eigenvalue equations. We find (1) $\phi' = C$, where C is a constant, and (2) $\phi' = \phi$.

Using Theorem 18, we obtain the solutions of the integral equations. In each case, denote by ϕ_0 the normed eigenfunction, and by λ_0 the corresponding eigenvalue.[25]

If g is orthogonal to ϕ_0, we have

$$f(x) = \begin{cases} -\frac{1}{\lambda} g(x), & \text{if } \lambda \neq \lambda_0; \\ C\phi_0 - \frac{1}{\lambda_0} g(x), & \text{if } \lambda = \lambda_0. \end{cases}$$

If g is not orthogonal to ϕ_0, we have

$$f(x) = -\frac{1}{\lambda}\big(g(x) - \langle \phi_0 \mid g \rangle \phi_0(x)\big) + \frac{1}{\lambda_0 - \lambda} \langle \phi_0 \mid g \rangle \phi_0(x),$$

if $\lambda \neq \lambda_0$, and no solution if $\lambda = \lambda_0$.

17. K being defined on $L^2_m([a,b])$, the kernel $(x,y) \mapsto K(x,y)$ belongs to $L^2_m([a,b]) \times L^2_m([a,b])$. Let (e_n) be a Hilbert basis in $L^2_m([a,b])$ that consists of the sequence (ϕ_n) of the eigenvectors of K completed by a basis in $\ker K$. Let I be a subset of indices such that $e_n = \phi_n$, if $n \in I$, and $e_n \in \ker K$, if $n \notin I$. The set of functions

$$(x,y) \mapsto e_m(x)\overline{e_n(y)} \quad ((m,n) \in \mathbf{N})$$

is a Hilbert basis in $L^2_m([a,b]) \times L^2_m([a,b])$, and we have

$$K(x,y) = \sum_{m,n} c_{mn} e_m(x) \overline{e_n(y)}.$$

Hence,

$$\int_a^b K(x,y) e_n(y)\, dy = \sum_m c_{mn} e_m(x).$$

If $n \in I$, the left-hand side is equal to $\lambda_n e_n(x)$, whereas if $n \notin I$, the right-hand side is equal to zero. The only coefficients c_{mn} that are not equal to zero are the coefficients c_{nn} where $n \in I$, and, in this case, $c_{nn} = \lambda_n$.

18. The kernel $K(x,y)$ is symmetric since it may be written $\min(x,y)$; the operator K is, therefore, compact and self-adjoint. Differentiating twice the eigenvalue equation $K\phi = \lambda\phi$ yields

$$\lambda\phi'' + \phi = 0.$$

---

[25] Note that with these notations, we have $K(x,y) = \lambda_0 \phi_0(x) \phi_0(y)$ (see Problem 17).

To solve this differential equation, we need to know the boundary conditions to be satisfied by ϕ. Since $(K\phi)(0) = 0$, $\phi(0)$ must be equal to zero, and from $(K\phi)'(x) = \int_x^1 \phi(y)\,dy$, it follows that $\phi'(1) = 0$.

The eigenfunctions satisfying these boundary conditions are

$$\phi_n : x \mapsto \sin \frac{(2n+1)\pi x}{2},$$

and the corresponding eigenvalues are

$$\lambda_n = \frac{4}{(2n+1)^2\pi^2},$$

where n is any integer. zero is not an eigenvalue, thus the sequence (ϕ_n) is a Hilbert basis in $L^2_m([0,1])$.

19. Proceeding as in Example 4.4, we can show that A is self-adjoint. Its spectrum is, therefore, real. A Fourier transform (see Example 5.4) shows that this spectrum is identical to the spectrum of the operator B on $L^2_m(\mathbf{R})$ defined by

$$(Bf)(x) = x^2 f(x),$$

and

$$D_B = \{f | f \in L^2_m(\mathbf{R}), x^2 f \in L^2_m(\mathbf{R})\}.$$

As in Example 5.2, it can then be shown that the only nonempty spectrum is the continuous spectrum, and that it coincides with the set of nonnegative reals.

If $\lambda \in \rho(A)$, the solution of

$$-f'' - \lambda f = g,$$

belonging to $L^2_m(\mathbf{R})$, is

$$f = R_\lambda(A)g,$$

which, after a Fourier transform, can be written

$$f(x) = \frac{1}{k}\int_{-\infty}^{\infty} e^{-k|x-\xi|} g(\xi)\,d\xi,$$

where k is the square root of $-\lambda$ whose real part is positive.

20. K is compact and self-adjoint. The eigenvalue equation is a convolution equation of the form

$$\frac{1}{2}k * f = \lambda f,$$

where k, defined on $[-\pi, \pi]$, is given by

$$k(\theta) = \frac{1}{1 - 2r\cos\theta + r^2}.$$

This equation can be solved using Theorem 10 of Chapter 4. The Fourier series expansion of the function k is easy to find if one considers the analytic function $z \mapsto (1+z)/(1-z)$. In the neighborhood of the origin, we have

$$\frac{1+z}{1-z} = 1 + 2\sum_{n=1}^{\infty} z^n.$$

This series is uniformly convergent for $|z| < 1$. Let $z = r\exp(i\theta)$, and equate the real part of the function with the real part of its expansion (Remark 2 of Example 3.8 of Chapter 3). This yields

$$\frac{1}{1 - 2r\cos\theta + r^2} = \frac{1}{1-r^2}\Big(1 + \sum_{n=1}^{\infty} r^n \cos n\theta\Big).$$

The expansion of the kernel K is, therefore,

$$K(x,y) = \frac{1}{2\pi}\frac{1}{1-r^2} + \sum_{n=1}^{\infty} \frac{2r^n}{1-r^2}\frac{\cos nx}{\sqrt{2\pi}}\frac{\cos ny}{\sqrt{2\pi}}$$
$$+ \sum_{n=1}^{\infty} \frac{2r^n}{1-r^2}\frac{\sin nx}{\sqrt{2\pi}}\frac{\sin ny}{\sqrt{2\pi}},$$

which shows (see Mercer's theorem, Problem 17) that the eigenvalues of K are $1/(1-r^2)$ associated to the normed eigenfunction $x \mapsto 1/\sqrt{2\pi}$, and $2r^n/(1-r^2)$ associated to the normed eigenfunctions $x \mapsto (1/\sqrt{2\pi})\cos nx$ and $x \mapsto (1/\sqrt{2\pi})\sin nx$. Since zero is not an eigenvalue of K, the set of all eigenfunctions is a Hilbert basis in $L^2_m([-\pi, \pi])$.

21. This is not a regular Sturm–Liouville operator, since $p(0) = 0$. The results obtained in the case of a regular operator might, therefore, not be true. In the present case, however, we can show that they remain valid.

Since the general solution of the equation $Af = 0$, which is $c_1 \log x + c_2$, does not satisfy the boundary condition at the origin, zero is not an eigenvalue of A. Let us, in this case, determine the resolvent $R_0(A) = A^{-1}$. From the equation:

$$(xf'(x))' = g(x)$$

it follows that

$$xf'(x) = \int_0^x g(y)\,dy,$$

since the right-hand side must be equal to zero if $x = 0$, and

$$f(x) = \int_0^x \Big(\frac{1}{t}\int_0^t g(y)\,dy\Big)\,dt + \text{const.},$$

or

$$f(x) = \int_0^x \Big(g(y)\int_y^x \frac{dt}{t}\Big)\,dy + \text{const.}$$

$$= \int_0^x (\log x - \log y) g(y)\,dy + \text{const.}$$

The constant is equal to $\int_0^1 \log y g(y)\,dy$ in order to satisfy the boundary condition $f(1) = 0$. Finally;

$$f(x) = \int_0^x \log x g(y)\,dy + \int_x^1 \log y g(y)\,dy.$$

The resolvent $R_0(A)$ is, therefore, an integral operator whose kernel is

$$K(x,y) = \begin{cases} \log x, & \text{if } 0 \le y \le x \le 1; \\ \log y, & \text{if } 0 \le x \le y \le 1. \end{cases}$$

This operator is compact ($\int_0^1\int_0^1 |K(x,y)|^2\,dx\,dy < \infty$) and self-adjoint ($K(x,y) = K(y,x)$). zero is not an eigenvalue of the operator K because the solution of the equation

$$\int_0^x \log x g(y)\,dy + \int_x^1 \log y g(y)\,dy = 0$$

is $g = 0$, since, if we differentiate the left-hand side of this equation, we obtain

$$(\forall x \in [0,1]) \qquad \int_0^1 g(y)\,dy = 0.$$

The spectrum of A is, therefore, a pure point spectrum.

For $0 < x < 1$, $\log x$ being negative, the eigenvalues of K (and of A) are all negative. To find the eigenvalues of A we have to find the solutions of the differential equation

$$\big(xf'(x)\big)' = \lambda f(x)$$

that satisfy the boundary conditions $f(0) < \infty$ and $f(1) = 0$. Let $u = 2(-\lambda x)^{1/2}$; then the differential equation takes the form

$$\frac{d^2 f}{du^2} + \frac{1}{u}\frac{df}{du} + f = 0.$$

This is a Bessel's equation, whose only solution finite for $u = 0$ is $J_0(u)$. The eigenvalues of A are, therefore, the solutions of the equation $J_0(2\sqrt{-\lambda}) = 0$.

References

On History

E. T. BELL, *Men of Mathematics* (2 volumes) Penguin Books, Harmondsworth, 1965.

T. HAWKINS, *Lebesgue's Theory of Integration. Its Origin and Development*, Chelsea, New York, 1979.

H. LEBESGUE, *Measure and the Integral*, Holden–Day, San Francisco, 1966.

J. LÜTZEN, *The Prehistory of the Theory of Distributions*, Springer–Verlag, Heidelberg, 1982.

F. RIESZ, *összegyűjtött munkái* (Collected Papers in Hungarian, French and German) (2 volumes), Akadémiai Kiadó, Budapest, 1960.

Textbooks

G. de BARRA, *Measure Theory and Integration*, Ellis Horwoood, Chichester, 1981.

S. B. CHAE, *Lebesgue Integration*, Marcel Dekker, New York, 1980.

J. W. DETTMAN, *Applied Complex Variables*, Dover, New York, 1984.

J. DIEUDONNÉ, *Foundations of Modern Analysis*, Academic Press, New York, 1969.

A. N. KOLMOGOROV and S. V. FOMIN, *Introductory Real Analysis*, translated and edited by R. A. Silverman, Dover, New York, 1975.

R. D. RICHTMYER, *Principles of Advanced Mathematical Physics* (2 volumes), Springer–Verlag, Heidelberg, 1981.

W. RUDIN, *Real and Complex Analysis*, McGraw–Hill, New York, 1970.

W. RUDIN, *Functional Analysis*, Tata McGraw–Hill, New Delhi, 1977.

L. SCHWARTZ, *Cours d'Analyse de l'Ecole Polytechnique* (in French) (2 volumes), Hermann, Paris, 1967.

Advanced or Specialized Books

N. I. AKHIEZER and I. M. GLAZMAN *Theory of Linear Operators in Hilbert Spaces* (2 volumes), Pitman, Boston, 1981.

K. CHANDRASEKHARAN *Classical Fourier Transforms*, Springer–Verlag, Heidelberg, 1989.

I. M. GELFAND and G. E. SHILOV *Generalized Functions* (4 volumes), Academic Press, New York, 1964.

J. R. HIGGINS *Completeness and Basic Properties of Sets of Special Functions*, Cambridge University Press, Cambridge, 1977.

M. REED and B. SIMON, *Methods of Modern Mathematical Physics* (4 volumes published), Academic Press, New York, 1972.

L. SCHWARTZ, *Théorie des Distributions* (in French), Hermann, Paris, 1978.

I. N. SNEDDON, *The Use of Integral Transforms*, Tata McGraw–Hill, New Delhi, 1974.

A. ZYGMUND, *Trigonometric Series*, Cambridge University Press, Cambridge, England, 1979.

Index

A

Abel's
    criterion, 34, 35, 53, 62
    integral equation, 234, 247
Abscissa of convergence, 94
Accumulation point, 50, 291, 299
Adjoint, 264, 273–274
Admittance, 203, 206
Almost everywhere, 13, 22, 88
Annihilation operator, 273, 304
Approximate
    eigenvector, 281
    eigenvalue, 281
Axiom of choice, 9

B

Banach
    algebra, 259–261
    space, 105
Bayes's formula, 12
Beppo Levi theorem, 21
Bessel
    functions, 33, 45, 57, 96, 107
    inequality, 131
Best approximation, 139, 163
Borel
    measurable mapping, 4
    set, 4
Boundary, 50
Bounded linear operator, 257–270, 287–296

C

Canonical representation, 291–292
Cantor
    diagonal process, 15
    function, 16
    sets, 14, 55, 59, 60
Cauchy
    integral, 52
    problem, 200–201
    sequence, 104, 122
    sum, 52
Cauchy–Schwarz inequality, 119–120, 124, 138, 257
Causality,
Central limit theorem, 92
Character, 85
Characteristic function
    of a random variable, 90

of a subset, 5
Chebyshev
inequality, 25
polynomials, 130
Closable operator, 275
Closed set, 50
Closure of an operator, 275
Commutator, 256, 260
Compact
operator, 270–272, 290–293
resolvent, 294–297
Complete
measure, 13
metric space, 76, 79, 104
orthonormal system, 131–138
Completion, 104
Completely continuous, 270
Conditional probability, 11
Conditionally convergent integral, 29
Conjugate exponents, 71, 109
Continuous mapping, 50
Convergence
almost everywhere, 76
factor, 90
in $\mathcal{D}$, 169
in $\mathcal{D}'$, 190–195
in the mean of order p, 76
Convex function, 69–70
Convolution 43, 196–204, 211, 218, 220, 228
algebra, 201
Cosine integral, 108, 114–115
Counting measure, 8
Creation operator, 274, 304

D

Dalzell's criterion, 143
Decreasing sequence, 12
Delta-function (δ-function), 167
Dense, 80–82, 104, 205
Densely defined, 273–276
Derivative
of a distribution, 173–190
normal, 187
Dirac
distribution,
function, 167
measure, 8
sequence, 190–192
Dirichlet
kernel, 145
function,28
Distribution, 168–172
δ, 171, 174, 177, 197, 209
$\log|x|$, 175–177
$\log(x \pm i0)$, 177

Pf x^{-1}, 177, 219, 230, 232, 237
Pf x^{-n}, 181, 232, 235, 237–238, 251–252
Pv x^{-1}, 177
$(x \pm i0)^{-n}$, 194, 219, 225, 232, 238
x_+^α, 177–180
x_-^α, 180
regular, 171
scalar, 184
singular, 171
tempered, 214–215, 218, 221, 226
tensorial, 184
vectorial, 184–190
Distribution function, 24
Dual space, 170

E

Eigenvector, 281
Eigenvalue, 281, 288–300
Electric
charge, 188–190
displacement, 188
field, 188
potential, 188
Elementary
events,
solution, 199
Elliptic integrals, 56
Error function, 56, 108
Essential supremum, 77
Essentially bounded, 78
Euler constant, 65, 229
Eulerian functions, 56, 74, 82, 99, 202
Extension of an operator, 256, 263, 275–276
Exterior, 50, 104

F

Fatou's lemma, 30
Fejér's theorem, 146–148
Finite-rank, 271–272, 311
Form
Hermitian,
linear, 170
sesquilinear, 119
Fourier
inversion theorem, 88, 214
series
of a distribution, 193, 207–212
of a function, 143–148, 155, 161–162
transform
of a distribution, 212–225
of a function, 83–94, 107, 113–114, 138
Fourier–Plancherel, 267–270
Fractals, 14, 59

Fredholm alternative, 293
 integral equation, 305
Fresnel's integrals, 42
Fubini's theorem, 40–47, 57, 98
Fubini–Tonelli theorem, 41, 67
Fundamental solution, 199

G

Gegenbauer polynomials, 130
Gibbs phenomenon, 149
Graph of an operator, 275
Green's function, 199, 203
Green's formula, 186

H

Hadamard's finite part, 182–184
Hardy's inequality, 113
Harmonic function, 92
Hausdorff
 dimension, 18, 60
 outer measure, 17
Heat equation, 93, 102, 200
Heaviside function, 95
Hermite polynomials, 129–130, 139–141
Hermitian operator, 264–265, 276, 304, 309–310
Hilbert
 basis, 131, 151
 space, 121, 125–127
 transform, 223
Hilbert–Schmidt operator, 258, 272, 299, 303
Hölder's inequality, 71–73, 75, 109

I

Idempotent operator, 265
Impedance, 203
Improper integral, 31
Increasing sequence, 12
Independent events, 11
Integration by parts, 301–302
Interior, 50, 81, 104
Isometry, 153, 266, 268, 309

J

Jacobi polynomials, 129–130
Jensen's inequality, 70

K

Koch
 curve, 54, 59
 island, 60

Kramers–Kronig relations, 223–224

L

Laguerre polynomials, 108, 129–130, 139–140
Laplace transform
    of a distribution, 226–231
    of a function, 94–103, 108, 114–116
inversion theorem, 99, 231
Lebesgue
    characterization theorem, 48
    dominated convergence theorem, 29–33, 61
    function, 17
    integrable, 27, 106
    integral, 27, 80
    measurable, 9
    measure, 8, 16
    space, 69
Legendre polynomials, 127–129, 154, 158, 165
Limit point, 50
Linear system, 205
Locally integrable, 170
Lower limit, 51

M

Mathematical expectation, 25
Matrix elements, 262, 265
Measurable
    function, 1–7
    space, 2, 8
    set, 2
Mercer's theorem, 305
Metric space, 104
Minkowski's inequality, 72–73, 75
Momentum operator, 256, 277–280, 293, 304, 309–310
Monotone class, 37
Monotone convergence theorem, 20, 30
Mu-integrable (μ-integrable), 19–23

N

Negligible, 13
Neighborhood, 104
Nonmeasurable subset, 9
Norm, 105
    of a partition, 52
    of uniform convergence, 52
Normed vector space, 105

O

Open set, 50
Orthogonal polynomials,129–130, 154, 160
Orthogonal projection theorem, 122

Orthogonal system, 127–139
Orthonormal system, 127
Oscillation of a function, 48

P

Parseval's
 equality, 133
 formula, 269
Partition, 52
Peano curve, 60
Plancherel transform
Poincaré formula, 54, 58
Poisson's
 equation, 188
 formula, 93
 summation formula, 216
Position operator, 256, 258, 277
Prehilbert space, 119, 121, 124
Probability
 distribution, 24
 measure, 11
 theory, 11
Product
 measure, 37
 of measurable spaces, 3
Projection operator, 265–266, 291–292, 303, 308

R

Rademacher functions, 156, 166
Random variable, 24, 90
Rectangle, 36
Regularization, 204
Regulated function, 52
Resolution of the identity, 292
Resolvent
 operator, 280
 set, 280, 287
Riemann
 integrable, 27, 48, 53, 80
 integral, 27
Riemann–Lebesgue lemma, 106, 110–111
Riesz's theorem, 124
Riesz–Fischer theorem, 76, 79
Rodrigues formula, 127, 129, 157–158

S

Scalar product, 118
Schmidt orthogonalization process, 135–136, 154
Schwartz
Section
 of functions,
 of sets, 36
Self-adjoint operator, 265, 276–277, 286, 290–297, 303, 307–308
Self-similar set, 18

Separable, 104
Serpiński carpet, 55, 61
Sigma-algebra (σ-algebra), 2–4, 51
Sigma-finite (σ-finite) measure, 11, 40, 43
Simple function, 6, 19
Sine integral, 108, 114–115
Spectral radius, 288
Spectrum, 281–297
Step function, 52
Stieljes measure, 10
Sturm–Liouville operator, 297–300, 305, 315–317
Support
    bounded (or compact), 80
    of a distribution, 196
    of a function, 80
Symmetric operator, 264, 276

T

Test function, 168–172, 213
Topological space, 50
Topology, 50, 104
Total system, 131
Translation invariance, 9, 84

U

Unbounded operator, 273–280
Unitary operator, 265, 266, 305, 312
Upper limit, 51

V

Vanish at infinity, 84
Variance, 25
Vitali's criterion, 142

W

Wave equation, 107, 233, 245–246
Weierstrass theorem, 206
Weight function, 129, 154

Z

Zorn axiom, 151